Solutions Manual for
Foundations of College Chemistry

11th Edition
Alternate 11th Edition

Morris Hein
Mount San Antonio College

Susan Arena
University of Illinois, Urbana-Champaign

JOHN WILEY & SONS, INC.

Cover Photo: Tim Davis/Corbis Images

ISBN 0-471-46812-6

Printed in the United States of America

10 9 8 7 6 5 4 3 2 1

Printed and bound by Von Hoffmann Graphics, Inc.

TO THE STUDENT

This Solutions Manual contains answers to all questions and solutions to all problems at the end of each chapter in the text.

This book will be valuable to you if you use it properly. It is not intended to be a substitute for answering the questions by yourself. It is important that you go through the process of writing down the answers to questions and the solutions to the problems before you see them in the solutions manual. Once you have seen the answer, much of the value of whether or not you have learned the material is lost. This does not mean that you cannot learn from the answers. If you absolutely cannot answer a question or problem, study the answer carefully to see where you are having difficulty. You should spend sufficient time answering each assigned exercise. Only after you have made a serious attempt to answer a question or problem should you resort to the solutions manual.

Chemistry is one of the most challenging courses you will encounter. The core of the study of chemistry is problem solving. Skill at solving problems is achieved through effective and consistent practice. Watching and listening to others solve problems may be useful, but will not result in facility with chemistry problems. The methods used for solving problems in this manual are essentially the same as those used in the text. The basic steps in problem solving are fairly universal:

1. Read the problem carefully and determine the type of problem.
2. Develop a plan for solving the problem.
3. Write down the given information in an organized fashion.
4. Write a complete set-up and solution for the problem.
5. Use the solution manual to check the answer and solution.

Your solution might be correct and yet will vary from the manual. Usually, these differences are the result of completing operations in a different order, or using separate steps instead of a single line approach.

If you make an error, take the time to analyze what went wrong. An error caused by an improper entry to the calculator can be frustrating, but it is not as serious as the inability to properly set up the problem. Do not become discouraged. Once you understand the steps in a problem, go back and rework it without looking at the answer.

We have made every attempt to produce a manual with as few errors as possible. Please let us know of errors that you encounter, so that they can be corrected. Allow this manual to help you have success as you begin the great adventure of studying chemistry.

MORRIS HEIN
SUSAN ARENA

CONTENTS

CHAPTER 2

STANDARDS FOR MEASUREMENT

1. $100 \text{ cm} = 1 \text{ m}$
   $1000 \text{ m} = 1 \text{ km}$
   $100{,}000 \text{ cm} = 1 \text{ km}$

   $$(1 \text{ km})\left(\frac{1000 \text{ m}}{\text{km}}\right)\left(\frac{100 \text{ cm}}{\text{m}}\right) = 100{,}000 \text{ cm}$$

2. 7.6 cm

3. The volumetric flask is a more precise measuring instrument than the graduated cylinder. This is because the narrow opening at the point of measurement (calibration mark in the neck of the flask) means that a small error made in not filling the flask exactly to the mark will be a smaller percentage error in total volume than will a similar error with the cylinder.

4. The three materials would sort out according to their densities with the most dense (mercury) at the bottom and the least dense (glycerin) at the top. In the cylinder, the solid magnesium would sink in the glycerin and float on the liquid mercury.

5. Order of increasing density: ethyl alcohol, vegetable oil, salt, lead.

6. The density of ice must be less than 0.91 g/mL and greater than 0.789 g/mL.

7. Heat is a form of energy, while temperature is a measure of the intensity of heat (how hot the system is).

8. Density is the ratio of the mass of a substance to the volume occupied by that mass. Density has the units of mass over volume. Specific gravity is the ratio (no units) of the density of a substance to the density of a reference substance (usually water at a specific temperature for solids and liquids). Specific gravity has no units.

9. Rule 1. When the first digit after those you want to retain is 4 or less, that digit and all others to its right are dropped. The last digit retained is not changed.

   Rule 2. When the first digit after those you want to retain is 5 or greater, that digit and all others to the right of it are dropped and the last digit retained is increased by one.

10. The number of degrees between the freezing and boiling point of water are
    Fahrenheit 180°F
    Celsius 100°C
    Kelvin 100 K

11. Weight is a measure of how much attraction the earth's gravity has for an object (or person). In this case, the farther the astronaut is from the earth the less gravitational force is pulling on him or her. Less gravitational attraction means the astronaut will weigh less. The mass of the astronaut is the amount of matter that makes up him or her. This does not change as the astronaut moves away from the earth.

12. The density of water is 1.0 g/mL at approximately 4 degrees Celsius. However, when water changes from a liquid to a solid at 0 degrees Celsius there is actually an increase in volume. The density of water at 0 degrees Celsius is 0.917 g/mL. Therefore, ice floats in water because solid water is less dense than liquid water.

Metric Abbreviations

13. (a) kilogram = 1000 grams
    (b) centimeter = 1/100 of a meter (0.01 m)
    (c) microliter = 1/1,000,000 of a liter (0.000001 L)
    (d) millimeter = 1/1000 of a meter (0.001 m)
    (e) deciliter = 1/10 of a liter (0.1 L).

14. (a) 1000 meters = a kilometer
    (b) 0.1 gram = a decigram
    (c) 0.000001 liter = a microliter
    (d) 0.01 meter = a centimeter
    (e) 0.001 liter = a milliliter

15. (a) gram = g
    (b) microgram = μg
    (c) centimeter = cm
    (d) micrometer = μm
    (e) milliliter = mL
    (f) deciliter = dL

16. (a) milligram = mg
    (b) kilogram = kg
    (c) meter = m
    (d) nanometer = nm
    (e) angstrom = Å
    (f) microliter = μL

17. (a) 503 zero is significant
    (b) 0.007 zeros are not significant
    (c) 4200 zeros are not significant
    (d) 3.0030 zeros are significant
    (e) 100.00 zeros are significant
    (f) 8.00×10^2 zeros are significant

18. (a) 63,000 zeros are not significant
    (b) 6.004 zeros are significant
    (c) 0.00543 zeros are not significant
    (d) 8.3090 zeros are significant

(e) 60. zero is significant
(f) 5.0×10^{-4} zero is significant

19. Significant figures
(a) 0.025 (2)
(b) 22.4 (3)
(c) 0.0404 (3)
(d) 5.50×10^3 (3)

20. Significant figures
(a) 40.0 (3)
(b) 0.081 (2)
(c) 129,042 (6)
(d) 4.090×10^{-3} (4)

21. Round to three significant figures
(a) 93.2
(b) 0.0286
(c) 4.64
(d) 34.3

22. Round to three significant figures
(a) 8.87
(b) 21.3
(c) 130. (1.30×10^2)
(d) 2.00×10^6

23. Exponential notation
(a) 2.9×10^6
(b) 5.87×10^{-1}
(c) 8.40×10^{-3}
(d) 5.5×10^{-6}

24. Exponential notation
(a) 4.56×10^{-2}
(b) 4.0822×10^3
(c) 4.030×10^1
(d) 1.2×10^7

25. (a)
$$\begin{array}{r} 12.62 \\ 1.5 \\ 0.25 \\ \hline 14.37 \end{array} = 14.4$$

(b) $(2.25 \times 10^3)(4.80 \times 10^4) = 10.8 \times 10^7 = 1.08 \times 10^8$

(c) $\dfrac{(452)(6.2)}{14.3} = 195.97 = 2.0 \times 10^2$

(d) $(0.0394)(12.8) = 0.504$

(e) $\dfrac{0.4278}{59.6} = 0.00718 = 7.18 \times 10^{-3}$

(f) $10.4 + (3.75)(1.5 \times 10^4) = 5.6 \times 10^4$

26. (a)

$$\begin{array}{r} 15.2 \\ -2.75 \\ 15.67 \\ \hline 28.1 \end{array}$$

(b) $(4.68)(12.5) = 58.5$

(c) $\dfrac{182.6}{4.6} = 4.0 \times 10^1$ or 40.

(d)

$$\begin{array}{r} 1986 \\ 23.84 \\ 0.012 \\ \hline 2009.852 \end{array} = 2010. = 2.010 \times 10^3$$

(e) $\dfrac{29.3}{(284)(415)} = 2.49 \times 10^{-4}$

(f) $(2.92 \times 10^{-3})(6.14 \times 10^5) = 1.79 \times 10^3$

27. Fractions to decimals (3 significant figures)

(a) $\dfrac{5}{6} = 0.833$

(b) $\dfrac{3}{7} = 0.429$

(c) $\dfrac{12}{16} = 0.750$

(d) $\dfrac{9}{18} = 0.500$

28. Decimals to fractions

(a) $0.25 = \dfrac{1}{4}$

(b) $0.625 = \dfrac{5}{8}$

(c) $1.67 = 1\dfrac{2}{3}$ or $\dfrac{5}{3}$

(d) $0.888 = \dfrac{8}{9}$

29. (a) $3.42x = 6.5$

$$\frac{3.42x}{3.42} = \frac{6.5}{3.42}$$

$$x = \frac{6.5}{3.42} = 1.9$$

(b) $\dfrac{x}{12.3} = 7.05$

$$x = (7.05)(12.3) = 86.7$$

(c) $\dfrac{0.525}{x} = 0.25$

$$0.525 = 0.25x$$

$$x = \frac{0.525}{0.25} = 2.1$$

30. (a) $x = \dfrac{212 - 32}{1.8}$

$x = 1.0 \times 10^2$

(b) $8.9\dfrac{\text{g}}{\text{mL}} = \dfrac{40.90\text{ g}}{x}$

$\left(8.9\dfrac{\text{g}}{\text{mL}}\right)x = 40.90\text{ g}$

$x = \dfrac{40.90\text{ g}}{8.9\dfrac{\text{g}}{\text{mL}}} = 4.6\text{ mL}$

(c) $72 = 1.8x + 32$

$72 - 32 = 1.8x$

$40. = 1.8x$

$\dfrac{40.}{1.8} = x$

$22 = x$

31. (a) $(28.0\text{ cm})\left(\dfrac{1\text{ m}}{100\text{ cm}}\right) = 0.280\text{ m}$

(b) $(1000.\text{ m})\left(\dfrac{1\text{ km}}{1000\text{ m}}\right) = 1.000\text{ km}$

(c) $(9.28\text{ cm})\left(\dfrac{10\text{ mm}}{1\text{ cm}}\right) = 92.8\text{ mm}$

(d) $(10.68\text{ g})\left(\dfrac{1000\text{ mg}}{1\text{ g}}\right) = 1.068 \times 10^4\text{ mg}$

(e) $(6.8 \times 10^4\text{ mg})\left(\dfrac{1\text{ g}}{1000\text{ mg}}\right)\left(\dfrac{1\text{ kg}}{1000\text{ g}}\right) = 6.8 \times 10^{-2}\text{ kg}$

(f) $(8.54\text{ g})\left(\dfrac{1\text{ kg}}{1000\text{ g}}\right) = 0.00854\text{ kg}$

(g) $(25.0\text{ mL})\left(\dfrac{1\text{ L}}{1000\text{ mL}}\right) = 2.50 \times 10^{-2}\text{ L}$

(h) $(22.4\text{ L})\left(\dfrac{10^6\ \mu\text{L}}{1\text{ L}}\right) = 2.24 \times 10^7\ \mu\text{L}$

32. (a) $(4.5\text{ cm})\left(\dfrac{1\text{ m}}{100\text{ cm}}\right)\left(\dfrac{1\text{ Å}}{10^{-10}\text{ m}}\right) = 4.5 \times 10^8\text{ Å}$

(b) $(12\text{ nm})\left(\dfrac{10^{-9}\text{ m}}{1\text{ nm}}\right)\left(\dfrac{100\text{ cm}}{1\text{ m}}\right) = 1.2 \times 10^{-6}\text{ cm}$

(c) $(8.0\text{ km})\left(\dfrac{1000\text{ m}}{1\text{ km}}\right)\left(\dfrac{1000\text{ mm}}{1\text{ m}}\right) = 8.0 \times 10^6\text{ mm}$

(d) $(164\ \text{mg})\left(\dfrac{1\ \text{g}}{1000\ \text{mg}}\right) = 0.164\ \text{g}$

(e) $(0.65\ \text{kg})\left(\dfrac{1000\ \text{g}}{1\ \text{kg}}\right)\left(\dfrac{1000\ \text{mg}}{1\ \text{g}}\right) = 6.5 \times 10^5\ \text{mg}$

(f) $(5.5\ \text{kg})\left(\dfrac{1000\ \text{g}}{1\ \text{kg}}\right) = 5.5 \times 10^3\ \text{g}$

(g) $(0.468\ \text{L})\left(\dfrac{1000\ \text{mL}}{1\ \text{L}}\right) = 468\ \text{mL}$

(h) $(9.0\ \mu\text{L})\left(\dfrac{1\ \text{L}}{10^6\ \mu\text{L}}\right)\left(\dfrac{1000\ \text{mL}}{1\ \text{L}}\right) = 9.0 \times 10^{-3}\ \text{mL}$

33. (a) $(42.2\ \text{in.})\left(\dfrac{2.54\ \text{cm}}{1\ \text{in.}}\right) = 107\ \text{cm}$

(b) $(0.64\ \text{mi})\left(\dfrac{5280\ \text{ft}}{1\ \text{mi}}\right)\left(\dfrac{12\ \text{in.}}{1\ \text{ft}}\right) = 4.1 \times 10^4\ \text{in.}$

(c) $(2.00\ \text{in.}^2)\left(\dfrac{2.54\ \text{cm}}{1\ \text{in.}}\right)^2 = 12.9\ \text{cm}^2$

(d) $(42.8\ \text{kg})\left(\dfrac{2.205\ \text{lb}}{\text{kg}}\right) = 94.4\ \text{lb}$

(e) $(3.5\ \text{qt})\left(\dfrac{946\ \text{mL}}{1\ \text{qt}}\right) = 3.3 \times 10^3\ \text{mL}$

(f) $(20.0\ \text{L})\left(\dfrac{1\ \text{qt}}{0.946\ \text{L}}\right)\left(\dfrac{1\ \text{gal}}{4\ \text{qt}}\right) = 5.29\ \text{gal}$

34. (a) The conversion is: m $\rightarrow$ cm $\rightarrow$ in. $\rightarrow$ ft

$$(35.6\ \text{m})\left(\dfrac{100\ \text{cm}}{1\ \text{m}}\right)\left(\dfrac{1\ \text{in.}}{2.54\ \text{cm}}\right)\left(\dfrac{1\ \text{ft}}{12\ \text{in.}}\right) = 117\ \text{ft}$$

(b) $(16.5\ \text{km})\left(\dfrac{1\ \text{mi}}{1.609\ \text{km}}\right) = 10.3\ \text{mi}$

(c) $(4.5\ \text{in.}^3)\left(\dfrac{2.54\ \text{cm}}{1\ \text{in.}}\right)^3\left(\dfrac{10\ \text{mm}}{1\ \text{cm}}\right)^3 = 7.4 \times 10^4\ \text{mm}^3$

(d) $(95\ \text{lb})\left(\dfrac{453.6\ \text{g}}{1\ \text{lb}}\right) = 4.3 \times 10^4\ \text{g}$

(e) $(20.0\text{ gal})\left(\frac{4\text{ qt}}{1\text{ gal}}\right)\left(\frac{0.946\text{ L}}{1\text{ qt}}\right) = 75.7\text{ L}$

(f) The conversion is: $\text{ft}^3 \rightarrow \text{in.}^3 \rightarrow \text{cm}^3 \rightarrow \text{m}^3$

$$(4.5 \times 10^4\text{ ft}^3)\left(\frac{12\text{ in.}}{1\text{ ft}}\right)^3\left(\frac{2.54\text{ cm}}{1\text{ in.}}\right)^3\left(\frac{1\text{ m}}{1000\text{ cm}}\right)^3 = 1.3\text{ m}^3$$

35. $\left(55\frac{\text{mi}}{\text{hr}}\right)\left(1.609\frac{\text{km}}{1\text{ mi}}\right) = 88\frac{\text{km}}{\text{hr}}$

36. The conversion is: $\frac{\text{km}}{\text{hr}} \rightarrow \frac{\text{mi}}{\text{hr}} \rightarrow \frac{\text{ft}}{\text{hr}} \rightarrow \frac{\text{ft}}{\text{s}}$

$$\left(55\frac{\text{km}}{\text{hr}}\right)\left(\frac{1\text{ mi}}{1.609\text{ km}}\right)\left(\frac{5280\text{ ft}}{1\text{ mi}}\right)\left(\frac{1\text{ hr}}{3600\text{ s}}\right) = 50.\frac{\text{ft}}{\text{s}}$$

37. The conversion is: $\frac{\text{m}}{\text{s}} \rightarrow \frac{\text{cm}}{\text{s}} \rightarrow \frac{\text{in.}}{\text{s}} \rightarrow \frac{\text{ft}}{\text{s}}$

$$\left(\frac{100.\text{ m}}{9.92\text{ s}}\right)\left(\frac{100\text{ cm}}{1\text{ m}}\right)\left(\frac{1\text{ in.}}{2.54\text{ cm}}\right)\left(\frac{1\text{ ft}}{12\text{ in.}}\right) = 33.1\frac{\text{ft}}{\text{s}}$$

38. The conversion is: $\frac{\text{mi}}{\text{hr}} \rightarrow \frac{\text{km}}{\text{hr}} \rightarrow \frac{\text{km}}{\text{s}}$

$$\left(\frac{229\text{ mi}}{1\text{ hr}}\right)\left(\frac{1.609\text{ km}}{\text{mi}}\right)\left(\frac{1\text{ hr}}{3600\text{ s}}\right) = 0.102\frac{\text{km}}{\text{s}}$$

39. The conversion is: $\frac{\text{mi}}{\text{hr}} \rightarrow \frac{\text{km}}{\text{hr}} \rightarrow \frac{\text{km}}{\text{s}}$

$$\left(\frac{27{,}000\text{ mi}}{\text{hr}}\right)\left(\frac{1.609\text{ km}}{\text{mi}}\right)\left(\frac{1\text{ hr}}{3600\text{ s}}\right) = 12\frac{\text{km}}{\text{s}}$$

40. The conversion is: $\text{mi} \rightarrow \text{km} \rightarrow \text{m} \rightarrow \text{s}$

93 million miles $= 9.3 \times 10^7$ mi

$$(9.3 \times 10^7\text{ mi})\left(\frac{1.609\text{ km}}{\text{mi}}\right)\left(\frac{1000\text{ m}}{1\text{ km}}\right)\left(\frac{1\text{ s}}{3.00 \times 10^8\text{ m}}\right) = 5.0 \times 10^2\text{ s}$$

41. $(176\text{ lb})\left(\frac{453.6\text{ g}}{1\text{ lb}}\right)\left(\frac{1\text{ kg}}{1000\text{ g}}\right) = 79.8\text{ kg}$

42. The conversion is: oz $\rightarrow$ lb $\rightarrow$ g $\rightarrow$ mg

$$(1\text{ oz})\left(\frac{1\text{ lb}}{16\text{ oz}}\right)\left(\frac{453.6\text{ g}}{1\text{ lb}}\right)\left(\frac{1000\text{ mg}}{1\text{ g}}\right) = 3 \times 10^4\text{ mg}$$

43. $$(5.0\text{ grains})\left(\frac{1\text{ lb}}{7000.\text{ grains}}\right)\left(\frac{453.6\text{ g}}{1\text{ lb}}\right) = 0.32\text{ g}$$

44. $$(21\text{ lb})\left(\frac{453.6\text{ g}}{1\text{ lb}}\right) = 9.5 \times 10^3\text{ g} = \text{mass condor}$$

$$\left(\frac{9.5 \times 10^3\text{ g/(condor)}}{3.2\text{ g/(hummingbird)}}\right) = 3.0 \times 10^3\text{ hummingbirds to equal the mass of one (1) condor}$$

45. $$\left(\frac{\$1.49}{283.5\text{ g}}\right)\left(\frac{453.6\text{ g}}{1\text{ lb}}\right) = \frac{\$2.38}{\text{lb}}$$

46. The conversion is: $\frac{\$}{\text{oz}} \rightarrow \frac{\$}{\text{lb}} \rightarrow \frac{\$}{\text{g}} \rightarrow \$$

$$\left(\frac{\$350}{1\text{ oz}}\right)\left(\frac{14.58\text{ oz}}{1\text{ lb}}\right)\left(\frac{1\text{ lb}}{453.6\text{ g}}\right)(250\text{ g}) = \$2800$$

47. The conversion is: $\frac{\$}{\text{L}} \rightarrow \frac{\$}{\text{qt}} \rightarrow \frac{\$}{\text{gal}} \rightarrow \$$

$$\left(\frac{\$0.35}{1\text{ L}}\right)\left(\frac{0.946\text{ L}}{1\text{ qt}}\right)\left(\frac{4\text{ qt}}{1\text{ gal}}\right)(15.8\text{ gal}) = \$21$$

48. The conversion is: mi $\rightarrow$ gal $\rightarrow$ qt $\rightarrow$ L

$$(525\text{ mi})\left(\frac{1\text{ gal}}{35\text{ mi}}\right)\left(\frac{4\text{ qt}}{1\text{ gal}}\right)\left(\frac{0.946\text{ L}}{1\text{ qt}}\right) = 57\text{ L}$$

49. The conversion is: $\frac{\text{drops}}{\text{mL}} \rightarrow \frac{\text{drops}}{\text{qt}} \rightarrow \frac{\text{drops}}{\text{gal}} \rightarrow \text{drops}$

$$\left(\frac{20.\text{ drops}}{\text{mL}}\right)\left(\frac{946\text{ mL}}{\text{qt}}\right)\left(\frac{4\text{ qt}}{\text{gal}}\right)(1.0\text{ gal}) = 7.6 \times 10^4\text{ drops}$$

50. $$(42\text{ gal})\left(\frac{4\text{ qt}}{\text{gal}}\right)\left(\frac{0.946\text{ L}}{\text{qt}}\right) = 160\text{ L}$$

51. The conversion is: $ft^3 \rightarrow in.^3 \rightarrow cm^3 \rightarrow mL$

$$(1.00\ ft^3)\left(\frac{12\ in.}{ft}\right)^3\left(\frac{2.54\ cm}{1\ in.}\right)^3\left(\frac{1\ mL}{1\ cm^3}\right) = 2.83 \times 10^4\ mL$$

52. $V = A \times h$ A = area h = height V = volume

The conversion is: $\frac{cm^3}{nm} \rightarrow \frac{cm^3}{m} \rightarrow m^2$

$$A = \frac{V}{h} = \left(\frac{200\ cm^3}{0.5\ nm}\right)\left(\frac{1\ nm}{10^{-9}\ m}\right)\left(\frac{1\ m}{100\ cm}\right)^3 = 4 \times 10^5\ m^2$$

53. (a) $(27\ cm)(21\ cm)(4.4\ cm) = 2.5 \times 10^3\ cm^3$

(b) $2.5 \times 10^3\ cm^3$ is $2.5 \times 10^3\ mL\left(\frac{1\ L}{1000\ mL}\right) = 2.5\ L$

(c) $(2.5 \times 10^3\ cm^3)\left(\frac{1\ in.}{2.54\ cm}\right)^3 = 1.5 \times 10^2\ in.^3$

54. $$(16\ in.)(8\ in.)(10\ in.)\left(\frac{2.54\ cm}{1\ in.}\right)^3\left(\frac{1\ L}{1000\ mL}\right)\left(\frac{1\ qt}{0.946\ L}\right)\left(\frac{1\ gal}{4\ qt}\right) = 6\ gal$$

55. $$°C = \frac{°F - 32}{1.8} = \frac{98.6 - 32}{1.8} = 37.0°C$$

56. $°F = 1.8°C + 32$ $(1.8)(45) + 32 = 113°F$ Summer!

57. (a) $\frac{162 - 32}{1.8} = 72.2°C$ Remember to express the answer to the same precision as the original measurement.

(b) $°C + 273 = K$ $\frac{0.0 - 32}{1.8} + 273 = 255.2\ K$ $(2.6 \times 10^2\ K)$

(c) $1.8(-18) + 32 = -0.40°F$

(d) $212 - 273 = -61°C$

58. (a) $1.8(32) + 32 = 90.°F$

(b) $\frac{-8.6 - 32}{1.8} = -22.6°C$

(c) $273 + 273 = 546\ K$

(d) $°C = 100 - 273 = -173°C$

$(-173)(1.8) + 32 = -279°F = -300°F$ (1 significant figure in 100 K)

59.
$$°F = °C$$
$$°F = 1.8(°C) + 32 \qquad \text{substitute °F for °C}$$
$$°F = 1.8(°F) + 32$$
$$-32 = 0.8(°F)$$
$$\frac{-32}{0.8} = °F$$
$$-40 = °F$$
$$-40°F = -40°C$$

60.
$$°F = -°C$$
$$°F = 1.8(°C) + 32 \qquad \text{substitute } -°C \text{ for °F}$$
$$-°C = 1.8(°C) + 32$$
$$2.8(°C) = -32$$
$$°C = \frac{-32}{2.8}$$
$$°C = -11.4$$
$$-11.4°C = 11.4°F$$

61. $d = \frac{m}{V} = \frac{78.26\text{ g}}{50.00\text{ mL}} = 1.565\frac{g}{mL}$

62. $d = \frac{m}{V} = \frac{39.9\text{ g}}{12.8\text{ mL}} = 3.12\frac{g}{mL}$

63. $29.6\text{ mL} - 25.0\text{ mL} = 4.6\text{ mL}$ (volume of chromium)

$$d = \frac{m}{V} = \frac{32.7\text{ g}}{4.6\text{ mL}} = 7.1\frac{g}{mL}$$

64. $106.773\text{ g} - 42.817\text{ g} = 63.956\text{ g}$ (mass of the liquid)

$$d = \frac{m}{V} = \frac{63.956\text{ g}}{50.0\text{ mL}} = 1.28\frac{g}{mL}$$

65. $d = \frac{m}{V}$

$$m = dV = \left(1.19\frac{g}{mL}\right)(250.0\text{ mL}) = 298\text{ g}$$

66. $d = \frac{m}{V}$

$$m = dV = \left(13.6\frac{g}{mL}\right)(25.0\text{ mL}) = 3.40 \times 10^2\text{ g}$$

67. (a) report 10.01 grams
(b) report 10.012 grams
(c) report 10.0124 grams

68. A graduated cylinder would be the best choice for adding 100 mL of solvent to a reaction. While the volumetric flask is also labeled 100 mL, volumetric flasks are typically used for doing dilutions. The other three pieces of glassware could also be used, but they hold smaller volumes so it would take a longer time to measure out 100 mL. Also, because you would have to repeat the measurement many times using the other glassware, there is a greater chance for error.

69. To add increments of a solution the best choice would be the buret. The stopcock allows you to dispense a desired amount without having to refill the buret each time. A pipet could also be used to accomplish the same task.

70. Density equals mass divided by volume. The volume is given as 50 mL. To find the mass of the chemical first weigh the beaker with the chemical in it. Then pour the chemical into another beaker and reweigh the empty beaker. Subtract the mass of the empty beaker from the mass of the beaker containing the chemical to obtain the mass of the chemical. To find the density simply divide the mass of the chemical by the volume.

71. $$d = \left(\frac{1032\text{ g}}{1\text{ L}}\right)\left(\frac{1\text{ L}}{1000\text{ mL}}\right) = 1.032\frac{g}{mL}$$

$$d = \left(\frac{1032\text{ g}}{1\text{ L}}\right)\left(\frac{1\text{ kg}}{1000\text{ g}}\right) = 1.032\frac{kg}{L}$$

72. The conversion is: $L \rightarrow cm^3 \rightarrow g \rightarrow lb$

$$(3.1\text{ L})\left(\frac{1000\text{ cm}^3}{1\text{ L}}\right)\left(1.03\frac{g}{cm^3}\right)\left(\frac{1\text{ lb}}{453.6\text{ g}}\right) = 7.0\text{ lbs}$$

73. $$\text{area per lane marker} = (2.5\text{ ft})(4.0\text{ in.})\left(\frac{1\text{ ft}}{12\text{ in.}}\right) = 0.83\text{ ft}^2$$

The conversion is: $\frac{\text{lane marker}}{\text{ft}^2} \rightarrow \frac{\text{lane}}{\text{qt}} \rightarrow \frac{\text{lane}}{\text{gal}} \rightarrow$ lane markers

$$\left(\frac{1\text{ lane marker}}{0.83\text{ ft}^2}\right)\left(\frac{43\text{ ft}^2}{1.0\text{ qt}}\right)\left(\frac{4\text{ qt}}{1.0\text{ gal}}\right)(15\text{ gal}) = 3100\text{ lane markers}$$

74. $V = \text{side}^3 = (0.50\ \text{m})^3 = 0.125\ \text{m}^3$ (volume of the cube)

$$(0.125\ \text{m}^3)\left(\frac{100.\ \text{cm}}{\text{m}}\right)^3\left(\frac{1\ \text{L}}{1000\ \text{cm}^3}\right) = 125\ \text{L} \quad \text{(volume of cube)}$$

Yes, the cube will hold the solution. $125\ \text{L} - 8.5\ \text{L} = 116.5\ \text{L}$ additional solution is necessary to fill the container.

75. The conversion is: $\frac{\mu g}{m^3} \rightarrow \frac{\mu g}{L} \rightarrow \frac{\mu g}{\text{day}}$

$$\left(\frac{180\ \mu g}{1\ \text{m}^3}\right)\left(\frac{1\ \text{m}^3}{1000\ \text{L}}\right)\left(2 \times 10^4 \frac{\text{L}}{\text{day}}\right) = 4000\ \mu g/\text{day ingested} \qquad \text{(1 sig. figure)}$$

Yes, the technician is at risk. This is well over the toxic limit.

76. Convert 4.5°F to °C

$$\frac{°F - 32}{1.8} = °C$$

$$\frac{(4.5 - 32)}{1.8} = -15.3°C \equiv 4.5°F$$

$-15°C > -15.3°C$

$-15°C > 4.5°F$

therefore, $-15°C$ is the higher temperature (just slightly)

77. Equal volumes of lead and aluminum have different masses because their densities are different. The density of lead is $11.24\ \text{g/cm}^3$ and the density of aluminum is $2.70\ \text{g/cm}^3$. The mass of $2.50\ \text{cm}^3$ of lead $= (2.50\ \text{cm}^3)(11.34\ \text{g/cm}^3) = 28.35\ \text{g}$. The mass of $2.50\ \text{cm}^3$ of aluminum $= (2.50\ \text{cm}^3)(2.70\ \text{g/cm}^3) = 6.75\ \text{g}$.

78. Based on their densities, the four substances would layer from largest density to smallest density, from bottom to top, in the cylinder. The order would be as follows; karo syrup ($d = 1.37\text{g/mL}$), water ($d = 1.000\ \text{g/mL}$), vegetable oil ($d = 0.91\ \text{g/mL}$) and ethyl alcohol ($d = 0.789\ \text{g/mL}$).

79. $m_{\text{pan 1}} = m_{\text{pan 2}}$ (when balanced) (m = mass)

$$m_{\text{pan 1}} = m_{\text{flask}} + m_{\text{alcohol}} = m_{\text{flask}} + (100.\ \text{mL})(0.789\ \text{g/mL})$$

$$= m_{\text{flask}} + 78.9\ \text{g}$$

let x = volume of turpentine

$$m_{\text{pan 2}} = (m_{\text{flask}} + 11.0\ \text{g}) + m_{\text{turpentine}} = (m_{\text{flask}} + 11.0\ \text{g}) + x\left(0.87\frac{\text{g}}{\text{mL}}\right)$$

Since

$$m_{pan\,1} = m_{pan\,2}$$

$$m_{flask} + 78.9\text{ g} = (m_{flask} + 11.0\text{ g}) + x\left(0.87\frac{\text{g}}{\text{mL}}\right)$$

$$78.9\text{ g} = 11.0\text{ g} + x\left(0.87\frac{\text{g}}{\text{mL}}\right)$$

$$67.9\text{ g} = x\left(0.87\frac{\text{g}}{\text{mL}}\right)$$

$$x = 78\text{ mL turpentine}$$

80. $d = \frac{\text{m}}{\text{V}}$ $\quad \text{V} = \frac{\text{m}}{d}$

$$\text{V}_\text{A} = \frac{25\text{ g}}{10.\frac{\text{g}}{\text{mL}}} = 2.5\text{ mL}$$

$$\text{V}_\text{B} = \frac{65\text{ g}}{4.0\frac{\text{g}}{\text{mL}}} = 16\text{ mL}$$

B occupies the larger volume

81.

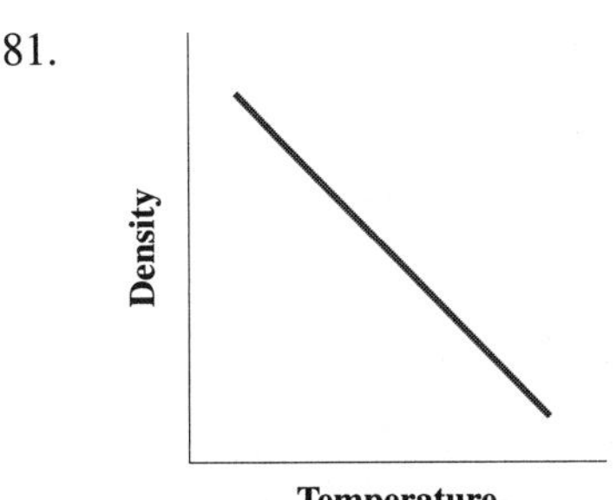

Since $d = \frac{\text{m}}{\text{V}}$, as the volume increases, the density decreases. As solids are heated the density decreases due to an increase in the volume of the solid.

82. $\text{m} = d\text{V} = (0.789\text{ g/mL})(35.0\text{ mL}) = 27.6\text{ g}$ ethyl alcohol
$27.6\text{ g} + 49.28\text{ g} = 76.9\text{ g}$ (mass of cylinder and alcohol)

83. $d = \frac{\text{m}}{\text{V}}$ The cube with the largest V has the lowest density. Use Table 2.5.

Cube A - lowest density	1.74 g/mL - magnesium
Cube B	2.70 g/mL - aluminum
Cube C - highest density	10.5 g/mL - silver

84. $(1.00\ \text{cm}^3)\left(\frac{2.54\ \text{cm}}{\text{in.}}\right)^3 = 16.4\ \text{cm}^3$ in 1.00 cubic inch

85. The volume of the aluminum cube is:

$$V = \frac{m}{d} = \frac{500.\ \text{g}}{2.70\frac{\text{g}}{\text{mL}}} = 185\ \text{mL} \qquad d(\text{Al}) \text{ is } 2.70\ \text{g/mL}$$

This is the same volume as the gold cube thus:

$m = dV = (185\ \text{mL})(19.3\ \text{g/mL}) = 3.57 \times 10^3$ g of gold   d(Au) is 19.3 g/mL

86. $d = \frac{m}{V} = \frac{24.12\ \text{g}}{25.0\ \text{mL}} = \frac{0.965\ \text{g}}{\text{mL}}$ (density of water at 90°C)

87. $150.50\ \text{g} - 88.25\ \text{g} = 62.25\ \text{g}$ (mass of liquid)

$$d = \frac{m}{V} \text{ thus } V = \frac{m}{d} = \frac{62.25\ \text{g}}{1.25\frac{\text{g}}{\text{mL}}} = 49.8\ \text{mL} \qquad \text{(volume of liquid)}$$

The container must hold at least 50 mL.

88. H_2O

$$\frac{50\ \text{g}}{1.0\frac{\text{g}}{\text{mL}}} = 50\ \text{mL}$$

alcohol

$$\frac{50\ \text{g}}{0.789\frac{\text{g}}{\text{mL}}} = 60\ \text{mL}$$

Ethyl alcohol has the greater volume due to its lower density.

89. The conversion is: g $\rightarrow$ lb $\rightarrow$ oz

$$(8.1\ \text{g})\left(\frac{1\ \text{lb}}{453.6\ \text{g}}\right)\left(\frac{16\ \text{oz}}{\text{lb}}\right) = 0.29\ \text{oz} \qquad \text{(mass of the coin)}$$

$$(0.29\ \text{oz})\left(\frac{5.3\%}{100}\right) = 0.010\ \text{oz Mn}$$

90. Volume of sulfuric acid

$$\left(\frac{1\ \text{mL}}{1.84\ \text{g}}\right)(100.\ \text{g}) = 54.3\ \text{mL}$$

91. $(1.00\text{ kg Pd})\left(\dfrac{1000\text{ g}}{\text{kg}}\right)\left(\dfrac{1.00\text{ mL}}{12.0\text{ g}}\right) = 83.3\text{ mL Pd at }20°\text{C}$

$(1.00\text{ kg Pd})\left(\dfrac{1000\text{ g}}{\text{kg}}\right)\left(\dfrac{1.00\text{ mL}}{11.0\text{ g}}\right) = 90.9\text{ mL Pd at }1550°\text{C}$

$90.9\text{ mL} - 83.3\text{ mL} = 7.6\text{ mL change in volume}$

92. $V = (2.00\text{ cm})(15.0\text{ cm})(6.00\text{ cm})\left(\dfrac{1\text{ mL}}{1\text{ cm}^3}\right) = 180.\text{mL}$ (volume of bar)

$d = \dfrac{m}{V} = 3300\text{ g}/180.\text{ mL} = 18.3\text{ g/mL}$

The density of pure gold is 19.3 g/mL (from Table 2.5), therefore, the gold bar is not pure gold, since its density is only 18.3 g/mL, or it is hollow inside.

93. $93.3\text{ kg} = 9.33 \times 10^4\text{ g}$ $\quad d(\text{gold}) = 19.3\text{ g/mL}$

$V = \left(\dfrac{9.33 \times 10^4\text{ g}}{19.3\text{ g/mL}}\right) = 4.83 \times 10^3\text{ mL} = 4.83 \times 10^3\text{ cm}^3$ (volume of nugget)

convert to ounces

$(9.33 \times 10^4\text{ g})\left(\dfrac{1\text{ lb}}{453.6\text{ g}}\right)\left(\dfrac{14.58\text{ oz}}{1\text{ lb}}\right) = 3.00 \times 10^3\text{ oz gold}$

$(3.00 \times 10^3\text{ oz})(\$345/\text{oz}) = \$1.04 \times 10^6$

Challenge Exercises

94. The density of lead is 11.34 g/mL. The density of aluminum is 2.70 g/mL. The density of silver is 10.5 g/mL. The density of the unknown piece of metal can be calculated from the mass (20.25 g) and the volume ($57.5\text{ mL} - 50\text{ mL} = 7.5\text{ mL}$) of the metal. Density of the unknown metal $= 20.25\text{ g}/7.5\text{ mL} = 2.7\text{ g/mL}$. The metal must be aluminum.

95. Volume of slug $\quad 30.7\text{ mL} - 25.0\text{ mL} = 5.7\text{ mL}$

Density of slug $\quad d = \dfrac{m}{V} = \dfrac{15.454\text{ g}}{5.7\text{ mL}} = 2.7\text{ g/mL}$

Mass of liquid, cylinder, and slug	125.934 g
Mass of slug (subtract)	−15.454 g
Mass of cylinder (subtract)	−89.450 g
Mass of the liquid	21.030 g

Density of liquid $d = \dfrac{m}{V} = \dfrac{21.030\text{ g}}{25.0\text{ mL}} = 0.841\text{ g/mL}$

CHAPTER 3

CLASSIFICATION OF MATTER

1. Answers will vary
   Solids: sugar, salt, copper, silver
   Liquids: water, sulfuric acid, ethyl alcohol, benzene
   Gases: hydrogen, oxygen, nitrogen, ammonia

2. (a) The attractive forces among the ultimate particles of a solid (atoms, ions, or molecules) are strong enough to hold these particles in a fixed position within the solid and thus maintain the solid in a definite shape. The attractive forces among the ultimate particles of a liquid (usually molecules) are sufficiently strong enough to hold them together (preventing the liquid from rapidly becoming gas) but are not strong enough to hold the particles in fixed positions (as in a solid).
   (b) The ultimate particles in a liquid are quite closely packed (essentially in contact with each other) and thus the volume of the liquid is fixed at a given temperature. But, the ultimate particles in a gas are relatively far apart and essentially independent of each other. Consequently, the gas does not have a definite volume.
   (c) In a gas the particles are relatively far apart and are easily compressed, but in a solid the particles are closely packed together and are virtually incompressible.

3. The water in the beaker does not fill the test tube. Since the tube is filled with air (a gas) and two objects cannot occupy the same space at the same time, the gas is shown to occupy space.

4. Mercury and water are the only liquids in the table which are not mixtures.

5. Air is the only gas mixture found in the table.

6. Three phases are present within the bottle; solid and liquid are observed visually, while gas is detected by the immediate odor.

7. The system is heterogeneous as multiple phases are present.

8. A system containing only one substance is not necessarily homogeneous. Two phases may be present. Example: ice in water.

9. A system containing two or more substances is not necessarily heterogeneous. In a solution only one phase is present. Examples: sugar dissolved in water, dilute sulfuric acid.

10. Silicon 25.67% Hydrogen 0.87%

In 100 g $\frac{25.67 \text{ g Si}}{0.87 \text{ g H}} = 30 \text{ g Si}/1 \text{ g H}$

Si is 28 times heavier than H, thus since 30 > 28, there are more Si atoms than H atoms.

11. (a) Ag (b) O (c) H (d) C (e) Fe (f) N (g) Mg (h) K

12. (a) Sodium (b) Fluorine (c) Nickel (d) Zinc (e) Neon (f) Helium (g) Calcium (h) Chlorine

13. The symbol of an element represents the element itself. It may stand for a single atom or a given quantity of the element.

14.

phosphorus	P	sodium	Na
aluminum	Al	nitrogen	N
hydrogen	H	nickel	Ni
potassium	K	silver	Ag
magnesium	Mg	plutonium	Pu

15. (a) Si—1 atom silicon SI—System International or 1 atom sulfur 1 atom iodine

(b) Pb—1 atom lead PB—1 atom phosphorus 1 atom boron

(c) 4P—4 atoms phosphorus P_4—1 molecule phosphorus (made of 4 phosphorus atoms)

16.

Na	sodium	Ag	silver
K	potassium	W	tungsten
Fe	iron	Au	gold
Sb	antimony	Hg	mercury
Sn	tin	Pb	lead

17.

H	hydrogen	S	sulfur
B	boron	K	potassium
C	carbon	V	vanadium
N	nitrogen	Y	yttrium
O	oxygen	I	iodine
F	fluorine	W	tungsten
P	phosphorus	U	uranium

18. In an element all atoms are alike, while a compound contains two or more elements (different atoms) which are chemically combined. Compounds may be decomposed into simpler substances while elements cannot.

19. 84 metals | 7 metalloids | 18 nonmetals (Based on 109 elements)

20. 7 metals | 1 metalloid | 2 nonmetals

21. 1 metal | 0 metalloids | 5 nonmetals

22. The symbol for gold is based upon the Latin word for gold, aurum.

23. (a) iodine (b) bromine

24. A compound is composed of two or more elements which are chemically combined in a definite proportion by mass. Its properties differ from those of its components. A mixture is the physical combining of two or more substances (not necessarily elements). The composition may vary, the substances retain their properties, and they may be separated by physical means.

25. Molecular compounds exist as molecules formed from two or more elements bonded together. Ionic compounds exist as cations and anions held together by electrical attractions.

26. Compounds are distinguished from one another by their characteristic physical and chemical properties.

27. (a) H_2–2 atoms (b) H_2O–3 atoms (c) H_2SO_4–7 atoms

28. Cations are positively charged, while anions are charged negatively.

29. H_2–hydrogen
N_2–nitrogen
O_2–oxygen
F_2–fluorine
Cl_2–chlorine
Br_2–bromine
I_2–iodine

30. Homogeneous mixtures contain only one phase, while heterogeneous mixtures contain two or more phases.

31.

Metals	Nonmetals
solid at room T (except Hg)	solids, liquids or gas at room T
luster	dull (no luster)
conduct heat & electricity	insulator (does not conduct electricity)
malleable	react with each other forming compounds.
react with nonmetals to form compounds	react with metals to form compounds

32. diatomic molecules (a) H_2, (c) HCl, (e) NO

33. (a) Potassium, iodine
(b) Sodium, carbon, oxygen
(c) Aluminum, oxygen
(d) Calcium, bromine
(e) Hydrogen, carbon, oxygen

34. (a) Magnesium, bromine
(b) Carbon, chlorine
(c) Hydrogen, nitrogen, oxygen
(d) Barium, sulfur, oxygen
(e) Aluminum, phosphorus, oxygen

35. (a) ZnO
(b) $KClO_3$
(c) NaOH
(d) C_2H_6O

36. (a) $AlBr_3$
(b) CaF_2
(c) $PbCrO_4$
(d) C_6H_6

37. (a) 2 atoms H, 1 atom O
(b) 2 atoms Na, 1 atom S, 4 atoms O
(c) 4 atoms H, 2 atoms C, 2 atoms O

38. (a) 1 atom Al, 3 atoms Br
(b) 1 atom Ni, 2 atoms N, 6 atoms O, 2 (NO_3) units
(c) 12 atoms C, 22 atoms H, 11 atoms O

39. (a) 2 atoms
(b) 5 atoms
(c) 11 atoms
(d) 8 atoms
(e) 16 atoms

40. (a) 2 atoms
(b) 2 atoms
(c) 9 atoms
(d) 5 atoms
(e) 17 atoms

41. (a) 1 atom O
(b) 4 atoms O
(c) 2 atoms O
(d) 3 atoms O
(e) 9 atoms O

42. (a) 2 atoms H
(b) 6 atoms H
(c) 12 atoms H
(d) 4 atoms H
(e) 8 atoms H

43. (a) mixture
(b) pure substance
(c) mixture
(d) mixture
(e) pure substance

44. (a) mixture
(b) pure substance
(c) pure substance
(d) mixture
(e) pure substance

45. (a) homogeneous
(b) element
(c) heterogeneous
(d) homogeneous
(e) compound

46. (a) homogeneous
(b) compound
(c) element
(d) heterogeneous
(e) compound

47. (a) mixture
(b) element
(c) compound
(d) mixture

48. (a) element
(b) compound
(c) elemen
(d) mixture

49. (a) mixture
(b) compound
(c) element
(d) mixture

50. (a) compound
(b) element
(c) mixture
(d) compound

51. (a) CH_2O (b) C_4H_9 (c) $C_{25}H_{52}$

52. (a) HO (b) C_2H_6O (c) $Na_2Cr_2O_7$

53. Yes. The gaseous elements are all found on the extreme right of the periodic table. They are the entire last column and in the upper right corner of the table. Hydrogen is the exceptions and located at the upper left of the table.

54. No. The only common liquid elements (at room temperature) are mercury and bromine.

55. $$\frac{18\text{ metals}}{36\text{ elements}} \times 100 = 50\%\text{ metals}$$

56. $$\frac{26\text{ solids}}{36\text{ elements}} \times 100 = 72\%\text{ solids}$$

57. Use a sifter to separate the sand from the rock. The sand would go through the holes in the sifter but the rock would not. If the rocks were large enough you could also remove them by hand.

58. A physical change is reversible. Therefore, boil the salt-water solution. The water will evaporate and leave the salt behind.

59. The atoms that make up ionic compounds are on opposite ends of the periodic table from one another. They are made up of metal-nonmetal combinations.

60. (a) 1 carbon atom and 1 oxygen atom, total number of atoms = 2
(b) 1 boron atom and 3 fluorine atoms, total number of atoms = 4
(c) 1 hydrogen atom, 1 nitrogen atom, 3 oxygen atoms, total number of atoms = 5
(d) 1 potassium atom, 1 manganese atom, 4 oxygen atoms, total number of atoms = 6
(e) 1 calcium atom, 2 nitrogen atoms, 6 oxygen atoms, total number of atoms = 9
(f) 3 iron atoms, 2 phosphorus atoms, 8 oxygen atoms, total number of atoms = 13

61. (a) 181 atoms/molecule

63 C
88 H
1 Co
14 N
14 O
1 P
―――――
181 atoms

(b) $\frac{63\ \text{C}}{181\text{atoms}} \times 100 = 35\%\ \text{C atoms}$

(c) $\frac{1\ \text{Co}}{181\ \text{atoms}} = \frac{1}{181}\ \text{metals}$

62. The conversion is: $\text{cm}^3 \rightarrow \text{L} \rightarrow \text{mg} \rightarrow \text{g} \rightarrow \$$

$$(1 \times 10^{15}\ \text{cm}^3)\left(\frac{1\ \text{L}}{1000\ \text{cm}^3}\right)\left(\frac{4 \times 10^{-4}\ \text{mg}}{\text{L}}\right)\left(\frac{1\ \text{g}}{1000\ \text{mg}}\right)\left(\frac{\$19.40}{\text{g}}\right) = \$8 \times 10^6$$

63. HNO_3 has 5 atoms/molecule

7 dozen = 84

(84 molecules)(5atoms/molecule) = 420 atoms

or $(7\ \text{dz})\left(\frac{12\ \text{molecules}}{\text{dz}}\right)\left(\frac{5\ \text{atoms}}{\text{molecule}}\right) = 420\ \text{atoms}$

64.

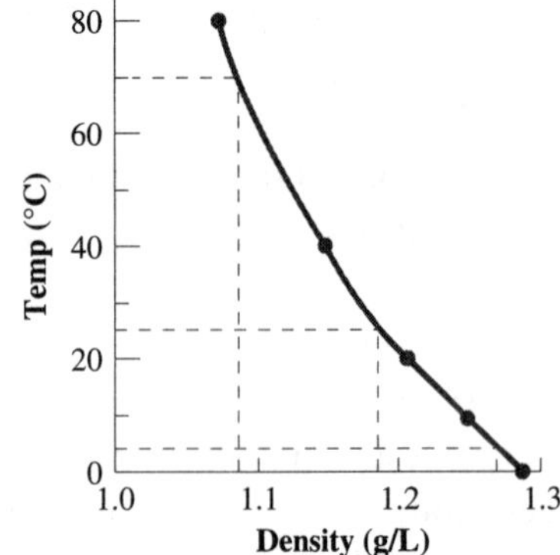

(a) As temperatures decreases, density increases.
(b) approximately 1.28 g/L 5°C
approximately 1.19 g/L 25°C
approximately 1.09 g/L 70°C

65. Each represents eight units of sulfur. In 8 S the atoms are separate and distinct. In S_8 the atoms are joined as a unit (molecule).

66. $Ca(H_2PO_4)_2$

$$(10 \text{ formula units})\left(\frac{4 \text{ atoms H}}{\text{formula unit}}\right) = 40 \text{ atoms H}$$

67. $C_{145}H_{293}O_{168}$

145 C
293 H
168 O
606 atoms/molecule

68. (a) magnesium, manganese, molybdenum, mendeleevium, mercury
(b) carbon, phosphorus, sulfur, selenium, iodine, astatine, boron
(c) sodium, potassium, iron, silver, tin, antimony

69. Add water to the mixture to dissolve the sugar. Filter the mixture to separate the sugar solution from the insoluble sand. Add another small amount of water to remove last traces of sugar. Filter. Allow the water to evaporate from the sugar solution to obtain crystals of sugar. Sand is the insoluble residue.

70. (a) NaCl (d) Fe_2S_3 (g) $C_6H_2O_6$
(b) H_2SO_4 (e) K_3PO_4 (h) C_2H_5OH
(c) K_2O (f) $Ca(CN)_2$ (i) $Cr(NO_3)_3$

71. Let X = grams sea water

$$\left(\frac{5.0 \times 10^{-8}\%\ I_2}{100}\right)(X) = 1.0\ g\ I_2$$

$$X = \frac{(1.0\ g\ I_2(100)}{(5.0 \times 10^{-8}\%\ I_2)} = 2.0 \times 10^9\ g\ \text{sea water}$$

$$(2.0 \times 10^9\ g)\left(\frac{1\ kg}{1000\ g}\right) = 2.0 \times 10^6\ kg\ \text{sea water}$$

72. (a) NH_4Cl (d) FeF_2
(b) H_2SO_4 (e) $Pb_3(PO_4)_2$
(c) MgI_2 (f) Al_2O_3

CHAPTER 4

PROPERTIES OF MATTER

1. Solid (melting point of acetic acid is 16.7°C)

2. Solid $\quad 102\ \text{K} = -171°\text{C} \quad$ melting point of chlorine is $-101.6°\text{C}$

3. Small bubbles appear at each electrode. Gas collects above the electrodes. The system now contains water and gas.

4. Water disappears. Gas appears above each electrode and as bubbles in solution.

5. Physical properties are characteristics which may be determined without altering the composition of the substance. Chemical properties describe the ability of a substance to form new substances by chemical reaction or decomposition.

6. A new substance is always formed during a chemical change, but never formed during physical changes.

7. Potential energy is the energy of position. By the position of an object, it has the potential of movement to a lower energy state. Kinetic energy is the energy matter possesses due to its motion.

8. (a) $118.0°\text{C} + 273 = 391.0\ \text{K}$
   (b) $(118.0°\text{C})(1.8) + 32 = 244.4°\text{F}$

9. Boiling water is a physical change so the chemical make-up of the water does not change. Therefore, the bubbles that you see in boiling water are . . . water!

10. Dissolving salt in water is a physical change. If the salt-water solution is boiled the water will evaporate away leaving the salt behind.

11. (a) physical
    (b) physical
    (c) physical
    (d) chemical
    (e) chemical
    (f) chemical

12. (a) chemical
    (b) physical
    (c) physical
    (d) chemical
    (e) chemical
    (f) physical

13. Although the appearance of the platinum wire changed during the heating, the original appearance was restored when the wire cooled. No change in the composition of the platinum could be detected.

14. The copper wire, like the platinum wire, changed to a glowing red color when heated (physical change). Upon cooling, the original appearance of the copper wire was not restored, but a new substance, black copper(II) oxide, had appeared (chemical change).

15. Reactants: copper, oxygen
    Product: copper(II) oxide

16. Reactant: water
    Product: hydrogen, oxygen

17. (a) potential energy
    (b) potential energy
    (c) kinetic energy
    (d) kinetic energy
    (e) kinetic energy

18. (a) potential energy
    (b) potential energy
    (c) kinetic energy
    (d) kinetic energy
    (e) potential energy

19. The kinetic energy is converted to thermal energy (heat), chiefly in the brake system, and eventually dissipated into the atmosphere.

20. The transformation of kinetic energy to thermal energy (heat) is responsible for the fiery reentry of a space vehicle.

21. (a) +
    (b) −
    (c) +
    (d) +
    (e) −

22. (a) +
    (b) −
    (c) −
    (d) +
    (e) −

23. $$\begin{aligned} E &= (m)(\text{specific heat})(\Delta t) \\ &= (75\text{ g})(4.184\text{ J/g°C})(70.0°\text{C} - 20.0°\text{C}) \\ &= 1.6 \times 10^4\text{ J} \end{aligned}$$

24. $$\begin{aligned} E &= (m)(\text{specific heat})(\Delta t) \\ &= (65\text{ g})(0.473\text{ J/g°C})(95°\text{C} - 25°\text{C}) \\ &= 2.2 \times 10^3\text{ J} \end{aligned}$$

25. $E = (m)(\text{specific heat})(\Delta t)$; change kJ to J

$$\text{specific heat} = \frac{E}{m(\Delta t)} = \frac{5.866 \times 10^3\ \text{J}}{(250.0\ \text{g})(100.0°\text{C} - 22°\text{C})} = 0.30\ \text{J/g°C}$$

26. $E = (m)(\text{specific heat})(\Delta t)$; change kg to g and kJ to J

$$1.00\ \text{kg} = 1.00 \times 10^3\ \text{g} \qquad (30.7\ \text{kJ})\left(\frac{1000\ \text{J}}{\text{kJ}}\right) = 3.07 \times 10^4\ \text{J}$$

$$\text{specific heat} = \frac{E}{m(\Delta t)} = \frac{3.07 \times 10^4\ \text{J}}{(1.00 \times 10^3\ \text{g})(630.0°\text{C} - 20.0°\text{C})} = 5.03 \times 10^{-2}\ \text{J/g°C}$$

27. heat lost by gold = heat gained by water $\quad x$ = final temperature

$$(m)(\text{specific heat})(\Delta t) = (m)(\text{specific heat})(\Delta t)$$
$$(325\ \text{g})(0.131\ \text{J/g°C})(427°\text{C} - x) = (200.0\ \text{g})(4.184\ \text{J/g°C})(x - 22.0°\text{C})$$
$$18180\ \text{J} - 42.575x\ \text{J/°C} = 836.8\ x\ \text{J/°C} - 18410\ \text{J}$$
$$18180\ \text{J} + 18410\ \text{J} = 836.8x + 42.575x$$
$$36590\ \text{J} = 879.4x\ \text{J/°C}$$
$$41.6°\text{C} = x$$

28. heat lost by iron = heat gained by water

(x = final temperature; Δt = change in temperature)

$$m = Vd = (2.0\ \text{L})\left(\frac{1000\ \text{mL}}{1\ \text{L}}\right)\left(1.0\frac{\text{g}}{\text{mL}}\right) = 2.0 \times 10^3\ \text{g}\ H_2O$$

$$(m)(\text{specific heat})(\Delta t) = (m)(\text{specific heat})(\Delta t)$$
$$(500.0\ \text{g})(0.473\ \text{J/g°C})(212°\text{C} - x) = (2.0 \times 10^3\ \text{g})(4.184\ \text{J/g°C})(x - 24.0°\text{C})$$
$$50138\ \text{J} - 236.5x\ \text{J/°C} = 8368x\ \text{J/°C} - 200832\ \text{J}$$
$$250970\ \text{J} = 8604.5x\ \frac{\text{J}}{°\text{C}}$$
$$x = 29°\text{C}$$
$$\Delta t = 29°\text{C} - 24°\text{C} = 5°\text{C}$$

29. When wood burns carbon dioxide is released into the air. If the wood has been chemically treated, those chemicals may be released as gases also. Therefore, the mass of the wood before burning will not equal the mass of the wood after burning. However, the law of conservation of mass still holds true. Some of the original mass of the wood is converted to gases that are released into the atmosphere.

30. heat lost by metal = heat gained by water + heat lost to environment
heat gained by water:

$$\begin{aligned} E &= (m)(\text{specific heat})(\Delta t) \\ &= (425\ g)(4.184\ J/g°C)(38.4°C - 26.0°C) \\ &= 22049.68\ J \end{aligned}$$

heat lost to environment: E = 1200 J
heat lost by metal = 22049.68 J + 1200 J = 23249.68 J
specific heat capacity of the metal:

$$\begin{aligned} \text{specific heat} &= E/(m \times \Delta t) \\ &= 23249.68\ J/(250.\ g)(130.°C - 38.4°C) \\ &= 1.015269869\ J/g°C = 1.02\ J/g°C \end{aligned}$$

(The final temperature of the metal is equal to the final temperature of the water.)

31. According to the law of conservation of mass the amount of nitrogen dioxide formed should equal the total amount of oxygen and nitrogen initially reacted. So, the amount of nitrogen that reacted = amount of nitrogen dioxide formed − amount of oxygen reacted.
mass of N_2 = 56 g − 18 g = 38 g

32. $$\begin{aligned} E &= (m)(\text{specific heat})(\Delta t) \\ &= (250.\ g)(0.096\ cal/g°C)(150.0°C - 24°C) \\ &= 3.0 \times 10^3\ cal \end{aligned}$$

33. E = (m)(specific heat)(Δt) change kJ to J x = final temperature

$4.00 \times 10^4\ J = (500.0\ g)(4.184\ J/g°C)(x - 10.0°C)$

$4.00 \times 10^4\ J = 2092x\ J/°C - 20920\ J$

$60{,}920\ J = 2092x\ J/°C$

$60{,}920°C = 2092x$

$29.1°C = x$

34. E = (m)(specific heat)(Δt)
heat lost by coal = heat gained by water x = mass of coal in g

$(5500\ cal/g)x = (500.0\ g)(1.00\ cal/g°C)(90.0°C - 20.0°C) = 35{,}000\ cal$

$$x = \frac{35{,}000\ cal}{5500\ cal/g} = 6.36\ g\ coal$$

35. (7000. cal)(4.184 J/cal) = 29290 J
heat lost by coal = heat gained by water x = mass of coal in g

$4.0\ L\ H_2O = 4.0 \times 10^3\ g\ H_2O$

$$\left(2.929 \times 10^4 \frac{J}{g}\right)x = (4.0 \times 10^3\ g)\left(4.184 \frac{J}{g°C}\right)(100.0°C - 20.0°C)$$

$$\left(2.929 \times 10^4 \frac{J}{g}\right)x = 1.3 \times 10^6\ J$$

$x = 44$ g coal

36. (a) E = (m)(specific heat)(Δt)
(100.0 g)(0.0921 cal/g°C)(100.0°C − 10.0°C) = 829 cal to heat Cu
(b) let x = temperature of A1 after adding 829 cal
829 cal = (100.0 g)(0.215 cal/g°C)(x − 10.0°C)
829 cal = (21.5 cal/°C)x − 215 cal
x = 48.6°C (final temperature for aluminum)
Therefore the copper gets hotter since it ended up at 100.0°C.
Note: You can figure this out without calculation if you consider the specific heats of the metals. Since the specific heat of copper is much less than aluminum the copper heats more easily.

37. heat lost by iron = heat gained by water $\qquad$ x = initial temperature of iron

$$(m)(\text{specific heat})\Delta(t) = (m)(\text{specific heat})(\Delta t)$$

$$(500.0\ g)(0.473\ J/g°C)(x - 90.0°C) = (400.\ g)(4.184\ J/g°C)(90.0°C - 10.0°C)$$

$$\left(237 \frac{J}{°C}\right)x - 2.13 \times 10^4\ J = 1.34 \times 10^3\ J$$

$$\left(237 \frac{J}{°C}\right)x = 1.55 \times 10^5\ J$$

$$x = \frac{1.55 \times 10^5\ J°C}{237\ J}$$

$$x = 654°C$$

38. heat lost by metal = heat gained by water $\qquad$ x = specific heat of metal

$$(m)(\text{specific heat})(\Delta t) = (m)(\text{specific heat})(\Delta t)$$

$$(20.0\ g)(x)(203.°C - 29.0°C) = (100.0\ g)(4.184\ J/g°C)(29.0°C - 25.0°C)$$

$$4060x\ g°C - 580x\ g°C = 12134\ J - 10460\ J$$

$$(3480\ g°C)x = 1674\ J$$

$$x = 0.481\ J/g°C$$

39. heat lost = heat gained $\qquad$ x = final temperature
(m)(specific heat)(Δt) = (m)(specific heat)(Δt) $\qquad$ (specific heats are the same)
Let x = final temperature

$$(10.0\text{ g})\left(\cancel{4.184}\frac{\text{J}}{\text{g°C}}\right)(50.0°\text{C} - x) = (50.0\text{ g})\left(\cancel{4.184}\frac{\text{J}}{\text{g°C}}\right)(x - 10.0°\text{C})$$

$$500.\text{g°C} - 10.0\text{ g}x = 50.0\text{ g}x - 500\text{ g°C}$$

$$1.00 \times 10^3\text{ g°C} = 60.0\text{ g}x$$

$$x = \frac{1.00 \times 10^3\text{ g°C}}{60.0\text{ g}} = 16.7°\text{C}$$

40. Specific heats for the metals are Fe: 0.473 J/g°C; Cu: 0.385 J/g°C; Al: 0.900 J/g°C. The metal with the lowest specific heat will warm most quickly, therefore, the copper pan heats fastest, and fries the egg fastest.

41. In order for the water to boil both the pan and water must reach 100.0°C. Specific heat for copper is 0.385 J/g°C.

$$(300.0\text{ g})\left(0.385\frac{\text{J}}{\text{g°C}}\right)(100.°\text{C} - 25°\text{C}) + (800.0\text{ g})\left(4.184\frac{\text{J}}{\text{g°C}}\right)(100.°\text{C} - 25°\text{C})$$

$$= 8.7 \times 10^3\text{ J} + 2.5 \times 10^5\text{ J}$$

$$= 2.6 \times 10^5\text{ J needed to heat the pan and water to 100°C}$$

$$(2.6 \times 10^5\text{ J})\left(\frac{1\text{ s}}{628\text{ cal}}\right) = 414\text{ s} = 6.9\text{ min} = 6\text{ min} + 54\text{ s}$$

The water will boil at 6:06 and 54 s p.m.

42. Heat is transferred from the molecules of water on the surface to the air above them. As you blow you move the warmed air molecules away from the surface replacing them with cooler ones which are warmed by the coffee and cool it in a repeating cycle. Inserting a spoon into hot coffee cools the coffee by heat transfer as well. Heat is transferred from the coffee to the spoon lowering the temperature of the coffee and raising the temperature of the spoon.

43. The potatoes will cook at the same rate whether the water boils vigorously or slowly. Once the boiling point is reached the water temperature remains constant. The energy available is the same so the cooking time should be equal.

44. $(250\text{ mL})(0.04) = 10\text{ mL fat}$
$(10\text{ mL})(0.8\text{ g/mL}) = 8\text{ g fat in a glass of milk}$

45. mercury + sulfur $\rightarrow$ compound
The mercury and sulfur react to form a compound since the properties of the product are different from the properties of either reactant.

$$(100.0\text{ mL})\left(13.6\frac{\text{g}}{\text{mL}}\right) = 1.36 \times 10^3 \text{ g mercury}$$

mercury	+	sulfur	→	compound
1360 g	+	100.0 g		1460 g

This supports the Law of Conservation of Matter since the mass of the product is equal to the mass of the reactants.

46. According to the law of conservation of energy the amount of potential energy the ball has initially should equal the amount of kinetic energy it has while in motion. If the hill that the ball rolled down was frictionless then the ball should role up the other side until it has reached the same level as where it started. However, no hill is really frictionless. Therefore, some of the ball's potential energy is converted to kinetic energy of motion and some is lost by way of heat friction between the ball and the surface of the hill.

CHAPTER 5

EARLY ATOMIC THEORY AND STRUCTURE

1.

	Element	Atomic number
(a)	copper	29
(b)	nitrogen	7
(c)	phosphorus	15
(d)	radium	88
(e)	zinc	30

2. The neutron is about 1840 times heavier than an electron.

3.

Particle	charge	mass
proton	+1	1 amu
neutron	0	1 amu
electron	−1	0

4. An atom is electrically neutral, containing equal numbers of protons and electrons. An ion has a charge resulting from an imbalance between the numbers of protons and electrons.

5. Isotopic notation $^{A}_{Z}X$

   Z represents the atomic number
   A represents the mass number

6. Isotopes contain the same number of protons and the same number of electrons. Isotopes have different numbers of neutrons and thus different atomic masses.

7. Each peak on the chromatogram represents one isotope of the atom being studied. In the example given in Figure 5.9 for copper, there are two isotopes. The numbers along the *x*-axis of the chromatogram give the mass in atomic mass units. For copper, there is a peak at approximately 63 atomic mass units and one at approximately 65 atomic mass units. The *y*-axis gives the relative abundance (percentages) of each isotope. For copper, there is approximately 69% of the 63 amu isotope and approximately 31% of the 65 amu isotope. Using the mass and the relative abundance of each copper isotope, the average atomic mass of copper can be calculated.

8. Pure vanilla extract is obtained from vanilla beans while imitation vanilla flavoring is made from lignin. Lignin is a waste product from the wood pulp industry that is

converted into vanillin, the same compound that is extracted from vanilla beans. There are two isotopes of carbon that are found in organic compounds, carbon-12 and carbon-13. A chemist can tell real vanilla flavoring from imitation vanilla flavoring by determining the ratio of carbon-12 to carbon-13 in the vanilla. The ratio of the two isotopes is different in the two types of vanilla flavoring.

9. The formula for water is H_2O. There is one atom of oxygen for every two atoms of hydrogen. The molar mass of oxygen is 16.00 g and the molar mass of hydrogen is 1.008 g. For H_2O the mass of two hydrogen atoms is 2.016 g and the mass of one oxygen atom is 16.00 g. The ratio of hydrogen to oxygen is approximately 2 : 16 or 1 : 8. Therefore, there is 1 gram of hydrogen for every 8 grams of oxygen.

10. The formula for hydrogen peroxide is H_2O_2. There are two atoms of oxygen for every two atoms of hydrogen. The molar mass of oxygen is 16.00 g and the molar mass of hydrogen is 1.01 g. For hydrogen peroxide the total mass of hydrogen is 2.016 g and the total mass of oxygen is 32.00 g for a ratio of hydrogen to oxygen of approximately 2 : 32 or 1 : 16. Therefore, there is 1 gram of hydrogen for every 16 grams of oxygen.

11. Gold nuclei are very massive (compared to an alpha particle) and have a large positive charge. As the positive alpha particles approach the atom, some are deflected by this positive charge. Alpha particles approaching a gold nucleus directly are deflected backwards by the massive positive nucleus.

12. (a) The nucleus of the atom contains most of the mass since only a collision with a very dense, massive object would cause an alpha particle to be deflected back towards the source.
    (b) The deflection of the positive alpha particles from their initial flight indicates the nucleus of the atom is also positively charged.
    (c) Most alpha particles pass through the gold foil undeflected leading to the conclusion that the atom is mostly empty space.

13. In the atom, protons and neutrons are found within the nucleus. Electrons occupy the remaining space within the atom outside the nucleus.

14. The nucleus of an atom contains nearly all of its mass.

15. (a) Dalton contributed the concept that each element is composed of atoms which are unique, and can combine in ratios of small whole numbers.
    (b) Thomson discovered the electron, determined its properties, and found that the mass of a proton is 1840 times the mass of the electron. He developed the Thomson model of the atom.
    (c) Rutherford devised the model of a nuclear atom with the positive charge and mass concentrated in the nucleus. Most of the atom is empty space.

16. Electrons: Dalton - electrons are not part of his model
Thomson - electrons are scattered throughout the positive mass of matter in the atom
Rutherford - electrons are located out in space away from the central positive nucleus

Positive matter: Dalton - no positive matter in his model
Thomson - positive matter is distributed throughout the atom
Rutherford - positive matter is concentrated in a small central nucleus

17. Atomic masses are not whole numbers because:

(a) the neutron and proton do not have identical masses and neither is exactly 1 amu.

(b) most elements exist in nature as a mixture of isotopes with different atomic masses due to different numbers of neutrons. The atomic mass given in the periodic table is the average mass of all these isotopes.

18. The isotope of C with a mass of 12 is an exact number by definition. The mass of other isotopes such as $^{63}_{29}Cu$ will not be an exact number for reasons given in Exercise 17.

19. The isotopes of hydrogen are protium, deuterium, and tritium.

20. All three isotopes of hydrogen have the same number of protons (1) and electrons (1). They differ in the number of neutrons (0, 1, and 2).

21. $^{52}_{24}Cr$ chromium-52

22. (a) $201 - 121 = 80$ protons; electrical charge of the nucleus is +80.
(b) Hg, mercury

23. All six isotopes have 20 protons and 20 electrons. The number of neutrons are

Isotope mass number	Neutrons
40	20
42	22
43	23
44	24
46	26
48	28

24. The most abundant Ca isotope has a mass number of 40. This is certain because 40 is about the average of all the isotopes and has the lowest mass number on the list. An arithmetic average would be between 40 and 48. Since the atomic mass given on the periodic table is 40.08, there must be only small amounts of the other isotopes.

25. (a) $^{55}_{26}Fe$ (b) $^{26}_{12}Mg$ (c) $^{6}_{3}Li$ (d) $^{188}_{79}Au$

26. (a) $^{59}_{27}Co$ Nucleus contains 27 protons and 32 neutrons
 (b) $^{31}_{15}P$ Nucleus contains 15 protons and 16 neutrons
 (c) $^{184}_{74}W$ Nucleus contains 74 protons and 110 neutrons
 (d) $^{235}_{92}U$ Nucleus contains 92 protons and 143 neutrons

27. For each isotope:
 (%)(amu) = that portion of the average atomic mass for that isotope.
 Add together to obtain the average atomic mass.
 (0.2360)(205.9745 amu) + (0.2260)(206.9759 amu) +
 (0.5230)(207.9766 amu) + (0.01480)(203.973 amu)
 = 48.61 amu + 46.78 amu + 108.8 amu + 3.019 amu
 = 207.2 amu = average atomic mass of Pb

28. For each isotope:
 (%)(amu) = that portion of the average atomic mass for that isotope.
 (0.7899)(23.985 amu) + (0.1000)(24.986) amu) + (0.1101)(25.983 amu)
 = 18.95 amu + 2.500 amu + 2.861 amu
 = 24.31 amu = average atomic mass Mg

29. (0.604)(68.9257 amu) + (1.00 − 0.604)(70.9249 amu)
 = 41.6 amu + 28.1 amu
 = 69.7 amu = average atomic mass
 The element is gallium (see periodic table).

30. (0.3000)(6.015 amu) + (0.7000)(7.016 amu)
 = 1.805 amu + 4.911 amu = 6.716 amu = average atomic mass of Li sample

31. $V_{sphere} = \frac{4}{3}\pi r^3$ r_A = radius of atom, r_N = radius of nucleus

$$\frac{V_{atom}}{V_{nucleus}} = \frac{\frac{4}{3}\pi r_A^3}{\frac{4}{3}\pi r_N^3} = \frac{r_A^3}{r_N^3} = \frac{(1.0 \times 10^{-8})^3}{(1.0 \times 10^{-13})^3} = 1.0 \times 10^{15} : 1.0$$

(ratio of atomic volume to nuclear volume)

32. $$\frac{3.0 \times 10^{-8}\ \text{cm}}{2.0 \times 10^{-13}\ \text{cm}} = 1.5 \times 10^5 : 1.0$$ (ratio of the diameter of an Al atom to its nucleus diameter)

33. (a) In Rutherford's experiment the majority of alpha particles passed through the gold foil without deflection. This shows that the atom is mostly empty space and the nucleus is very small.

(b) In Thomson's experiments with the cathode ray tube, rays were observed coming from both the anode and the cathode.

(c) In Rutherford's experiment an alpha particle was occasionally dramatically deflected by the nucleus of a gold atom. The direction of deflection showed the nucleus to be positive. Positive charges repel each other.

34. (a) These two atoms are isotopes.

(b) These two atoms follow one another on the periodic table. The atoms have about the same mass.

35. The nucleus of the atoms will have 2 less protons (lowering the nuclear charge by 2) and 2 less neutrons. The nuclear mass will be reduced by 4 amu.

36. $\frac{1.5\text{ cm}}{0.77 \times 10^{-8}\text{ cm}} = 1.9 \times 10^8 : 1.0$ $\quad (1.9 \times 10^8$ enlargement)

37. The properties of an element are related to the number of protons and electrons. If the number of neutrons differs, isotopes result. Isotopes of the same element are still the same element even though the nuclear composition of the atoms are different.

38. ^{210}Bi has $210 - 83 = 127$ neutrons $\rightarrow$ largest number of neutrons/atom
^{210}Po has $210 - 84 = 126$ neutrons
^{210}At has $210 - 85 = 125$ neutrons
^{211}At has $211 - 85 = 126$ neutrons

39. percent of sample $^{60}\text{Q} = x$
percent of sample $^{63}\text{Q} = 1 - x$

$$\begin{aligned}(x)(60.\text{ amu}) + (1 - x)(63\text{ amu}) &= 61.5\text{ amu}\\ 60.x\text{ amu} + 63\text{ amu} - 63x\text{ amu} &= 61.5\text{ amu}\\ 63\text{ amu} - 61.5\text{ amu} &= 63x\text{ amu} - 60x\text{ amu}\\ 1.5 &= 3x\\ 0.50 &= x\\ ^{60}\text{Q} &= 50\%\\ ^{63}\text{Q} &= 50\%\end{aligned}$$

On a simpler basis: The average atomic mass (61.5 amu) is exactly half way between the masses of the two isotopes. Therefore, the relative percentages are 50% for each isotope.

40. Compare the mass of the unknown atom to the mass of carbon-12 $(1.9927 \times 10^{-23}\text{ g})$

$$\left(\frac{2.18 \times 10^{-22}\text{ g}}{1.9927 \times 10^{-23}\text{ g C}}\right)(12.0\text{ g C}) = 131\text{ g} \quad \text{(atomic mass of unknown element)}$$

41. $(40.0\text{ g})\left(\dfrac{1\text{ atom}}{6.63 \times 10^{-24}\text{ g}}\right) = 6.03 \times 10^{24}\text{ atoms Ar}$

42.

	protons	neutrons	electrons
He	2	2	2
C	6	6	6
N	7	7	7
O	8	8	8
Ne	10	10	10
Mg	12	12	12
Si	14	14	14
S	16	16	16
Ca	20	20	20

43.

	Atomic Number	Mass Number	Symbol	Protons	Neutrons
(a)	8	16	O	8	8
(b)	28	58	Ni	28	30
(c)	80	199	Hg	80	119

44.

	Element	Symbol	Atomic #	Protons	Neutrons	Electrons
(a)	platinum	^{195}Pt	78	78	117	78
(b)	phosphorus	^{30}P	15	15	15	15
(c)	iodine	^{127}I	53	53	74	53
(d)	krypton	^{84}Kr	36	36	48	36
(e)	selenium	^{79}Se	34	34	45	34
(f)	calcium	^{40}Ca	20	20	20	20

45.

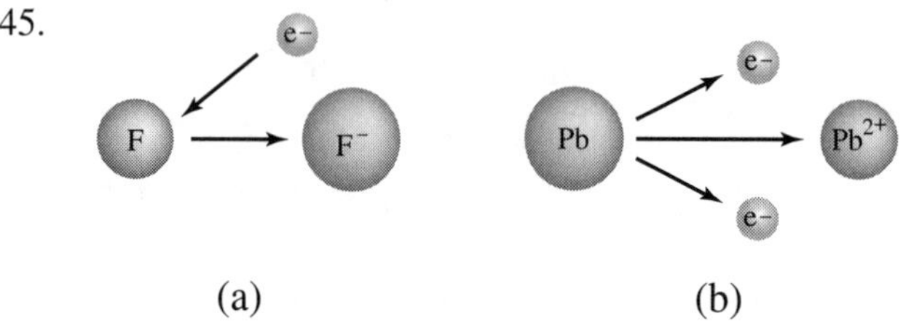

(a) (b)

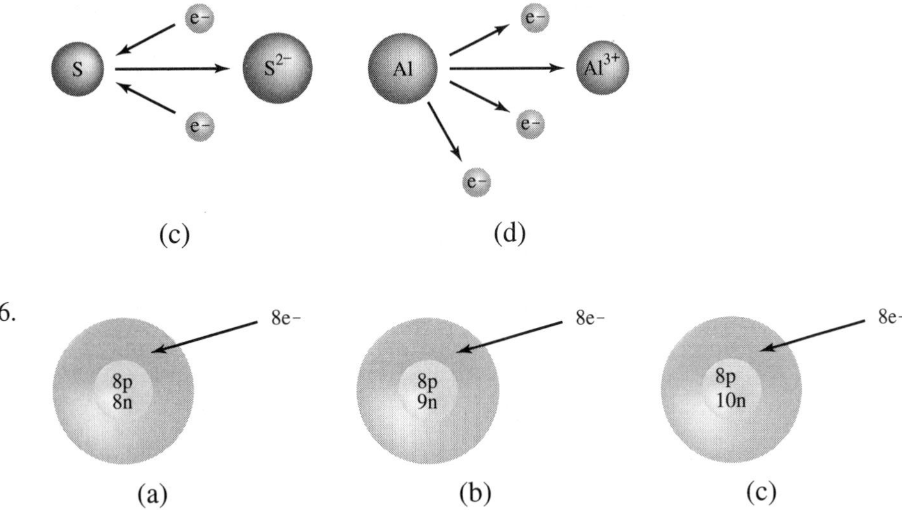

46. (a), (b), (c) as shown above.

47. The mass of one electron is 9.110×10^{-28} grams.

(a) Iron has 26 electrons. $\dfrac{(26)(9.110 \times 10^{-28}\text{ g})}{9.274 \times 10^{-23}\text{ g}}(100) = 0.02554\%$

(b) Nitrogen has 7 electrons. $\dfrac{(7)(9.110 \times 10^{-28}\text{ g})}{2.326 \times 10^{-23}\text{ g}}(100) = 0.02742\%$

(c) Carbon has 6 electrons. $\dfrac{(6)(9.110 \times 10^{-28}\text{ g})}{1.994 \times 10^{-23}\text{ g}}(100) = 0.02741\%$

(d) Potassium has 19 electrons. $\dfrac{(19)(9.110 \times 10^{-28}\text{ g})}{6.493 \times 10^{-23}\text{ g}}(100) = 0.02666\%$

48. The mass of one proton is 1.673×10^{-24} grams.

(a) Sodium has 11 protons. $\dfrac{(11)(1.673 \times 10^{-24}\text{ g})}{3.818 \times 10^{-23}\text{ g}}(100) = 48.20\%$

(b) Oxygen has 8 protons. $\dfrac{(8)(1.673 \times 10^{-24}\text{ g})}{2.657 \times 10^{-23}\text{ g}}(100) = 50.37\%$

(c) Mercury has 80 protons. $\dfrac{(80)(1.673 \times 10^{-24}\text{ g})}{3.331 \times 10^{-22}\text{ g}}(100) = 40.18\%$

(d) Fluorine has 9 protons. $\dfrac{(9)(1.673 \times 10^{-24}\text{ g})}{3.155 \times 10^{-23}\text{ g}}(100) = 47.72\%$

49. The electron region is the area around the nucleus where electrons are most likely to be located.

Challenge Exercises

50.

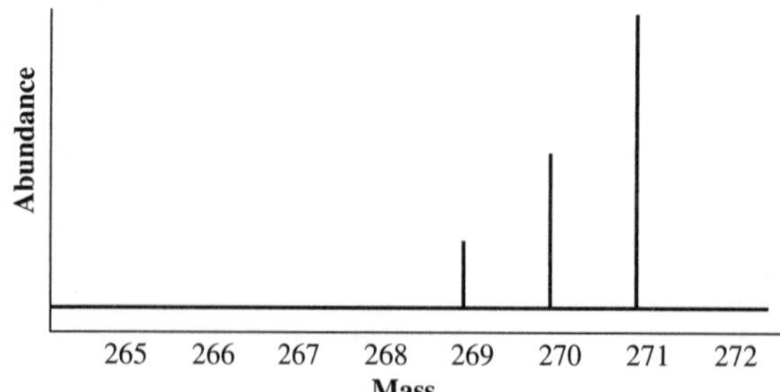

average atomic mass:

$$\begin{aligned} (0.1081)(269.14\text{ amu}) &= 29.09\text{ amu} \\ (0.3407)(270.51\text{ amu}) &= 92.16\text{ amu} \\ (0.5512)(271.23\text{ amu}) &= \underline{149.50}\text{ amu} \\ \text{Total} &= 270.75\text{ amu} \end{aligned}$$

An atomic mass of 270.75 amu would come somewhere after Rutherfordium (mass = 261.1 amu). So, the atomic number of this new element would be greater than 105.

CHAPTER 6

NOMENCLATURE OF INORGANIC COMPOUNDS

1. (a) $NaClO_3$
   (b) H_2SO_4
   (c) $Sn(C_2H_3O_2)_2$
   (d) Cu_2O
   (e) $Zn(HCO_3)_2$
   (f) $Fe_2(CO_3)_3$

2. No, if elements combine in a one-to-one ratio the charges on their ions must be equal and opposite in sign. They could be +1, −1, or +2, −2 or +3, −3 etc.

3. (a)

HBrO	hypobromous acid
$HBrO_2$	bromous acid
$HBrO_3$	bromic acid
$HBrO_4$	perbromic acid

   (b)

HIO	hypoiodous acid
HIO_2	iodous acid
HIO_3	iodic acid
HIO_4	periodic acid

4. The system for naming binary compounds composed of two nonmetals uses the stem of the second element in the formula plus the suffix -ide. A prefix is attached to each element indicating the number of atoms of that element in the formula. Thus N_2O_5 is named dinitrogen pentoxide and PCl_3 is named phosphorus trichloride.

5. Chromium(III) compounds

   (a) $Cr(OH)_3$
   (b) $Cr(NO_3)_3$
   (c) $Cr(NO_2)_3$
   (d) $Cr(HCO_3)_3$
   (e) $Cr_2(CO_3)_3$
   (f) $Cr_2(Cr_2O_7)_3$
   (g) $CrPO_4$
   (h) $Cr_2(C_2O_4)_3$
   (i) Cr_2O_3
   (j) CrF_3

6. Magnesium forms one series of compounds in which the cation is Mg^{2+}. Thus the name for $MgCl_2$ (magnesium chloride) does not need to be distinguished from any other compound. Copper forms two series of compounds in which the copper ion is Cu^+ and Cu^{2+}. Thus the name copper chloride does not indicate which compound is in question. Therefore, $CuCl_2$ is called copper(II) chloride to indicate that the compound contains the Cu^{2+} ion.

7. (a) Metals are located in groups IA (except for hydrogen), IIA, IIIB–IIB and atomic numbers 13, 31, 49, 50, 81, 82, 83, lanthanides and actinides.
   (b) The nonmetals include hydrogen, group VIIA, the noble gases and atomic numbers 6–9, 15–17, 34, 35, 53 and 85.
   (c) The transition metals are located in groups IIIB–IIB in the center of the periodic table. The lanthanides and actinides are located below the main body of the periodic table.

8. water (H_2O)–dihydrogen monoxide
ammonia (NH_3) - nitrogen trihydride

9. Formulas of compounds.

(a)	Na and I	NaI	(d)	K and S	K_2S
(b)	Ba and F	BaF_2	(e)	Cs and Cl	CsCl
(c)	Al and O	Al_2O_3	(f)	Sr and Br	$SrBr_2$

10. Formulas of compounds.

(a)	Ba and O	BaO	(d)	Be and Br	$BeBr_2$
(b)	H and S	H_2S	(e)	Li and Si	Li_4Si
(c)	Al and Cl	$AlCl_3$	(f)	Mg and P	Mg_3P_2

11.

sodium	Na^+	cobalt(II)	Co^{2+}
magnesium	Mg^{2+}	barium	Ba^{2+}
aluminum	Al^{3+}	hydrogen	H^+
copper(II)	Cu^{2+}	mercury(II)	Hg^{2+}
iron(II)	Fe^{2+}	tin(II)	Sn^{2+}
iron(III)	Fe^{3+}	chromium(III)	Cr^{3+}
lead(II)	Pb^{2+}	tin(IV)	Sn^{4+}
silver	Ag^+	manganese(II)	Mn^{2+}
		bismuth(III)	Bi^{3+}

12.

chloride	Cl^-	hydrogen sulfate	HSO_4^-
bromide	Br^-	hydrogen sulfite	HSO_3^-
fluoride	F^-	chromate	CrO_4^{2-}
iodide	I^-	carbonate	CO_3^{2-}
cyanide	CN^-	hydrogen carbonate	HCO_3^-
oxide	O^{2-}	acetate	$C_2H_3O_2^-$
hydroxide	OH^-	chlorate	ClO_3^-
sulfide	S^{2-}	permanganate	MnO_4^-
sulfate	SO_4^{2-}	oxalate	$C_2O_4^{2-}$

Paired Exercises

13. (a) dinitrogen monoxide
(b) calcium carbonate
(c) aluminum oxide
(d) lead (II) sulfide
(e) hydrochloric acid
(f) sodium chloride

14. (a) calcium hydroxide
(b) sodium nitrate
(c) sulfur
(d) sodium hydrogen carbonate
(e) iron (II) sulfide
(f) potassium carbonate

15.

Ion	Br^-	O^{2-}	NO_3^-	PO_4^{3-}	CO_3^{2-}
K^+	KBr	K_2O	KNO_3	K_3PO_4	K_2CO_3
Mg^{2+}	$MgBr_2$	MgO	$Mg(NO_3)_2$	$Mg_3(PO_4)_2$	$MgCO_3$
Al^{3+}	$AlBr_3$	Al_2O_3	$Al(NO_3)_3$	$AlPO_4$	$Al_2(CO_3)_3$
Zn^{2+}	$ZnBr_2$	ZnO	$Zn(NO_3)_2$	$Zn_3(PO_4)_2$	$ZnCO_3$
H^+	HBr	H_2O	HNO_3	H_3PO_4	H_2CO_3

16.

Ion	SO_4^{2-}	Cl^-	AsO_4^{3-}	$C_2H_3O_2^-$	CrO_4^{2-}
NH_4^+	$(NH_4)_2SO_4$	NH_4Cl	$(NH_4)_3AsO_4$	$NH_4C_2H_3O_2$	$(NH_4)_2CrO_4$
Ca^{2+}	$CaSO_4$	$CaCl_2$	$Ca_3(AsO_4)_2$	$Ca(C_2H_3O_2)_2$	$CaCrO_4$
Fe^{3+}	$Fe_2(SO_4)_3$	$FeCl_3$	$FeAsO_4$	$Fe(C_2H_3O_2)_3$	$Fe_2(CrO_4)_3$
Ag^+	Ag_2SO_4	$AgCl$	Ag_3AsO_4	$AgC_2H_3O_2$	Ag_2CrO_4
Cu^{2+}	$CuSO_4$	$CuCl_2$	$Cu_3(AsO_4)_2$	$Cu(C_2H_3O_2)_2$	$CuCrO_4$

17. Nonmetal binary compound formulas

(a) carbon monoxide, CO
(b) sulfur trioxide, SO_3
(c) carbon tetrabromide, CBr_4
(d) Phosphorus trichloride, PCl_3
(e) nitrogen dioxide, NO_2
(f) dinitrogen pentoxide, N_2O_5
(g) iodine monobromide, IBr
(h) silicon tetrachloride, $SiCl_4$
(i) phosphorus pentiodide, PI_5
(j) diboron trioxide, B_2O_3

18. Naming binary nonmetal compounds:

(a) CO_2 carbon dioxide
(b) N_2O dinitrogen oxide
(c) PCl_5 phosphorus pentachloride
(d) CCl_4 carbon tetrachloride
(e) SO_2 sulfur dioxide
(f) N_2O_4 dinitrogen tetroxide
(g) P_2O_5 diphosphorus pentoxide
(h) OF_2 oxygen difluoride
(i) NF_3 nitrogen trifluoride
(j) CS_2 carbon disulfide

19. (a) sodium nitrate, $NaNO_3$
(b) magnesium fluoride, MgF_2
(c) barium hydroxide, $Ba(OH)_2$
(d) ammonium sulfate, $(NH_4)_2SO_4$
(e) silver carbonate, Ag_2CO_3
(f) calcium phosphate, $Ca_3(PO_4)_2$
(g) potassium nitrite, KNO_2
(h) strontium oxide, SrO

20. (a) K_2O, potassium oxide
(b) NH_4Br, ammonium bromide
(c) CaI_2, calcium iodide
(d) $BaCO_3$, barium carbonate
(e) Na_3PO_4, sodium phosphate
(f) Al_2O_3, aluminum oxide
(g) $Zn(NO_3)_2$, zinc nitrate
(h) Ag_2SO_4, silver sulfate

21. (a) $CuCl_2$ copper(II) chloride
(b) $CuBr$ copper(I) bromide
(c) $Fe(NO_3)_2$ iron(II) nitrate
(d) $FeCl_3$ iron(III) chloride
(e) SnF_2 tin(II) fluoride
(f) $HgCO_3$ mercury(II) carbonate

22. Formulas:

(a) tin(IV) bromide $SnBr_4$
(b) copper(I) sulfate Cu_2SO_4
(c) iron(III) carbonate $Fe_2(CO_3)_3$
(d) mercury(II) nitrite $Hg(NO_2)_2$
(e) titanium(IV) sulfide TiS_2
(f) iron(II) acetate $Fe(C_2H_3O_2)_2$

23. Acid formulas:

(a) hydrochloric acid, HCl
(b) chloric acid, $HClO_3$
(c) nitric acid, HNO_3
(d) carbonic acid, H_2CO_3
(e) sulfurous acid, H_2SO_3
(f) phosphoric acid, H_3PO_4

24. Formulas of acids:

(a) acetic acid, $HC_2H_3O_2$
(b) hydrofluoric acid, HF
(c) hypochlorous acid, $HClO$
(d) boric acid, H_3BO_3
(e) nitrous acid, HNO_2
(f) hydrosulfuric acid, H_2S

25. Naming acids:

(a) HNO_2, nitrous acid
(b) H_2SO_4, sulfuric acid
(c) $H_2C_2O_4$, oxalic acid
(d) HBr, hydrobromic acid
(e) H_3PO_3, phosphorous acid
(f) $HC_2H_3O_2$, acetic acid
(g) HF, hydrofluoric acid
(h) $HBrO_3$, bromic acid

26. Naming acids:

(a) H_3PO_4, phosphoric acid
(b) H_2CO_3 carbonic acid
(c) HIO_3, iodic acid
(d) HCl, hydrochloric acid
(e) $HClO$, hypochlorous acid
(f) HNO_3, nitric acid
(g) HI, hydroiodic acid
(h) $HClO_4$ perchloric acid

27. Formulas for:

(a) silver sulfite Ag_2SO_3
(b) cobalt(II) bromide $CoBr_2$

(c)	tin(II) hydroxide	$Sn(OH)_2$
(d)	aluminum sulfate	$Al_2(SO_4)_3$
(e)	manganese(II) fluoride	MnF_2
(f)	ammonium carbonate	$(NH_4)_2CO_3$
(g)	chromium(III) oxide	Cr_2O_3
(h)	cupric chloride	$CuCl_2$
(i)	potassium permanganate	$KMnO_4$
(j)	barium nitrite	$Ba(NO_2)_2$
(k)	sodium peroxide	Na_2O_2
(l)	iron(II) sulfate	$FeSO_4$
(m)	potassium dichromate	$K_2Cr_2O_7$
(n)	bismuth(III) chromate	$Bi_2(CrO_4)_3$

28. Formulas for:

(a)	sodium chromate	Na_2CrO_4
(b)	magnesium hydride	MgH_2
(c)	nickel(II) acetate	$Ni(C_2H_3O_2)_2$
(d)	calcium chlorate	$Ca(ClO_3)_2$
(e)	lead(II) nitrate	$Pb(NO_3)_2$
(f)	potassium dihydrogen phosphate	KH_2PO_4
(g)	manganese(II) hydroxide	$Mn(OH)_2$
(h)	cobalt(II) hydrogen carbonate	$Co(HCO_3)_2$
(i)	sodium hypochlorite	$NaClO$
(j)	arsenic(V) carbonate	$As_2(CO_3)_5$
(k)	chromium(III) sulfite	$Cr_2(SO_3)_3$
(l)	antimony(III) sulfate	$Sb_2(SO_4)_3$
(m)	sodium oxalate	$Na_2C_2O_4$
(n)	potassium thiocyanate	KSCN

29.

	Formula	Name
(a)	$ZnSO_4$	zinc sulfate
(b)	$HgCl_2$	mercury(II) chloride
(c)	$CuCO_3$	copper(II) carbonate
(d)	$Cd(NO_3)_2$	cadmium nitrate
(e)	$Al(C_2H_3O_2)_3$	aluminum acetate
(f)	CoF_2	cobalt(II) fluoride
(g)	$Cr(ClO_3)_3$	chromium(III) chlorate
(h)	Ag_3PO_4	silver phosphate
(i)	NiS	nickel(II) sulfide
(j)	$BaCrO_4$	barium chromate

30.

	Formula	Name
(a)	$Ca(HSO_4)_2$	calcium hydrogen sulfate
(b)	$As_2(SO_3)_3$	arsenic(III) sulfite
(c)	$Sn(NO_2)_2$	tin(II) nitrite
(d)	$FeBr_3$	iron(III) bromide
(e)	$KHCO_3$	potassium hydrogen carbonate
(f)	$BiAsO_4$	bismuth(III) arsenate
(g)	$Fe(BrO_3)_2$	iron(II) bromate
(h)	$(NH_4)_2HPO_4$	ammonium monohydrogen phosphate
(i)	$NaClO$	sodium hypochlorite
(j)	$KMnO_4$	potassium permanganate

31. Formulas for:

(a) baking soda $NaHCO_3$
(b) lime CaO
(c) Epsom salts $MgSO_4 \cdot 7\,H_2O$
(d) muriatic acid HCl
(e) vinegar $HC_2H_3O_2$
(f) potash K_2CO_3
(g) lye $NaOH$

32. Formulas for:

(a) fool's gold FeS_2
(b) saltpeter $NaNO_3$
(c) limestone $CaCO_3$
(d) cane sugar $C_{12}H_{22}O_{11}$
(e) milk of magnesia $Mg(OH)_2$
(f) washing soda $Na_2CO_3 \cdot 10\,H_2O$
(g) grain alcohol C_2H_5OH

33. (a) $K \longrightarrow K^+ + e^-$
(b) $I + e^- \longrightarrow I^-$
(c) $Br + e^- \longrightarrow Br^-$
(d) $Fe \longrightarrow Fe^{2+} + 2e^-$
(e) $Ca \longrightarrow Ca^{2+} + 2e^-$
(f) $O + 2e^- \longrightarrow O^{2-}$

34. (a) cation
(b) anion
(c) anion
(d) cation
(e) cation
(f) anion

35. (a) not an acid, sodium chloride
(b) an acid, nitrous acid
(c) an acid, hydrofluoric acid
(d) an acid, phosphoric acid
(e) not an acid, lead(II) iodide
(f) not an acid, potassium hydride

Most often the formulas for acids begin with hydrogen.

36. Possibilities for A include: Al^{3+}, Fe^{3+}, Ti^{3+}, Cr^{3+}, As^{3+}, Bi^{3+} or Sb^{3+}
Example unbalanced reaction: $Al^{3+}(aq) + Br^{-}(aq) \longrightarrow AlBr_3(aq)$
The reactions for the other cations listed above are similar.
To form ABr_3 the A cation must have a +3 charge.

37. Possibilities for Q include: Be^{2+}, Mg^{2+}, Ca^{2+}, Sr^{2+} or Ba^{2+}
Example unbalanced reaction: $Be^{2+}(aq) + PO_4^{3-}(aq) \longrightarrow Be_3(PO_4)_2(aq)$
Example unbalanced reaction: $Be^{2+}(aq) + PO_4^{3-}(aq) \longrightarrow Be_3(PO_4)_2(aq)$
The reactions for the other cations listed above are similar.

38. (a) Name the cation first by using the parent name (the name of the atom) and name the anion of the compound by dropping the *-ine* ending and replacing it with an *-ide* ending. (example: LiBr–lithium bromide)
(b) Using the Stock System, name the cation first by using the parent name and specifying the charge on the cation with a Roman numeral in parentheses after the name. Name the anion as explained in 38(a). (example: $FeCl_3$ – iron(III) chloride)
(c) Name the first element in the compound using the parent name and the second element by using the *-ide* ending. Then attach a prefix to each atom name (mono-, di-, tri- etc. . .) specifying the number of atoms in the compound. The prefix mono- is not used in the name of the first element. (example: CCl_4–carbon tetrachloride)

39. Naming compounds
(a) $Ba(NO_3)_2$, barium nitrate
(b) $NaC_2H_3O_2$, sodium acetate
(c) PbI_2, lead(II) iodide
(d) $MgSO_4$, magnesium sulfate
(e) $CdCrO_4$, cadmium chromate
(f) $BiCl_3$, bismuth(III) chloride
(g) NiS, nickel(II) suflide
(h) $Sn(NO_3)_4$, tin(IV) nitrate
(i) $Ca(OH)_2$, calcium hydroxide

40. ide: suffix is used to indicate a binary compound except for hydroxides, cyanides, and ammonium compounds.

ous: used as a suffix to name an acid that has a lower oxygen content than the *-ic* acid (e.g. HNO_2, nitrous acid and HNO_3, nitric acid); also used as a suffix to name the lower ionic charge of a multivalent metal (e.g. Fe^{2+}, ferrous and Fe^{3+}, ferric).

hypo: used as a prefix in naming an acid that has a lower oxygen content that the *-ous* acid when there are more than two oxyacids with the same elements (e.g. HClO, hypochlorous acid and $HClO_2$, chlorous acid).

per: used as a prefix in naming an acid that has a higher oxygen content than the *-ic* acid when there are more than two oxyacids with the same elements (e.g. $HClO_4$, perchloric acid and $HClO_3$, chloric acid).

ite: the suffix used in naming a salt derived from an *-ous* acid.
For example, HNO_2 (nitrous acid); $NaNO_2$ (sodium nitrite).

ate: the suffix used in naming a salt derived from an *-ic* acid.
For example, H_2SO_4 (sulfuric acid); $CaSO_4$ (calcium sulfate).

Roman numerals: In the Stock System Roman numerals are used in naming compounds that contain metals that may exist as more than one type of cation. The charge of a metal is indicated by a Roman numeral written in parenthesis immediately after the name of the metal.

41. (a) $AgNO_3 + NaCl \longrightarrow AgCl + NaNO_3$
(b) $Fe_2(SO_4)_3 + Ca(OH)_2 \longrightarrow Fe(OH)_3 + CaSO_4$
(c) $KOH + H_2SO_4 \longrightarrow K_2SO_4 + H_2O$

42. (a) 50 e^-, 50 p (b) 48e^-, 50 p (c) 46 e^-, 50 p

43. The formula for a compound must be electrically neutral. Therefore $X = +3$ and $Y = -2$ since in X_2Y_3 this would give $2(+3) + 3(-2) = 0$.

44. $Li_3Fe(CN)_6$
$AlFe(CN)_6$
$Zn_3[Fe(CN)_6]_2$

45. (a) N^{3-} nitride; NO_2^- nitrite — One has oxygen the other does not, charges on the ions differ.
(b) NO_2^- nitrite; NO_3^- nitrate — The number of oxygens differ, but the charge is the same.
(c) HNO_2 nitrous acid; HNO_3 nitric acid — The number of oxygens in the compounds differ but they both have only one hydrogen and one nitrogen atom.

46.

$(NH_4)_2O$	ammonium oxide (does not actually exist)
$(NH_4)_2CO_3$	ammonium carbonate
NH_4Cl	ammonium chloride
$NH_4C_2H_3O_2$	ammonium acetate
ZnO	zinc oxide
$ZnCO_3$	zinc carbonate
$ZnCl_2$	zinc chloride
$Zn(C_2H_3O_2)_2$	zinc acetate
H_2CO_3	carbonic acid
$HC_2H_3O_2$	acetic acid
HCl	hydrochloric acid
H_2O	water

Challenge Exercises

47. Let x = molar mass of Y

$2.44x + x = 110.27$

$x = 32.06$, Y = sulfur (from periodic table)

$2.44x = (110.27 - 32.06) = 78.21$

The compound formula is X_2Y, so the mass of $2X_2 = 78.21$

X = 39.11 which is potassium (from the periodic table)

$X_2Y = K_2S$

$$\text{Ratio} = \frac{78.21}{32.06} = 2.44$$

CHAPTER 7

QUANTITATIVE COMPOSITION OF COMPOUNDS

1. A mole is an amount of substance containing the same number of particles as there are atoms in exactly 12 g of carbon-12.

   It is Avogadro's number (6.022×10^{23}) of anything (atoms, molecules, ping-pong balls, etc.).

2. A mole of gold (197.0 g) has a higher mass than a mole of potassium (39.10 g).

3. Both samples (Au and K) contain the same number of atoms. (6.022×10^{23}).

4. A mole of gold atoms contains more electrons than a mole of potassium atoms, as each Au atom has 79 e^-, while each K atom has only 19 e^-.

5. The molar mass of an element is the mass of one mole (or 6.022×10^{23} atoms) of that element.

6. No. Avogadro's number is a constant. The mole is defined as Avogadro's number of C-12 atoms. Changing the atomic mass to 50 amu would change only the size of the atomic mass unit, not Avogadro's number.

7. 6.022×10^{23}

8. There are Avogadro's number of particles in one mole of substance.

9. (a) A mole of oxygen atoms (O) contains $\mathbf{6.022 \times 10^{23}}$ atoms.
   (b) A mole of oxygen molecules (O_2) contains $\mathbf{6.022 \times 10^{23}}$ molecules.
   (c) A mole of oxygen molecules (O_2) contains $\mathbf{1.204 \times 10^{24}}$ atoms.
   (d) A mole of oxygen atoms (O) has a mass of **16.00** g.
   (e) A mole of oxygen molecules (O_2) has a mass of **32.00** g.

10. 6.022×10^{23} molecules in one molar mass of H_2SO_4.
    4.215×10^{24} atoms in one molar mass of H_2SO_4.

11. The molecular formula represents the total number of atoms of each element in a molecule. The empirical formula represents the lowest number ratio of atoms of each element in a molecule.

12. Choosing 100.0 g of a compound allows us to simply drop the % sign and use grams for percent.

13. Molar masses

(a)	KBr	1	K	39.10 g
		1	Br	79.90 g
				119.0 g

	Formula	Atoms	Element	Mass
(b)	Na_2SO_4	2	Na	45.98 g
		1	S	32.07 g
		4	O	64.00 g
				142.1 g
(c)	$Pb(NO_3)_2$	1	Pb	207.2 g
		2	N	28.02 g
		6	O	96.00 g
				331.2 g
(d)	C_2H_5OH	2	C	24.02 g
		6	H	6.048 g
		1	O	16.00 g
				46.07 g
(e)	$HC_2H_3O_2$	4	H	4.032 g
		2	C	24.02 g
		2	O	32.00 g
				60.05 g
(f)	Fe_3O_4	3	Fe	167.6 g
		4	O	64.00 g
				231.6 g
(g)	$C_{12}H_{22}O_{11}$	12	C	144.1 g
		22	H	22.18 g
		11	O	176.0 g
				342.3 g
(h)	$Al_2(SO_4)_3$	2	Al	53.96 g
		3	S	96.21 g
		12	O	192.0 g
				342.2 g
(i)	$(NH_4)_2HPO_4$	9	H	9.072 g
		2	N	28.02 g
		1	P	30.97 g
		4	O	64.00 g
				132.1 g

14. Molar masses

(a) NaOH

1	Na	22.99 g
1	O	16.00 g
1	H	1.008 g
		40.00 g

(b) Ag_2CO_3

2	Ag	215.8 g
1	C	12.01 g
3	O	48.00 g
		275.8 g

(c) Cr_2O_3

2	Cr	104.0 g
3	O	48.00 g
		152.0 g

(d) $(NH_4)_2CO_3$

2	N	28.02 g
8	H	8.064 g
1	C	12.01 g
3	O	48.00 g
		96.09 g

(e) $Mg(HCO_3)_2$

1	Mg	24.31 g
2	H	2.016 g
2	C	24.02 g
6	O	96.00 g
		146.3 g

(f) C_6H_5COOH

7	C	84.07 g
6	H	6.048 g
2	O	32.00 g
		122.1 g

(g) $C_6H_{12}O_6$

6	C	72.06 g
12	H	12.10 g
6	O	96.00 g
		180.2 g

(h)	$K_4Fe(CN)_6$	4	K	156.4 g
		1	Fe	55.85 g
		6	C	72.06 g
		6	N	84.06 g
				368.4 g
(i)	$BaCl_2 \cdot 2\,H_2O$	1	Ba	137.3 g
		2	Cl	70.90 g
		4	H	4.032 g
		2	O	32.00 g
				244.2 g

15. Moles of atoms.

(a) $(22.5 \text{ g Zn})\left(\dfrac{1 \text{ mol}}{65.39 \text{ g}}\right) = 0.344 \text{ mol Zn}$

(b) $(0.688 \text{ g Mg})\left(\dfrac{1 \text{ mol}}{24.31 \text{ g}}\right) = 2.83 \times 10^{-2} \text{ mol Mg}$

(c) $(4.5 \times 10^{22} \text{ atoms Cu})\left(\dfrac{1 \text{ mol}}{6.022 \times 10^{23} \text{ atoms}}\right) = 7.5 \times 10^{-2} \text{ mol Cu}$

(d) $(382 \text{ g Co})\left(\dfrac{1 \text{ mol}}{58.93 \text{ g}}\right) = 6.48 \text{ mol Co}$

(e) $(0.055 \text{ g Sn})\left(\dfrac{1 \text{ mol}}{118.7 \text{ g}}\right) = 4.6 \times 10^{-4} \text{ mol Sn}$

(f) $(8.5 \times 10^{24} \text{ molecules } N_2)\left(\dfrac{2 \text{ atoms N}}{1 \text{ molecule } N_2}\right)\left(\dfrac{1 \text{ mol N atoms}}{6.022 \times 10^{23} \text{ atoms N}}\right)$

$= 28 \text{ mol N atoms}$

16. Number of moles.

(a) $(25.0 \text{ g NaOH})\left(\dfrac{1 \text{ mol}}{40.00 \text{ g}}\right) = 0.625 \text{ mol NaOH}$

(b) $(44.0 \text{ g } Br_2)\left(\dfrac{1 \text{ mol}}{159.8 \text{ g}}\right) = 0.275 \text{ mol } Br_2$

(c) $(0.684 \text{ g } MgCl_2)\left(\dfrac{1 \text{ mol}}{95.21 \text{ g}}\right) = 7.18 \times 10^{-3} \text{ mol } MgCl_2$

(d) $(14.8\text{ g CH}_3\text{OH})\left(\dfrac{1\text{ mol}}{32.04\text{ g}}\right) = 0.462\text{ mol CH}_3\text{OH}$

(e) $(2.88\text{ g Na}_2\text{SO}_4)\left(\dfrac{1\text{ mol}}{142.1\text{ g}}\right) = 2.03 \times 10^{-2}\text{ mol Na}_2\text{SO}_4$

(f) $(4.20\text{ lb ZnI}_2)\left(\dfrac{453.6\text{ g}}{1\text{ lb}}\right)\left(\dfrac{1\text{ mol}}{319.2\text{ g}}\right) = 5.97\text{ mol ZnI}_2$

17. Number of grams.

(a) $(0.550\text{ mol Au})\left(\dfrac{197.0\text{ g}}{1\text{ mol}}\right) = 108\text{ g Au}$

(b) $(15.8\text{ mol H}_2\text{O})\left(\dfrac{18.02\text{ g}}{\text{mol}}\right) = 285\text{ g H}_2\text{O}$

(c) $(12.5\text{ mol Cl}_2)\left(\dfrac{70.90\text{ g}}{\text{mol}}\right) = 886\text{ g Cl}_2$

(d) $(3.15\text{ mol NH}_4\text{NO}_3)\left(\dfrac{80.05\text{ g}}{\text{mol}}\right) = 252\text{ g NH}_4\text{NO}_3$

18. Number of grams.

(a) $(4.25 \times 10^{-4}\text{ mol H}_2\text{SO}_4)\left(\dfrac{98.09\text{ g}}{\text{mol}}\right) = 0.0417\text{ g H}_2\text{SO}_4$

(b) $(4.5 \times 10^{22}\text{ molecules CCl}_4)\left(\dfrac{1\text{ mol}}{6.022 \times 10^{23}\text{ molecules}}\right)\left(\dfrac{153.8\text{ g}}{\text{mol}}\right) = 11\text{ g CCl}_4$

(c) $(0.00255\text{ mol Ti})\left(\dfrac{47.87\text{ g}}{\text{mol}}\right) = 0.122\text{ g Ti}$

(d) $(1.5 \times 10^{16}\text{ atoms S})\left(\dfrac{32.07\text{ g}}{6.022 \times 10^{23}\text{ atoms}}\right) = 8.0 \times 10^{-7}\text{ g S}$

19. Number of molecules.

(a) $(1.26\text{ mol O}_2)\left(\dfrac{6.022 \times 10^{23}\text{ molecules}}{\text{mol}}\right) = 7.59 \times 10^{23}\text{ molecules O}_2$

(b) $(0.56\text{ mol C}_6\text{H}_6)\left(\dfrac{6.022 \times 10^{23}\text{ molecules}}{\text{mol}}\right) = 3.4 \times 10^{23}\text{ molecules C}_6\text{H}_6$

(c) $(16.0 \text{ g } CH_4)\left(\frac{6.022 \times 10^{23} \text{ molecules}}{16.04 \text{ g}}\right) = 6.01 \times 10^{23} \text{ molecules } CH_4$

(d) $(1000. \text{ g HCl})\left(\frac{6.022 \times 10^{23} \text{ molecules}}{36.46 \text{ g}}\right) = 1.652 \times 10^{25} \text{ molecules HCl}$

20. (a) $(1.75 \text{ mol } Cl_2)\left(\frac{6.022 \times 10^{23} \text{ molecules}}{\text{mol}}\right) = 1.05 \times 10^{24} \text{ molecules } Cl_2$

(b) $(0.27 \text{ mol } C_2H_6O)\left(\frac{6.022 \times 10^{23} \text{ molecules}}{\text{mol}}\right) = 1.6 \times 10^{23} \text{ molecules } C_2H_6O$

(c) $(12.0 \text{ g } CO_2)\left(\frac{6.022 \times 10^{23} \text{ molecules}}{44.01 \text{ g}}\right) = 1.64 \times 10^{23} \text{ molecules } CO_2$

(d) $(100. \text{ g } CH_4)\left(\frac{6.022 \times 10^{23} \text{ molecules}}{16.04 \text{ g}}\right) = 3.75 \times 10^{24} \text{ molecules } CH_4$

21. Number of grams.

(a) $(1 \text{ atom Pb})\left(\frac{207.2 \text{ g}}{6.022 \times 10^{23} \text{ atoms}}\right) = 3.441 \times 10^{-22} \text{ g Pb}$

(b) $(1 \text{ atom Ag})\left(\frac{107.9 \text{ g}}{6.022 \times 10^{23} \text{ atoms}}\right) = 1.792 \times 10^{-22} \text{ g Ag}$

(c) $(1 \text{ molecule } H_2O)\left(\frac{18.02 \text{ g}}{6.022 \times 10^{23} \text{ molecules}}\right) = 2.992 \times 10^{-23} \text{ g } H_2O$

(d) $(1 \text{ molecule } C_3H_5(NO_3)_3)\left(\frac{227.1 \text{ g}}{6.022 \times 10^{23} \text{ molecules}}\right) = 3.771 \times 10^{-22} \text{ g } C_3H_5(NO_3)_3$

22. (a) $(1 \text{ atom Au})\left(\frac{197.0 \text{ g}}{6.022 \times 10^{23} \text{ atoms}}\right) = 3.271 \times 10^{-22} \text{ g Au}$

(b) $(1 \text{ atom U})\left(\frac{238.0 \text{ g}}{6.022 \times 10^{23} \text{ atoms}}\right) = 3.952 \times 10^{-22} \text{ g U}$

(c) $(1 \text{ molecule } NH_3)\left(\frac{17.03 \text{ g}}{6.022 \times 10^{23} \text{ molecules}}\right) = 2.828 \times 10^{-23} \text{ g } NH_3$

(d) $(1 \text{ molecule } C_6H_4(NH_2)_2)\left(\frac{108.1 \text{ g}}{6.022 \times 10^{23} \text{ molecules}}\right) = 1.795 \times 10^{-22} \text{ g } C_6H_4(NH_2)_2$

23. (a) $(8.66 \text{ mol Cu})\left(\dfrac{63.55 \text{ g}}{\text{mol}}\right) = 550. \text{ g Cu}$

(b) $(125 \text{ mol Au})\left(\dfrac{197.0 \text{ g}}{\text{mol}}\right)\left(\dfrac{1 \text{ kg}}{1000 \text{ g}}\right) = 24.6 \text{ kg Au}$

(c) $(10. \text{ atoms C})\left(\dfrac{1 \text{ mol}}{6.022 \times 10^{23} \text{ atoms}}\right) = 1.7 \times 10^{-23} \text{ mol C}$

(d) $(5000 \text{ molecules } CO_2)\left(\dfrac{1 \text{ mol}}{6.022 \times 10^{23} \text{ molecules}}\right) = 8 \times 10^{-21} \text{ mol } CO_2$

24. (a) $(28.4 \text{ g S})\left(\dfrac{1 \text{ mol}}{32.07 \text{ g}}\right) = 0.886 \text{ mol S}$

(b) $(2.50 \text{ kg NaCl})\left(\dfrac{1000 \text{ g}}{\text{kg}}\right)\left(\dfrac{1 \text{ mol}}{58.44 \text{ g}}\right) = 42.8 \text{ mol NaCl}$

(c) $(42.4 \text{ g Mg})\left(\dfrac{6.022 \times 10^{23} \text{ atoms}}{24.31 \text{ g}}\right) = 1.05 \times 10^{24} \text{ atoms Mg}$

(d) $(485 \text{ mL } Br_2)\left(\dfrac{3.12 \text{ g}}{\text{mL}}\right)\left(\dfrac{1 \text{ mol}}{159.8 \text{ g}}\right) = 9.47 \text{ mol } Br_2$

25. One mole of carbon disulfide (CS_2) contains:

(a) 6.022×10^{23} molecules of CS_2

(b) $(6.022 \times 10^{23} \text{ molecules of } CS_2)\left(\dfrac{1 \text{ C atom}}{1 \text{ molecule } CS_2}\right) = 6.022 \times 10^{23} \text{ C atoms}$

(c) $(6.022 \times 10^{23} \text{ molecules of } CS_2)\left(\dfrac{2 \text{ S atoms}}{1 \text{ molecule } CS_2}\right) = 1.204 \times 10^{24} \text{ S atoms}$

(d) $(6.022 \times 10^{23} \text{ atoms}) + (1.204 \times 10^{24} \text{ atoms}) = 1.806 \times 10^{24} \text{ total atoms}$

26. One mole of ammonia (NH_3) contains

(a) 6.022×10^{23} molecules of NH_3

(b) $(6.022 \times 10^{23} \text{ molecules of } NH_3)\left(\dfrac{1 \text{ N atom}}{\text{molecule } NH_3}\right) = 6.022 \times 10^{23} \text{ N atoms}$

(c) $(6.022 \times 10^{23} \text{ molecules of } NH_3)\left(\dfrac{3 \text{ H atoms}}{\text{molecule } NH_3}\right) = 1.807 \times 10^{24} \text{ H atoms}$

(d) $(6.022 \times 10^{23} \text{ atoms}) + (1.807 \times 10^{24} \text{ atoms}) = 2.409 \times 10^{24} \text{ total atoms}$

27. Atoms of oxygen in:

(a) $(16.0 \text{ g } O_2)\left(\dfrac{1 \text{ mol}}{32.00 \text{ g}}\right)\left(\dfrac{2 \text{ mol O}}{1 \text{ mol } O_2}\right)\left(\dfrac{6.022 \times 10^{23} \text{ atoms}}{\text{mol}}\right) = 6.02 \times 10^{23} \text{ atoms O}$

(b) $(0.622 \text{ mol MgO})\left(\dfrac{1 \text{ mol O}}{\text{mol MgO}}\right)\left(\dfrac{6.022 \times 10^{23} \text{ atoms}}{\text{mol}}\right) = 3.75 \times 10^{23} \text{ atoms O}$

(c) $(6.00 \times 10^{22} \text{ molecules } C_6H_{12}O_6)\left(\dfrac{6 \text{ atoms O}}{\text{molecule } C_6H_{12}O_6}\right) = 3.60 \times 10^{23} \text{ atoms O}$

28. Atoms of oxygen in:

(a) $(5.0 \text{ mol } MnO_2)\left(\dfrac{2 \text{ mol O}}{\text{mol } MnO_2}\right)\left(\dfrac{6.022 \times 10^{23} \text{ atoms}}{\text{mol}}\right) = 6.0 \times 10^{24} \text{ atoms O}$

(b) $(255 \text{ g } MgCO_3)\left(\dfrac{1 \text{ mol}}{84.32 \text{ g}}\right)\left(\dfrac{3 \text{ mol O}}{\text{mol } MgCO_3}\right)\left(\dfrac{6.022 \times 10^{23} \text{ atoms}}{\text{mol}}\right)$

$= 5.46 \times 10^{24} \text{ atoms O}$

(c) $(5.0 \times 10^{18} \text{ molecules } H_2O)\left(\dfrac{1 \text{ atom O}}{\text{molecule } H_2O}\right) = 5.0 \times 10^{18} \text{ atoms O}$

29. The number of grams of:

(a) silver in 25.0 g AgBr

$$(25.0 \text{ g AgBr})\left(\dfrac{107.9 \text{ g Ag}}{187.8 \text{ g AgBr}}\right) = 14.4 \text{ g Ag}$$

(b) nitrogen in 6.34 mol $(NH_4)_3PO_4$

$$(6.34 \text{ mol } (NH_4)_3PO_4)\left(\dfrac{42.03 \text{ g N}}{\text{mol } (NH_4)_3PO_4}\right) = 266 \text{ g N}$$

(c) oxygen in 8.45×10^{22} molecules SO_3

The conversion is: molecules $SO_3 \longrightarrow$ mol $SO_3 \longrightarrow$ g O

$$(8.45 \times 10^{22} \text{ molecules } SO_3)\left(\frac{1 \text{ mol}}{6.022 \times 10^{23} \text{ molecules}}\right)\left(\frac{48.00 \text{ g O}}{\text{mol } SO_3}\right) = 6.74 \text{ g O}$$

30. The number of grams of:

(a) chlorine in 5.00 g $PbCl_2$

$$(5.00 \text{ g } PbCl_2)\left(\frac{70.90 \text{ g Cl}}{278.1 \text{ g } PbCl_2}\right) = 1.27 \text{ g Cl}$$

(b) hydrogen in 4.50 g H_2SO_4

$$(4.50 \text{ g } H_2SO_4)\left(\frac{2.016 \text{ g H}}{98.09 \text{ g } H_2SO_4}\right) = 9.25 \times 10^{-2} \text{ g H}$$

(c) hydrogen in 5.45×10^{22} molecules NH_3

The conversion is: molecules $NH_3 \longrightarrow$ moles $NH_3 \longrightarrow$ g H

$$(5.45 \times 10^{22} \text{ molecules } NH_3)\left(\frac{1 \text{ mol}}{6.022 \times 10^{23} \text{ molecules}}\right)\left(\frac{3.024 \text{ g H}}{\text{mol } NH_3}\right) = 0.274 \text{ g H}$$

31. Percent composition

(a) NaBr

Na	22.99 g	
Br	79.90 g	
	102.9 g	

$$\left(\frac{22.99 \text{ g}}{102.9 \text{ g}}\right)(100) = 22.34\% \text{ Na}$$

$$\left(\frac{79.90 \text{ g}}{102.9 \text{ g}}\right)(100) = 77.65\% \text{ Br}$$

(b) $KHCO_3$

K	39.10 g
H	1.008 g
3 O	48.00 g
C	12.01 g
	100.1 g

$$\left(\frac{39.10 \text{ g}}{100.1 \text{ g}}\right)(100) = 39.06\% \text{ K}$$

$$\left(\frac{1.008 \text{ g}}{100.1 \text{ g}}\right)(100) = 1.007\% \text{ H}$$

$$\left(\frac{12.01 \text{ g}}{100.1 \text{ g}}\right)(100) = 12.00\% \text{ C}$$

$$\left(\frac{48.00 \text{ g}}{100.1 \text{ g}}\right)(100) = 47.95\% \text{ O}$$

(c) $FeCl_3$

Fe	55.85 g
3 Cl	106.4 g
	162.3 g

$$\left(\frac{55.85\ \text{g}}{162.3\ \text{g}}\right)(100) = 34.41\%\ \text{Fe}$$

$$\left(\frac{106.4\ \text{g}}{162.3\ \text{g}}\right)(100) = 65.56\%\ \text{Cl}$$

(d) $SiCl_4$

Si	28.09 g
4 Cl	141.8 g
	169.9 g

$$\left(\frac{28.09\ \text{g}}{169.9\ \text{g}}\right)(100) = 16.53\%\ \text{Si}$$

$$\left(\frac{141.8\ \text{g}}{169.9\ \text{g}}\right)(100) = 83.46\%\ \text{Cl}$$

(e) $Al_2(SO_4)_3$

2 Al	53.96 g
3 S	96.21 g
12 O	192.0 g
	342.2 g

$$\left(\frac{53.96\ \text{g}}{342.2\ \text{g}}\right)(100) = 15.77\%\ \text{Al}$$

$$\left(\frac{96.21\ \text{g}}{342.2\ \text{g}}\right)(100) = 28.12\%\ \text{S}$$

$$\left(\frac{192.0\ \text{g}}{342.2\ \text{g}}\right)(100) = 56.11\%\ \text{O}$$

(f) $AgNO_3$

Ag	107.9 g
N	14.01 g
3 O	48.00 g
	169.9 g

$$\left(\frac{107.9\ \text{g}}{169.9\ \text{g}}\right)(100) = 63.51\%\ \text{Ag}$$

$$\left(\frac{14.01\ \text{g}}{169.9\ \text{g}}\right)(100) = 8.246\%\ \text{N}$$

$$\left(\frac{48.00\ \text{g}}{169.9\ \text{g}}\right)(100) = 28.25\%\ \text{O}$$

32. Percent composition

(a) $ZnCl_2$

Zn	65.39 g
2 Cl	70.90 g
	136.3 g

$$\left(\frac{65.39\ \text{g}}{136.3\ \text{g}}\right)(100) = 47.98\%\ \text{Zn}$$

$$\left(\frac{70.90\ \text{g}}{136.3\ \text{g}}\right)(100) = 52.02\%\ \text{Cl}$$

(b) $NH_4C_2H_3O_2$

N	14.01 g
7 H	7.056 g
2 C	24.02 g
2 O	32.00 g
	77.09 g

$$\left(\frac{14.01\text{ g}}{77.09\text{ g}}\right)(100) = 18.17\%\text{ N}$$

$$\left(\frac{7.056\text{ g}}{77.09\text{ g}}\right)(100) = 9.153\%\text{ H}$$

$$\left(\frac{24.02\text{ g}}{77.09\text{ g}}\right)(100) = 31.16\%\text{ C}$$

$$\left(\frac{32.00\text{ g}}{77.09\text{ g}}\right)(100) = 41.51\%\text{ O}$$

(c) MgP_2O_7

Mg	24.31 g
2 P	61.94 g
7 O	112.0 g
	198.3 g

$$\left(\frac{24.31\text{ g}}{198.3\text{ g}}\right)(100) = 12.26\%\text{ Mg}$$

$$\left(\frac{61.94\text{ g}}{198.3\text{ g}}\right)(100) = 31.24\%\text{ P}$$

$$\left(\frac{112.0\text{ g}}{198.3\text{ g}}\right)(100) = 56.48\%\text{ O}$$

(d) $(NH_4)_2SO_4$

2 N	28.02 g
8 H	8.064 g
S	32.07 g
4 O	64.00 g
	132.2 g

$$\left(\frac{28.02\text{ g}}{132.2\text{ g}}\right)(100) = 21.20\%\text{ N}$$

$$\left(\frac{8.064\text{ g}}{132.2\text{ g}}\right)(100) = 6.100\%\text{ H}$$

$$\left(\frac{32.07\text{ g}}{132.2\text{ g}}\right)(100) = 24.26\%\text{ S}$$

$$\left(\frac{64.00\text{ g}}{132.2\text{ g}}\right)(100) = 48.41\%\text{ O}$$

(e) $Fe(NO_3)_3$

Fe	55.85 g
3 N	42.03 g
9 O	144.0 g
	241.9 g

$$\left(\frac{55.85\text{ g}}{241.9\text{ g}}\right)(100) = 23.09\%\text{ Fe}$$

$$\left(\frac{42.03\text{ g}}{241.9\text{ g}}\right)(100) = 17.37\%\text{ N}$$

$$\left(\frac{144.0\text{ g}}{241.9\text{ g}}\right)(100) = 59.53\%\text{ O}$$

(f) ICl_3

I	126.9 g
3 Cl	106.4 g
	233.3 g

$$\left(\frac{126.9\text{ g}}{233.3\text{ g}}\right)(100) = 54.39\%\text{ I}$$

$$\left(\frac{106.4\text{ g}}{233.3\text{ g}}\right)(100) = 45.61\%\text{ Cl}$$

33. Percent of iron

(a) FeO

Fe	55.85 g
O	16.00 g
	71.85 g

$$\left(\frac{55.85\text{ g}}{71.85\text{ g}}\right)(100) = 77.73\%\text{ Fe}$$

(b) Fe_2O_3

2 Fe	111.7 g
3 O	48.00 g
	159.7 g

$$\left(\frac{111.7\text{ g}}{159.7\text{ g}}\right)(100) = 69.94\%\text{ Fe}$$

(c) Fe_3O_4

3 Fe	167.6 g
4 O	64.00 g
	231.6 g

$$\left(\frac{167.6\text{ g}}{231.6\text{ g}}\right)(100) = 72.37\%\text{ Fe}$$

(d) $K_4Fe(CN)_6$

Fe	55.85 g
4 K	156.4 g
6 C	72.06 g
6 N	84.06 g
	368.4 g

$$\left(\frac{55.85\text{ g}}{368.4\text{ g}}\right)(100) = 15.16\%\text{ Fe}$$

34. Percent chlorine

(a) KCl

K	39.10 g
Cl	35.45 g
	74.55 g

$$\left(\frac{35.45\text{ g}}{74.55\text{ g}}\right)(100) = 47.55\%\text{ Cl}$$

(b) $BaCl_2$

Ba	137.3 g	
2 Cl	70.90 g	$\left(\frac{70.90\text{ g}}{208.2\text{ g}}\right)(100) = 34.05\%\text{ Cl}$
	208.2 g	

(c) $SiCl_4$

Si	28.09 g	
4 Cl	141.8 g	$\left(\frac{141.8\text{ g}}{169.9\text{ g}}\right)(100) = 83.46\%\text{ Cl}$
	169.9 g	

(d) LiCl

Li	6.941 g	
Cl	35.45 g	$\left(\frac{35.45\text{ g}}{42.39\text{ g}}\right)(100) = 83.63\%\text{ Cl}$
	42.39 g	

Highest % Cl is in LiCl; lowest % Cl is in $BaCl_2$

35. Percent composition of an oxide

14.20 g oxide
−6.20 g P
8.00 g oxygen

$$\left(\frac{6.20\text{ g}}{14.20\text{ g}}\right)(100) = 43.7\%\text{ P}$$

$$\left(\frac{8.00\text{ g}}{14.20\text{ g}}\right)(100) = 56.3\%\text{ O}$$

36. Percent composition of ethylene chloride

6.00 g C
1.00 g H
17.75 g Cl
24.75 g total

$$\left(\frac{6.00\text{ g}}{24.75\text{ g}}\right)(100) = 24.2\%\text{ C}$$

$$\left(\frac{1.00\text{ g}}{24.75\text{ g}}\right)(100) = 4.04\%\text{ H}$$

$$\left(\frac{17.75\text{ g}}{24.75\text{ g}}\right)(100) = 71.72\%\text{ Cl}$$

37. (a) H_2O (by inspection of formulas)
(b) N_2O_3 (by inspection of formulas)
(c) equal (by inspection of formulas)

38. (a) $KClO_3$ (by inspection of formulas)
(b) $KHSO_4$ (by inspection of formulas)
(c) Na_2CrO_4 (by inspection of formulas)

39. Empirical Formulas. Change all percents to grams.

(a) $\dfrac{5.94 \text{ gH}}{1.008 \text{ g/mol}} = 5.89 \text{ mol H} \qquad \dfrac{94.06 \text{ g O}}{16.00 \text{ g/mol}} = 5.879 \text{ mol O}$

This is a ratio of 1 mol H to 1 mol O.

The empirical formula is HO; molar mass 34.02 $= H_2O_2$

(b) $\dfrac{80.34 \text{ g Zn}}{65.39 \text{ g/mol}} = 1.228 \text{ mol Zn} \qquad \dfrac{19.66 \text{ g O}}{16.00 \text{ g/mol}} = 1.229 \text{ mol O}$

This is a ratio of 1 mol Zn to 1 mol O.

The empirical formula is ZnO; molar mass 81.39

(c) $\dfrac{35.18 \text{ g Fe}}{55.85 \text{ g/mol}} = 0.6299 \text{ mol Fe} \qquad \dfrac{44.66 \text{ g Cl}}{35.45 \text{ g/mol}} = 1.260 \text{ mol Cl}$

$\dfrac{20.16 \text{ g O}}{16.00 \text{ g O/mol}} = 1.260 \text{ mol O}$

$\text{mol ratios} = \dfrac{0.6299 \text{ Fe}}{0.6299} = 1.000 \text{ mol Fe} \qquad \dfrac{1.260 \text{ mol O}}{0.6299} = 2.000 \text{ mol O}$

$\dfrac{1.260 \text{ mol Cl}}{0.6299} = 2.000 \text{ mol Cl}$

The empirical formula is $Fe(ClO)_2$; molar mass 158.75

(d) $\dfrac{26.19 \text{ g N}}{14.01 \text{ g/mol}} = 1.869 \text{ mol N} \qquad \dfrac{7.55 \text{ g H}}{1.008 \text{ g/mol}} = 7.49 \text{ mol H}$

$\dfrac{66.26 \text{ g Cl}}{35.45 \text{ g/mol}} = 1.869 \text{ mol Cl}$

$\text{mol ratios:} \dfrac{1.869 \text{ mol N}}{1.869} = 1 \text{ mol N} \qquad \dfrac{7.49 \text{ mol H}}{1.869} = 4.00 \text{ mol H}$

$\dfrac{1.869 \text{ mol Cl}}{1.869} = 1.000 \text{ mol Cl}$

The empirical formula is NH_4Cl; molar mass 59.49

40. Empirical formulas: Change all percents to grams.

(a) $\dfrac{32.86 \text{ g K}}{39.10 \text{ g/mol}} = 0.8404 \text{ mol K} \qquad \dfrac{67.14 \text{ g Br}}{79.90 \text{ g/mol}} = 0.8403 \text{ mol Br}$

This is a ratio of 1 mol K to 1 mol Br.

The empirical formula is KBr; molar mass 119.0 g

(b) $\dfrac{54.09 \text{ g Ca}}{40.08 \text{ g/mol}} = 1.350 \text{ mol Ca}$ $\qquad$ $\dfrac{2.72 \text{ g H}}{1.008 \text{ g/mol}} = 2.70 \text{ mol H}$

$\dfrac{43.18 \text{ g O}}{16.00 \text{ g/mol}} = 2.699 \text{ mol O}$

mole ratios: $\dfrac{1.350 \text{ mol Ca}}{1.350} = 1.000 \text{ mol Ca}$ $\qquad$ $\dfrac{2.70 \text{ mol H}}{1.350} = 2.00 \text{ mol H}$

$\dfrac{2.699 \text{ mol O}}{1.35} = 1.999 \text{ mol O}$

The empirical formula is CaO_2H_2 or $Ca(OH)_2$; molar mass 74.10 g

(c) $\dfrac{63.50 \text{ g Ag}}{107.9 \text{ g/mol}} = 0.5885 \text{ mol Ag}$ $\qquad$ $\dfrac{8.25 \text{ g N}}{14.01 \text{ g/mol}} = 0.5889 \text{ mol N}$

$\dfrac{28.25 \text{ g O}}{16.00 \text{ g/mol}} = 1.766 \text{ g mol O}$

mole ratios: $\dfrac{0.5885 \text{ mol Ag}}{0.5885} = 1.000 \text{ mol Ag}$ $\qquad$ $\dfrac{0.5889 \text{ mol N}}{0.5885} = 1.001 \text{ mol N}$

$\dfrac{1.766 \text{ mol O}}{0.5885} = 3.000 \text{ mol O}$

The empirical formula is $Ag\,NO_3$; molar mass 169.9 g

(d) $\dfrac{2.06 \text{ g H}}{1.008 \text{ g/mol}} = 2.07 \text{ mol H}$ $\qquad$ $\dfrac{32.69 \text{ g S}}{32.07 \text{ g/mol}} = 1.019 \text{ mol S}$

$\dfrac{65.25 \text{ g O}}{16.00 \text{ g/mol}} = 4.078 \text{ mol O}$

mole ratios: $\dfrac{2.04 \text{ mol H}}{1.019} = 2.00 \text{ mol H}$ $\qquad$ $\dfrac{1.019 \text{ mol S}}{1.019} = 1.000 \text{ mol S}$

$\dfrac{4.078 \text{ mol O}}{1.019} = 4.002 \text{ mol O}$

The empirical formula is H_2SO_4; molar mass 98.09

41. Empirical formulas from percent composition.

(a) Step 1. Express each element as grams/100 g material.

$$63.6\% \text{ N} = 63.6 \text{ g N}/100 \text{ g material}$$
$$36.4\% \text{ O} = 36.4 \text{ g O}/100 \text{ g material}$$

Step 2. Calculate the relative moles of each element.

$$(63.6\ \text{g N})\left(\frac{1\ \text{mol}}{14.01\ \text{g}}\right) = 4.54\ \text{mol N}$$

$$(36.4\ \text{g O})\left(\frac{1\ \text{mol}}{16.00\ \text{g}}\right) = 2.28\ \text{mol O}$$

Step 3. Change these moles to whole numbers by dividing each by the smaller number.

$$\frac{4.54\ \text{mol N}}{2.28} = 1.99\ \text{mol N}$$

$$\frac{2.28\ \text{mol O}}{2.28} = 1.00\ \text{mol O}$$

The simplest ratio of N:O is 2:1. The empirical formula, therefore, is N_2O.

(b) 46.7% N, 53.3% O

$$(46.7\ \text{g N})\left(\frac{1\ \text{mol}}{14.01\ \text{g}}\right) = 3.33\ \text{mol N} \qquad \frac{3.33}{3.33} = 1.00\ \text{mol N}$$

$$(53.3\ \text{g O})\left(\frac{1\ \text{mol}}{16.00\ \text{g}}\right) = 3.33\ \text{mol O} \qquad \frac{3.33}{3.33} = 1.00\ \text{mol O}$$

The empirical formula is NO.

(c) 25.9% N, 71.4% O

$$(25.9\ \text{g N})\left(\frac{1\ \text{mol}}{14.01\ \text{g}}\right) = 1.85\ \text{mol N} \qquad \frac{1.85}{1.85} = 1.00\ \text{mol N}$$

$$(74.1\ \text{g O})\left(\frac{1\ \text{mol}}{16.00\ \text{g}}\right) = 4.63\ \text{mol O} \qquad \frac{4.63}{1.85} = 2.5\ \text{mol O}$$

Since these values are not whole numbers, multiply each by 2 to change them to whole numbers.

$(1.00\ \text{mol N})(2) = 2.00\ \text{mol N}$; $(2.5\ \text{mol O})(2) = 5.00\ \text{mol O}$

The empirical formula is N_2O_5.

(d) 43.4% Na, 11.3% C, 45.3% O

$$(43.4\ \text{g Na})\left(\frac{1\ \text{mol}}{22.99\ \text{g}}\right) = 1.89\ \text{mol Na} \qquad \frac{1.89}{0.941} = 2.01\ \text{mol Na}$$

$$(11.3\ \text{g C})\left(\frac{1\ \text{mol}}{12.01\ \text{g}}\right) = 0.941\ \text{mol C} \qquad \frac{0.941}{0.941} = 1.00\ \text{mol C}$$

$$(45.3\text{ g O})\left(\frac{1\text{ mol}}{16.00\text{ g}}\right) = 2.83\text{ mol O} \qquad \frac{2.83}{0.941} = 3.00\text{ mol O}$$

The empirical formula is Na_2CO_3.

(e) 18.8% Na, 29.0% Cl, 52.3% O

$$(18.8\text{ g Na})\left(\frac{1\text{ mol}}{22.99\text{ g}}\right) = 0.818\text{ mol Na} \qquad \frac{0.818}{0.818} = 1.00\text{ mol Na}$$

$$(29.0\text{ g Cl})\left(\frac{1\text{ mol}}{35.45\text{ g}}\right) = 0.818\text{ mol Cl} \qquad \frac{0.818}{0.818} = 1.00\text{ mol Cl}$$

$$(52.3\text{ g O})\left(\frac{1\text{ mol}}{16.00\text{ g}}\right) = 3.27\text{ mol O} \qquad \frac{3.27}{0.818} = 4.00\text{ mol O}$$

The empirical formula is $NaClO_4$.

(f) 72.02% Mn, 27.98% O

$$(72.02\text{ g Mn})\left(\frac{1\text{ mol}}{54.94\text{ g}}\right) = 1.311\text{ mol Mn} \qquad \frac{1.311}{1.311} = 1.000\text{ mol Mn}$$

$$(27.98\text{ g O})\left(\frac{1\text{ mol}}{16.00\text{ g}}\right) = 1.749\text{ mol O} \qquad \frac{1.749}{1.311} = 1.334\text{ mol O}$$

Multiply both values by 3 to give whole numbers.

$(1.000\text{ mol Mn})(3) = 3.000\text{ mol Mn}$; $(1.334\text{ mol O})(3) = 4.002\text{ mol O}$

The empirical formula is Mn_3O_4.

42. Empirical formulas from percent composition.

(a) 64.1% Cu, 35.9% Cl

$$(64.1\text{ g Cu})\left(\frac{1\text{ mol}}{63.55\text{ g}}\right) = 1.01\text{ mol Cu} \qquad \frac{1.01}{1.01} = 1.00\text{ mol Cu}$$

$$(35.9\text{ g Cl})\left(\frac{1\text{ mol}}{35.45\text{ g}}\right) = 1.01\text{ mol Cl} \qquad \frac{1.01}{1.01} = 1.00\text{ mol Cl}$$

The empirical formula is CuCl.

(b) 47.2% Cu, 52.8% Cl

$$(47.2\text{ g Cu})\left(\frac{1\text{ mol}}{63.55\text{ g}}\right) = 0.743\text{ mol Cu} \qquad \frac{0.743}{0.743} = 1.00\text{ mol Cu}$$

$$(52.8\text{ g Cl})\left(\frac{1\text{ mol}}{35.45\text{ g}}\right) = 1.49\text{ mol Cl} \qquad \frac{1.49}{0.743} = 2.01\text{ mol Cl}$$

The empirical formula is $CuCl_2$.

(c) 51.9% Cr, 48.1% S

$$(51.9\text{ g Cr})\left(\frac{1\text{ mol}}{52.00\text{ g}}\right) = 0.998\text{ mol Cr} \qquad \frac{0.998}{0.998} = 1.00\text{ mol Cr}$$

$$(48.1\text{ g S})\left(\frac{1\text{ mol}}{32.07\text{ g}}\right) = 1.50\text{ mol S} \qquad \frac{1.50}{0.998} = 1.50\text{ mol S}$$

Multiply both values by 2 to give whole numbers.

$(1.00\text{ mol Cr})(2) = 2.00\text{ mol Cr}$; $(1.50\text{ mol S})(2) = 3.00\text{ mol S}$

The empirical formula is Cr_2S_3.

(d) 55.3% K, 14.6% P, 30.1% O

$$(55.3\text{ g K})\left(\frac{1\text{ mol}}{39.10\text{ g}}\right) = 1.41\text{ mol K} \qquad \frac{1.41}{0.471} = 2.99\text{ mol K}$$

$$(14.6\text{ g P})\left(\frac{1\text{ mol}}{30.97\text{ g}}\right) = 0.471\text{ mol P} \qquad \frac{0.471}{0.471} = 1.00\text{ mol P}$$

$$(30.1\text{ g O})\left(\frac{1\text{ mol}}{16.00\text{ g}}\right) = 1.88\text{ mol O} \qquad \frac{1.88}{0.471} = 3.99\text{ mol O}$$

The empirical formula is K_3PO_4.

(e) 38.9% Ba, 29.4% Cr, 31.7% O

$$(38.9\text{ g Ba})\left(\frac{1\text{ mol}}{137.3\text{ g}}\right) = 0.283\text{ mol Ba} \qquad \frac{0.283}{0.283} = 1.00\text{ mol Ba}$$

$$(29.4\text{ g Cr})\left(\frac{1\text{ mol}}{52.00\text{ g}}\right) = 0.565\text{ mol Cr} \qquad \frac{0.565}{0.283} = 2.00\text{ mol Cr}$$

$$(31.7\text{ g O})\left(\frac{1\text{ mol}}{16.00\text{ g}}\right) = 1.98\text{ mol O} \qquad \frac{1.98}{0.283} = 7.00\text{ mol O}$$

The empirical formula is $BaCr_2O_7$.

(f) 3.99% P, 82.3% Br, 13.7% Cl

$$(3.99\text{ g P})\left(\frac{1\text{ mol}}{30.97\text{ g}}\right) = 0.129\text{ mol P} \qquad \frac{0.129}{0.129} = 1.00\text{ mol P}$$

$$(82.3\text{ g Br})\left(\frac{1\text{ mol}}{79.90\text{ g}}\right) = 1.03\text{ mol Br} \qquad \frac{1.03}{0.129} = 7.98\text{ mol Br}$$

$$(13.7\text{ g Cl})\left(\frac{1\text{ mol}}{35.45\text{ g}}\right) = 0.386\text{ mol Cl} \qquad \frac{0.386}{0.129} = 2.99\text{ mol Cl}$$

The empirical formula is PBr_8Cl_3.

43. Empirical formula

$$(3.996 \text{ g Sn})\left(\frac{1 \text{ mol}}{118.7 \text{ g}}\right) = 0.0337 \text{ mol Sn} \qquad \frac{0.0337}{0.0337} = 1.00 \text{ mol Sn}$$

$$(1.077 \text{ g O})\left(\frac{1 \text{ mol}}{16.00 \text{ g}}\right) = 0.0673 \text{ mol O} \qquad \frac{0.0673}{0.0337} = 2.00 \text{ mol O}$$

The empirical formula is SnO_2.

44. Empirical formula

5.454 g product − 3.054 g V = 2.400 g O

$$(3.054 \text{ g V})\left(\frac{1 \text{ mol}}{50.94 \text{ g}}\right) = 0.0600 \text{ mol V} \qquad \frac{0.0600}{0.0600} = 1.00 \text{ mol V}$$

$$(2.400 \text{ g O})\left(\frac{1 \text{ mol}}{16.00 \text{ g}}\right) = 0.1500 \text{ mol O} \qquad \frac{0.1500}{0.0600} = 2.50 \text{ mol O}$$

Multiplying both by 2 gives the empirical formula V_2O_5.

45. Molecular formula of hydroquinone

65.45% C, 5.45% H, 29.09% O; molar mass = 110.1

$$(65.45 \text{ g C})\left(\frac{1 \text{ mol}}{12.01 \text{ g}}\right) = 5.450 \text{ mol C} \qquad \frac{5.450}{1.818} = 2.998 \text{ mol C}$$

$$(5.45 \text{ g H})\left(\frac{1 \text{ mol}}{1.008 \text{ g}}\right) = 5.41 \text{ mol H} \qquad \frac{5.41}{1.818} = 2.98 \text{ mol H}$$

$$(29.09 \text{ g O})\left(\frac{1 \text{ mol}}{16.00 \text{ g}}\right) = 1.818 \text{ mol O} \qquad \frac{1.818}{1.818} = 1.000 \text{ mol O}$$

The empirical formula is C_3H_3O making the empirical mass 55.05.

$$\frac{\text{molar mass}}{\text{empirical mass}} = \frac{110.1}{55.05} = 2$$

The molecular formula is twice that of the empirical formula.

Molecular formula = $(C_3H_3O)_2 = C_6H_6O_2$

46. Molecular formula of fructose

40.0% C, 6.7% H, 53.3% O; molar mass = 180.1

$$(40.0 \text{ g C})\left(\frac{1 \text{ mol}}{12.01 \text{ g}}\right) = 3.33 \text{ mol C} \qquad \frac{3.33}{3.33} = 1.00 \text{ mol C}$$

$$(6.7 \text{ g H})\left(\frac{1 \text{ mol}}{1.008 \text{ g}}\right) = 6.6 \text{ mol H} \qquad \frac{6.6}{3.33} = 2.0 \text{ mol H}$$

$$(53.3 \text{ g O})\left(\frac{1 \text{ mol}}{16.00 \text{ g}}\right) = 3.33 \text{ mol O} \qquad \frac{3.33}{3.33} = 1.00 \text{ mol O}$$

The empirical formula is CH_2O making the empirical mass 33.03.

$$\frac{\text{molar mass}}{\text{empirical mass}} = \frac{180.1}{33.03} = 5.994$$

The molecular formula is six times that of the empirical formula.

Molecular formula $= (CH_2O)_6 = C_6H_{12}O_6$

47. $\%\ \text{nitrogen} = \dfrac{12.04\ \text{g}}{39.54\ \text{g}}(100) = 30.45\%$

$$\%\ \text{oxygen} = \frac{39.54\ \text{g} - 12.04\ \text{g}}{39.54\ \text{g}}(100) = 69.55\%$$

$$\text{empirical formula: moles of nitrogen} = \frac{12.04\ \text{g}}{14.01\ \text{g/mol}} = 0.8594\ \text{mol}$$

$$\text{moles of oxygen} = \frac{27.50\ \text{g}}{16.00\ \text{g/mol}} = 1.7\ \text{mol}$$

$$\text{relative number of nitrogen atoms} = \frac{0.8594\ \text{mol}}{0.8594\ \text{mol}} = 1$$

$$\text{relative number of oxygen atoms} = \frac{1.7\ \text{mol}}{0.8594\ \text{mol}} = 1.98 = 2$$

empirical formula $= NO_2$

molecular formula: (molar mass of NO_2)x = 92.02 g, $46.01x = 92.02$, $x = 2$

The molecular formula is twice the empirical formula.

molecular formula $= N_2O_4$

48. Total mass of C + H + O = 30.21 g + 40.24 g + 5.08 g = 75.53 g

$$\%\ \text{carbon} = \frac{30.21\ \text{g}}{75.53\ \text{g}}(100) = 40.0\%$$

$$\%\ \text{hydrogen} = \frac{5.08\ \text{g}}{75.53\ \text{g}}(100) = 6.73\%$$

$$\%\ \text{oxygen} = \frac{40.24\ \text{g}}{75.53\ \text{g}}(100) = 53.3\%$$

$$\text{empirical formula: moles of carbon} = \frac{30.21\ \text{g}}{12.01\ \text{g/mol}} = 2.515\ \text{mol}$$

$$\text{moles of hydrogen} = \frac{5.080\ \text{g}}{1.01\ \text{g/mol}} = 5.03\ \text{mol}$$

$$\text{moles of oxygen} = \frac{40.24\ \text{g}}{16.00\ \text{g/mol}} = 2.515\ \text{mol}$$

$$\text{relative number of carbon atoms} = \frac{2.515 \text{ mol}}{2.515 \text{ mol}} = 1$$

$$\text{relative number of hydrogen atoms} = \frac{5.03 \text{ mol}}{2.515 \text{ mol}} = 2$$

$$\text{relative number of oxygen atoms} = \frac{2.515 \text{ mol}}{2.515 \text{ mol}} = 1$$

empirical formula $= CH_2O$

molecular formula: (molar mass of $CH_2O)x = 180.18$ g/mol,

$(30.03 \text{ g/mol})x = 180.18$ g/mol,

$$x = \frac{180.18 \text{ g/mol}}{30.03 \text{ g/mol}} = 6$$

The molecular formula is six times the empirical formula.

molecular formula $= C_6H_{12}O_6$

49. What is compound XYZ_3

X: $(0.4004)(100.09 \text{ g}) = 40.08$ g (calcium)

Y: $(0.1200)(100.09 \text{ g}) = 12.01$ g (carbon)

Z: $(0.4796)(100.09 \text{ g}) = 48.00 \text{ g}; \dfrac{48.00 \text{ g}}{3} = 16.00$ g (oxygen)

Elements determined from atomic masses in the periodic table.

$XYZ_3 = CaCO_3$

50. What is compound $X_2(YZ_3)_3$

X: $(0.1912)(282.23 \text{ g}) = \dfrac{53.96 \text{ g}}{2} = 26.98$ g (aluminum)

Y: $(0.2986)(282.23 \text{ g}) = \dfrac{84.27 \text{ g}}{3} = 28.09$ g (silicon)

Z: $(0.5102)(282.23 \text{ g}) = \dfrac{143.99 \text{ g}}{9} = 16.00$ g (oxygen)

Elements determined from atomic masses in the periodic table.

$X_2(YZ_3)_3 = Al_2(SiO_3)_3$

51. $$(0.350 \text{ mol } P_4)\left(\frac{6.022 \times 10^{23} \text{ molecules}}{\text{mol}}\right)\left(\frac{4 \text{ atoms P}}{\text{molecule } P_4}\right) = 8.43 \times 10^{23} \text{ atoms P}$$

52. $$(10.0 \text{ g K})\left(\frac{1 \text{ mol}}{39.10 \text{ g}}\right)\left(\frac{1 \text{ mol Na}}{1 \text{ mol K}}\right)\left(\frac{22.99 \text{ g}}{\text{mol}}\right) = 5.88 \text{ g Na}$$

53. $(1.79 \times 10^{-23}\ \text{g/atom})(6.022 \times 10^{23}\ \text{atoms/molar mass}) = 10.8\ \text{g/molar mass}$

54. $(6.022 \times 10^{23}\ \text{sheets})\left(\dfrac{4.60\ \text{cm}}{500\ \text{sheets}}\right)\left(\dfrac{1\ \text{m}}{100\ \text{cm}}\right) = 5.54 \times 10^{19}\ \text{m}$

55. $\left(\dfrac{6.022 \times 10^{23}\ \text{dollars}}{6.1 \times 10^{9}\ \text{people}}\right) = 9.9 \times 10^{13}\ \text{dollars/person}$

56. The conversion is: $\text{mi}^3 \longrightarrow \text{ft}^3 \longrightarrow \text{in.}^3 \longrightarrow \text{cm}^3 \longrightarrow \text{drops}$

(a) $(1\ \text{mi}^3)\left(\dfrac{5280\ \text{ft}}{\text{mile}}\right)^3\left(\dfrac{12.0\ \text{in.}}{\text{ft}}\right)^3\left(\dfrac{2.54\ \text{cm}}{\text{inch}}\right)^3\left(\dfrac{20\ \text{drops}}{1.0\ \text{cm}^3}\right) = 8 \times 10^{16}\ \text{drops}$

(b) $(6.022 \times 10^{23}\ \text{drops})\left(\dfrac{1\ \text{mi}^3}{8 \times 10^{16}\ \text{drops}}\right) = 8 \times 10^{16}\ \text{mi}^3$

57. $1\ \text{mol Ag} = 107.9\ \text{g Ag}$

(a) $(107.9\ \text{g Ag})\left(\dfrac{1\ \text{cm}^3}{10.5\ \text{g}}\right) = 10.3\ \text{cm}^3$ (volume of cube)

(b) $10.3\ \text{cm}^3 = \text{volume of cube} = (\text{one side})^3$

side $= \sqrt[3]{10.3\ \text{cm}^3} = 2.18\ \text{cm}$

58. The conversion is: $\text{L sol.} \longrightarrow \text{mL sol.} \longrightarrow \text{g sol.} \longrightarrow \text{g } H_2SO_4 \longrightarrow \text{mol } H_2SO_4$

$$(1.00\text{L})\left(\dfrac{1000\ \text{mL}}{1\ \text{L}}\right)\left(\dfrac{1.55\ \text{g}}{1.00\ \text{mL}}\right)\left(\dfrac{0.650\ \text{g } H_2SO_4}{1.00\ \text{g}}\right)\left(\dfrac{1\ \text{mol}}{98.09\ \text{g}}\right) = 10.3\ \text{mol } H_2SO_4$$

59. The conversion is: $\text{mL sol.} \longrightarrow \text{g sol.} \longrightarrow \text{g } HNO_3 \longrightarrow \text{mol } HNO_3$

$$(100.\ \text{mL})\left(\dfrac{1.42\ \text{g}}{1.00\ \text{mL}}\right)\left(\dfrac{0.720\ \text{g } HNO_3}{1.000\ \text{g}}\right)\left(\dfrac{1\ \text{mol}}{63.02\ \text{g}}\right) = 1.62\ \text{mol } HNO_3$$

60. (a) Determine the molar mass of each compound.
CO_2, 44.01 g; O_2, 32.00 g; H_2O, 18.02 g; CH_3OH, 32.04 g. The 1.00 gram sample with the lowest molar mass will contain the most molecules. Thus, H_2O will contain the most molecules.

(b) $(1.00\ \text{g } H_2O)\left(\dfrac{1\ \text{mol}}{18.02\ \text{g}}\right)\left(\dfrac{(3)(6.022 \times 10^{23}\ \text{atoms})}{\text{mol}}\right) = 1.00 \times 10^{23}\ \text{atoms}$

$$(1.00 \text{ g CH}_3\text{OH})\left(\frac{1 \text{ mol}}{32.04 \text{ g}}\right)\left(\frac{(6)(6.022 \times 10^{23} \text{ atoms})}{\text{mol}}\right) = 1.13 \times 10^{23} \text{ atoms}$$

$$(1.00 \text{ g CO}_2)\left(\frac{1 \text{ mol}}{44.01 \text{ g}}\right)\left(\frac{(3)(6.022 \times 10^{23} \text{ atoms})}{\text{mol}}\right) = 4.10 \times 10^{22} \text{ atoms}$$

$$(1.00 \text{ g O}_2)\left(\frac{1 \text{ mol}}{32.00 \text{ g}}\right)\left(\frac{(2)(6.022 \times 10^{23} \text{ atoms})}{\text{mol}}\right) = 3.76 \times 10^{22} \text{ atoms}$$

The 1.00 g sample of CH_3OH contains the most atoms

61. 1 mol Fe_2S_3 = 207.9 g Fe_2S_3 = 6.022×10^{23} formula units

$$(6.022 \times 10^{23} \text{ atoms})\left(\frac{1 \text{ formula unit}}{5 \text{ atoms}}\right)\left(\frac{207.9 \text{ g Fe}_2\text{S}_3}{6.022 \times 10^{23} \text{ formula units}}\right) = 41.58 \text{ g Fe}_2\text{S}_3$$

62. $3\,Ca + 2\,P \longrightarrow Ca_3P_2$

The conversion is g P $\longrightarrow$ mol P $\longrightarrow$ mol Ca $\longrightarrow$ g Ca

$$(1.00 \text{ g P})\left(\frac{1 \text{ mol P}}{30.97 \text{ g P}}\right)\left(\frac{3 \text{ mol Ca}}{2 \text{ mol P}}\right)\left(\frac{40.08 \text{ g Ca}}{1 \text{ mol Ca}}\right) = 1.94 \text{ g Ca}$$

1.94 g Ca combines with 1.00 g P.

63. Grams of Fe per ton of ore that contains 5% $FeSO_4$.

The conversion is: ton $\longrightarrow$ lb $\longrightarrow$ g $\longrightarrow$ g $FeSO_4$ $\longrightarrow$ g Fe

$$(1.0 \text{ ton})\left(\frac{2000 \text{ lb}}{\text{ton}}\right)\left(\frac{453.6 \text{ g}}{\text{lb}}\right)(0.05 \text{ FeSO}_4)\left(\frac{55.85 \text{ Fe}}{151.9 \text{ g FeSO}_4}\right) = 2 \times 10^4 \text{ g Fe}$$

1.0 ton of iron ore contains 2×10^4 g Fe.

64. From the formula, 2 Li (13.88 g) combine with 1 S (32.07 g).

$$\left(\frac{13.88 \text{ g Li}}{32.07 \text{ g S}}\right)(20.0 \text{ g S}) = 8.66 \text{ g Li}$$

65. (a) $HgCO_3$

Hg	200.6 g	
C	12.01 g	
3 O	48.00 g	
	260.6 g	

$$\left(\frac{200.6 \text{ g}}{260.6 \text{ g}}\right)(100) = 76.98\% \text{ Hg}$$

(b) $Ca(ClO_3)_2$

6 O	96.00 g	$\left(\frac{96.00\text{ g}}{207.0\text{ g}}\right)(100) = 46.38\%\text{ O}$
2 Cl	70.90 g	
Ca	40.08 g	
	207.0 g	

(c) $C_{10}H_{14}N_2$

2 N	28.02 g	$\left(\frac{28.02\text{ g}}{162.2\text{ g}}\right)(100) = 17.27\%\text{ N}$
10 C	120.1 g	
14 H	14.11 g	
	162.2 g	

(d) $C_{55}H_{72}MgN_4O_5$

Mg	24.31 g	$\left(\frac{24.31\text{ g}}{893.5\text{ g}}\right)(100) = 2.721\%\text{ Mg}$
55 C	660.55 g	
72 H	72.58 g	
4 N	56.04 g	
5 O	80.00 g	
	893.5 g	

66. According to the formula, 1 mol (65.39 g) Zn combines with 1 mol (32.07 g) S.

$$(19.5\text{ g Zn})\left(\frac{32.07\text{ g S}}{65.39\text{ g Zn}}\right) = 9.56\text{ g S}$$

19.5 g Zn require 9.56 g S for complete reaction. Therefore, there is not sufficient S present (9.40 g) to react with the Zn.

67. Molecular formula of aspirin

60.0% C, 4.48% H, 35.5% O; molar mass of aspirin = 180.2

$$(60.0\text{ g C})\left(\frac{1\text{ mol}}{12.01\text{ g}}\right) = 5.00\text{ mol C} \qquad \frac{5.00}{2.22} = 2.25\text{ mol C}$$

$$(4.48\text{ g H})\left(\frac{1\text{ mol}}{1.008\text{ g}}\right) = 4.44\text{ mol H} \qquad \frac{4.44}{2.22} = 2.00\text{ mol H}$$

$$(35.5\text{ g O})\left(\frac{1\text{ mol}}{16.00\text{ g}}\right) = 2.22\text{ mol O} \qquad \frac{2.22}{2.22} = 1.00\text{ mol O}$$

Multiplying each by 4 give the empirical formula $C_9H_8O_4$. The empirical mass is 180.2. Since the empirical mass equals the molar mass, the molecular formula is the same as the empirical formula, $C_9H_8O_4$.

68. Calculate the percent oxygen in $Al_2(SO_4)_3$.

2 Al	53.96 g
3 S	96.21 g
12 O	192.0 g
	342.2 g

$$\left(\frac{192.0\text{ g}}{342.2\text{ g}}\right)(100) = 56.11\%\text{ O}$$

Now take 56.11% of 8.50 g

$$(8.50\text{ g O})(0.5611) = 4.77\text{ g O}$$

69. Empirical formula of gallium arsenide; 48.2% Ga, 51.8% As

$$(48.2\text{ g Ga})\left(\frac{1\text{ mol}}{69.72\text{ g}}\right) = 0.691\text{ mol Ga} \qquad \frac{0.691}{0.691} = 1.00\text{ mol Ga}$$

$$(51.8\text{ g As})\left(\frac{1\text{ mol}}{74.92\text{ g}}\right) = 0.691\text{ mol As} \qquad \frac{0.691}{0.691} = 1.00\text{ mol As}$$

The empirical formula is GaAs.

70. (a) 7.79% C, 92.21% Cl

$$(7.79\text{ g C})\left(\frac{1\text{ mol}}{12.01\text{ g}}\right) = 0.649\text{ mol C} \qquad \frac{0.649}{0.649} = 1.00\text{ mol C}$$

$$(92.21\text{ g Cl})\left(\frac{1\text{ mol}}{35.45\text{ g}}\right) = 2.601\text{ mol Cl} \qquad \frac{2.601}{0.649} = 4.01\text{ mol Cl}$$

The empirical formula is CCl_4. The empirical mass is 153.8 which equals the molar mass, therefore the molecular formula is CCl_4.

(b) 10.13% C, 89.87% Cl

$$(10.13\text{ g C})\left(\frac{1\text{ mol}}{12.01\text{ g}}\right) = 0.8435\text{ mol C} \qquad \frac{0.8435}{0.8435} = 1.000\text{ mol C}$$

$$(89.87\text{ g Cl})\left(\frac{1\text{ mol}}{35.45\text{ g}}\right) = 2.535\text{ mol Cl} \qquad \frac{2.535}{0.8435} = 3.005\text{ mol Cl}$$

The empirical formula is CCl_3. The empirical mass is 118.4.

$$\frac{\text{molar mass}}{\text{empirical mass}} = \frac{236.7}{118.4} = 1.999$$

The molecular formula is twice that of the empirical formula.
Molecular formula = C_2Cl_6.

(c) 25.26% C, 74.74% Cl

$$(25.26\text{ g C})\left(\frac{1\text{ mol}}{12.01\text{ g}}\right) = 2.103\text{ mol C} \qquad \frac{2.103}{2.103} = 1.000\text{ mol C}$$

$$(74.74\text{ g Cl})\left(\frac{1\text{ mol}}{35.45\text{ g}}\right) = 2.108\text{ mol Cl} \qquad \frac{2.103}{2.108} = 1.002\text{ mol Cl}$$

The empirical formula is CCl. The empirical mass is 47.46.

$$\frac{\text{molar mass}}{\text{empirical mass}} = \frac{284.8}{47.46} = 6.000$$

The molecular formula is six times that of the empirical formula.
Molecular formula = C_6Cl_6.

(d) 11.25% C, 88.75% Cl

$$(11.25\text{ g C})\left(\frac{1\text{ mol}}{12.01\text{ g}}\right) = 0.9367\text{ mol C} \qquad \frac{0.9367}{0.9367} = 1.000\text{ mol C}$$

$$(88.75\text{ g Cl})\left(\frac{1\text{ mol}}{35.45\text{ g}}\right) = 2.504\text{ mol Cl} \qquad \frac{2.504}{0.9367} = 2.673\text{ mol Cl}$$

Multiplying each by 3 give the empirical formula C_3Cl_8. The empirical mass is 319.6. Since the molar mass is also 319.6 the molecular formula is C_3Cl_8.

71. The conversion is: s $\longrightarrow$ min $\longrightarrow$ hr $\longrightarrow$ day $\longrightarrow$ yr

$$(6.022 \times 10^{23}\text{ s})\left(\frac{1\text{ min}}{60\text{ s}}\right)\left(\frac{1\text{ hr}}{60\text{ min}}\right)\left(\frac{1\text{ day}}{24\text{ hr}}\right)\left(\frac{1\text{ year}}{365\text{ days}}\right) = 1.910 \times 10^{16}\text{ years}$$

72. The conversion is: g Cu $\longrightarrow$ mol $\longrightarrow$ atom

$$(2.5\text{ g Cu})\left(\frac{1\text{ mol}}{63.55\text{ g}}\right)\left(\frac{6.022 \times 10^{23}\text{ atoms}}{\text{mol}}\right) = 2.4 \times 10^{22}\text{ atoms Cu}$$

73. The conversion is: molecules $\longrightarrow$ mol $\longrightarrow$ g 1 trillion = 10^{12}

$$(1000. \times 10^{12}\text{ molecules } C_3H_8O_3)\left(\frac{1\text{ mol}}{6.022 \times 10^{23}\text{ molecules}}\right)\left(\frac{92.09\text{ g}}{\text{mol}}\right)$$

$$= 1.529 \times 10^{-7}\text{ g } C_3H_8O_3$$

74. $$(6.1 \times 10^9\text{ people})\left(\frac{1\text{ mol people}}{6.022 \times 10^{23}\text{ people}}\right) = 1.0 \times 10^{-14}\text{ mol of people}$$

75. Empirical formula

23.3% Co, 25.3% Mo, 51.4% Cl

$$(23.3 \text{ g Co})\left(\frac{1 \text{ mol}}{58.93 \text{ g}}\right) = 0.395 \text{ mol Co} \qquad \frac{0.395}{0.264} = 1.50$$

$$(25.3 \text{ g Mo})\left(\frac{1 \text{ mol}}{95.94 \text{ g}}\right) = 0.264 \text{ mol Mo} \qquad \frac{0.264}{0.264} = 1.00$$

$$(51.4 \text{ g Cl})\left(\frac{1 \text{ mol}}{35.45 \text{ g}}\right) = 1.45 \text{ mol Cl} \qquad \frac{1.45}{0.264} = 5.49$$

Multiplying by 2 gives the empirical formula $Co_3Mo_2Cl_{11}$.

76. The conversion is: g Al $\longrightarrow$ mol Al $\longrightarrow$ mol Mg $\longrightarrow$ g Mg

$$(18 \text{ g Al})\left(\frac{1 \text{ mol}}{26.98 \text{ g}}\right)\left(\frac{2 \text{ mol Mg}}{1 \text{ mol Al}}\right)\left(\frac{24.31 \text{ g}}{\text{mol}}\right) = 32 \text{ g Mg}$$

77. $(10.0 \text{ g N})(0.177) = 1.77 \text{ g N}$

$$(1.77 \text{ g N})\left(\frac{1 \text{ mol}}{14.01 \text{ g}}\right) = 0.126 \text{ mol N}$$

$$(3.8 \times 10^{23} \text{ atoms H})\left(\frac{1 \text{ mol}}{6.022 \times 10^{23} \text{ atoms}}\right) = 0.63 \text{ mol H}$$

To determine the mol C, first find grams H and subtract the grams of H and N from the grams of the sample.

$$(0.63 \text{ mol H})\left(\frac{1.008 \text{ g}}{\text{mol}}\right) = 0.64 \text{ g H}$$

10.0 g	sample
−1.77 g	N
−0.64 g	H
7.6 g	C

$$(7.6 \text{ g C})\left(\frac{1 \text{ mol}}{12.01 \text{ g}}\right) = 0.63 \text{ mol C}$$

Now determine the empirical formula from the moles of C, H, and N.

$$\text{N} \quad \frac{0.126}{0.126} = 1.00$$

$$\text{H} \quad \frac{0.63}{0.126} = 5.0$$

$$\text{C} \quad \frac{0.63}{0.126} = 5.0$$

The formula is C_5H_5N

78. Let x = molar mass of A_2O

$$0.400x = 16.00 \text{ g O (Since } A_2O \text{ has only one mol of O atoms)}$$

$$x = 40.0 \text{ g O/mol } A_2O$$

$$40.0 = 16.00 + 2y \qquad y = \text{molar mass of A}$$

$$40.0 - 16.00 = 2y$$

$$12.0\frac{\text{g}}{\text{mol}} = y$$

Look in the periodic table for the element that has 12.0 g/mol.
The element is carbon. The mystery element is carbon.

79. (a) CH_2O (divide by 6)
(b) C_4H_9 (divide by 2)
(c) CH_2O (divide by 3)
(d) $C_{25}H_{52}$ (divide by 1)
(e) $C_6H_2Cl_2O$ (divide by 2)

80. First determine the elements in compound $A(BC)_3$:

A: $(0.3459)(78.01 \text{ g}) = 26.98 \text{ g (aluminum)}$

B: $(0.6153)(78.01 \text{ g}) = \frac{48.00 \text{ g}}{3} = 16.00 \text{ g (oxygen)}$

C: $(0.0388)(78.01 \text{ g}) = \frac{3.03 \text{ g}}{3} = 1.01 \text{ g (hydrogen)}$

Elements determined from atomic masses in the periodic table.

$A(BC)_3 = Al(OH)_3$

Then compound $A_2B_3 = Al_2O_3$ with a molar mass of
$2(26.98 \text{ g}) + 3(16.00 \text{ g}) = 101.96 \text{ g}$

$$\% \text{ Al} = \frac{2(26.98 \text{ g})}{101.96 \text{ g}}(100) = 52.92\%$$

$$\% \text{ O} = \frac{3(16.00)}{101.96}(100) = 47.08\%$$

CHAPTER 8

CHEMICAL EQUATIONS

1. The purpose of balancing chemical equations is to conform to the Law of Conservation of Mass. Ratios of reactants and products can then be easily determined.

2. The coefficients in a balanced chemical equation represent the number of moles (or molecules or formula units) of each of the chemical species in the reaction.

3. (a) Yes. It is necessary to conserve atoms to follow the Law of Conservation of Mass.
   (b) No. Molecules can be taken apart and rearranged to form different molecules in reactions.
   (c) Moles of molecules are not conserved (b). Moles of atoms are conserved (a).

4. A chemical change that absorbs heat energy is said to be an *endothermic* reaction. The products are at a higher energy level than the reactants. A chemical change that liberates heat energy is said to be an *exothermic* reaction. The products are at a lower energy level than the reactants.

5. The charts are one way of keeping track of the number of atoms of each element on the reactant side of a chemical equation and on the product side of an equation. The top row in a chart gives the number and types of atoms on the reactant side and the bottom row gives the number and types of atoms on the product side of a chemical equation. Using a chart may make it easier to see where coefficients are needed in a reaction and what number that coefficient should be.

6. The symbols indicate whether a substance is a solid, a liquid, a gas or is in an aqueous solution. A solid is indicated by (*s*), a liquid by (*l*), a gas by (*g*) and an aqueous solution by (*aq*).

7. The activity series given in Table 8.2 shows the relative activity of certain metals and halogens. As you move up the table starting with gold (Au) and ending with potassium (K) the activity increases. The same is true as you move up from iodine (I_2) to fluorine (F_2). The table is useful for predicting the products of some reactions because an element in the series will replace any element given below it. For example, hydrogen can replace copper, silver, mercury or gold in a chemical reaction.

8. A combustion reaction is an exothermic process (usually burning) done in the presence of oxygen.

9. (a) exothermic
   (b) endothermic
   (c) exothermic
   (d) exothermic
   (e) endothermic

10. (a) endothermic
    (b) exothermic
    (c) endothermic
    (d) exothermic
    (e) exothermic

11. (a) $2\,H_2 + O_2 \longrightarrow 2\,H_2O$
    (b) $3\,C + Fe_2O_3 \longrightarrow 2\,Fe + 3\,CO$
    (c) $H_2SO_4 + 2\,NaOH \longrightarrow 2\,H_2O + Na_2SO_4$
    (d) $Al_2(CO_3)_3 \xrightarrow{\Delta} Al_2O_3 + 3\,CO_2$
    (e) $2\,NH_4I + Cl_2 \longrightarrow 2\,NH_4Cl + I_2$

12. (a) $H_2 + Br_2 \longrightarrow 2\,HBr$
    (b) $4\,Al + 3\,C \xrightarrow{\Delta} Al_4C_3$
    (c) $Ba(ClO_3)_2 \xrightarrow{\Delta} BaCl_2 + 3\,O_2$
    (d) $CrCl_3 + 3\,AgNO_3 \longrightarrow Cr(NO_3)_3 + 3\,AgCl$
    (e) $2\,H_2O_2 \longrightarrow 2\,H_2O + O_2$

13. (a) combination
    (b) single displacement
    (c) double displacement
    (d) decomposition
    (e) single displacement

14. (a) combination
    (b) combination
    (c) decomposition
    (d) double displacement
    (e) decomposition

15. There are many ways to form oxides. For example: (1) some metals plus oxygen, (2) some nonmetals plus oxygen, (3) some metals plus water (steam), (4) combustion of hydrocarbons

16. A metal and a nonmetal can react to form a salt; also an acid plus a base.

17. (a) $2\,MnO_2 + CO \longrightarrow Mn_2O_3 + CO_2$
    (b) $Mg_3N_2 + 6\,H_2O \longrightarrow 3\,Mg(OH)_2 + 2\,NH_3$
    (c) $4\,C_3H_5(NO_3)_3 \longrightarrow 12\,CO_2 + 10\,H_2O + 6\,N_2 + O_2$
    (d) $4\,FeS + 7\,O_2 \longrightarrow 2\,Fe_2O_3 + 4\,SO_2$

(e) $2\,Cu(NO_3)_2 \longrightarrow 2\,CuO + 4\,NO_2 + O_2$

(f) $3\,NO_2 + H_2O \longrightarrow 2\,HNO_3 + NO$

(g) $2\,Al + 3\,H_2SO_4 \longrightarrow Al_2(SO_4)_3 + 3\,H_2$

(h) $4\,HCN + 5\,O_2 \longrightarrow 2\,N_2 + 4\,CO_2 + 2\,H_2O$

(i) $2\,B_5H_9 + 12\,O_2 \longrightarrow 5\,B_2O_3 + 9\,H_2O$

18. (a) $2\,SO_2 + O_2 \longrightarrow 2\,SO_3$

(b) $4\,Al + 3\,MnO_2 \xrightarrow{\Delta} 3\,Mn + 2\,Al_2O_3$

(c) $2\,Na + 2\,H_2O \longrightarrow 2\,NaOH + H_2$

(d) $2\,AgNO_3 + Ni \longrightarrow Ni(NO_3)_2 + 2\,Ag$

(e) $Bi_2S_3 + 6\,HCl \longrightarrow 2\,BiCl_3 + 3\,H_2S$

(f) $2\,PbO_2 \xrightarrow{\Delta} 2\,PbO + O_2$

(g) $2\,LiAlH_4 \xrightarrow{\Delta} 2\,LiH + 2\,Al + 3\,H_2$

(h) $2\,KI + Br_2 \longrightarrow 2\,KBr + I_2$

(i) $2\,K_3PO_4 + 3\,BaCl_2 \longrightarrow 6\,KCl + Ba_3(PO_4)_2$

19. (a) $2\,H_2O \longrightarrow 2\,H_2 + O_2$

(b) $HC_2H_3O_2 + KOH \longrightarrow KC_2H_3O_2 + H_2O$

(c) $2\,P + 3\,I_2 \longrightarrow 2\,PI_3$

(d) $2\,Al + 3\,CuSO_4 \longrightarrow 3\,Cu + Al_2(SO_4)_3$

(e) $(NH_4)_2SO_4 + BaCl_2 \longrightarrow 2\,NH_4Cl + BaSO_4$

(f) $SF_4 + 2\,H_2O \longrightarrow SO_2 + 4\,HF$

(g) $Cr_2(CO_3)_3 \xrightarrow{\Delta} Cr_2O_3 + 3\,CO_2$

20. (a) $2\,Cu + S \xrightarrow{\Delta} Cu_2S$

(b) $2\,H_3PO_4 + 3\,Ca(OH)_2 \longrightarrow Ca_3(PO_4)_2 + 6\,H_2O$

(c) $2\,Ag_2O \xrightarrow{\Delta} 4\,Ag + O_2$

(d) $FeCl_3 + 3\,NaOH \longrightarrow Fe(OH)_3 + 3\,NaCl$

(e) $Ni_3(PO_4)_2 + 3\,H_2SO_4 \longrightarrow 3\,NiSO_4 + 2\,H_3PO_4$

(f) $ZnCO_3 + 2\,HCl \longrightarrow ZnCl_2 + H_2O + CO_2$

(g) $3\,AgNO_3 + AlCl_3 \longrightarrow 3\,AgCl + Al(NO_3)_3$

21. (a) $Ag(s) + H_2SO_4(aq) \longrightarrow$ no reaction
    (b) $Cl_2(g) + 2\,NaBr(aq) \longrightarrow Br_2(l) + 2\,NaCl(aq)$
    (c) $Mg(s) + ZnCl_2(aq) \longrightarrow Zn(s) + MgCl_2(aq)$
    (d) $Pb(s) + 2\,AgNO_3(aq) \longrightarrow 2\,Ag(s) + Pb(NO_3)_2(aq)$

22. (a) $Cu(s) + FeCl_3(aq) \longrightarrow$ no reaction
    (b) $H_2(g) + Al_2O_3(s) \xrightarrow{\Delta}$ no reaction
    (c) $2\,Al(s) + 6\,HBr(aq) \longrightarrow 3\,H_2(g) + 2\,AlBr_3(aq)$
    (d) $I_2(s) + HCl(aq) \longrightarrow$ no reaction

23. (a) $H_2 + I_2 \longrightarrow 2\,HI$
    (b) $CaCO_3 \xrightarrow{\Delta} CaO + CO_2$
    (c) $Mg + H_2SO_4 \longrightarrow H_2 + MgSO_4$
    (d) $FeCl_2 + 2\,NaOH \longrightarrow Fe(OH)_2 + 2\,NaCl$

24. (a) $SO_2 + H_2O \longrightarrow H_2SO_3$
    (b) $SO_3 + H_2O \longrightarrow H_2SO_4$
    (c) $Ca + 2\,H_2O \longrightarrow Ca(OH)_2 + H_2$
    (d) $2\,Bi(NO_3)_3 + 3\,H_2S \longrightarrow Bi_2S_3 + 6\,HNO_3$

25. (a) $2\,Ba + O_2 \longrightarrow 2\,BaO$
    (b) $2\,NaHCO_3 \xrightarrow{\Delta} Na_2CO_3 + H_2O + CO_2$
    (c) $Ni + CuSO_4 \longrightarrow NiSO_4 + Cu$
    (d) $MgO + 2\,HCl \longrightarrow MgCl_2 + H_2O$
    (e) $H_3PO_4 + 3\,KOH \longrightarrow K_3PO_4 + 3\,H_2O$

26. (a) $C + O_2 \xrightarrow{\Delta} CO_2$
    (b) $2\,Al(ClO_3)_3 \xrightarrow{\Delta} 9\,O_2 + 2\,AlCl_3$
    (c) $CuBr_2 + Cl_2 \longrightarrow CuCl_2 + Br_2$
    (d) $2\,SbCl_3 + 3(NH_4)_2S \longrightarrow Sb_2S_3 + 6\,NH_4Cl$
    (e) $2\,NaNO_3 \xrightarrow{\Delta} 2\,NaNO_2 + O_2$

27. (a) One mole of $MgBr_2$ reacts with two moles of $AgNO_3$ to yield one mole of $Mg(NO_3)_2$ and two moles of AgBr.
    (b) One mole of N_2 reacts with three moles of H_2 to produce two moles of NH_3.

(c) Two moles of C_3H_7OH react with nine moles of O_2 to form six moles of CO_2 and eight moles of H_2O.

28. (a) Two moles of Na react with one mole of Cl_2 to produce two moles of NaCl and release 822 kJ of energy. The reaction is exothermic.
    (b) One mole of PCl_5 absorbs 92.9 kJ of energy to produce one mole of PCl_3 and one mole of Cl_2. The reaction is endothermic.

29. (a) $CaO + H_2O \longrightarrow Ca(OH)_2 + 65.3\ kJ$
    (b) $2\,Al_2O_3 + 3260\ kJ \longrightarrow 4\,Al + 3\,O_2$

30. (a) $2\,Al + 3\,I_2 \longrightarrow 2\,AlI_3 + \text{heat}$
    (b) $4\,CuO + CH_4 + \text{heat} \longrightarrow 4\,Cu + CO_2 + 2\,H_2O$

31. (a) decomposition reaction, $2\,AgClO_3(s) \longrightarrow 2\,AgCl(s) + 3\,O_2(g)$
    (b) single-displacement, $Fe(s) + H_2SO_4(aq) \longrightarrow H_2(g) + FeSO_4(aq)$
    (c) combination reaction, $Zn(s) + Cl_2(g) \longrightarrow ZnCl_2(s)$
    (d) double-displacement, $HBr(aq) + KOH(aq) \longrightarrow KBr(aq) + H_2O(l)$

32. (a) single-displacement, $Ni(s) + Pb(NO_3)_2(aq) \longrightarrow Pb(s) + Ni(NO_3)_2(aq)$
    (b) combination, $MgO(s) + H_2O(l) \longrightarrow Mg(OH)_2(s)$
    (c) decomposition, $2\,HgO(s) \longrightarrow 2\,Hg(l) + O_2(g)$
    (d) double-displacement,
    $PbCl_2(aq) + (NH_4)_2CO_3(aq) \longrightarrow PbCO_3(s) + 2\,NH_4Cl(aq)$

33. (a) change in color and texture of the bread
    (b) change in texture of the white and the yoke
    (c) the flame (combustion), change in matchhead, odor

34. $P_4O_{10} + 12\,HClO_4 \longrightarrow 6\,Cl_2O_7 + 4\,H_3PO_4$

$$\begin{array}{ccc} 10\,O + 12\,(4\,O) & & 6\,(7\,O) + 4\,(4\,O) \\ 10\,O + 48\,O & & 42\,O + 16\,O \\ 58\,O & & 58\,O \end{array}$$

35. In $7\,Al_2(SO_4)_3$ there are:
    (a) 14 atoms of Al
    (b) 21 atoms of S
    (c) 84 atoms of O
    (d) 119 total atoms

36. A balanced equation tells us:
    (1) the types of atoms/molecules involved in the reaction
    (2) the relationship between quantities of the substances in the reaction

    A balanced equation gives no information about
    (1) the time required for the reaction
    (2) odors or colors which may result

37.

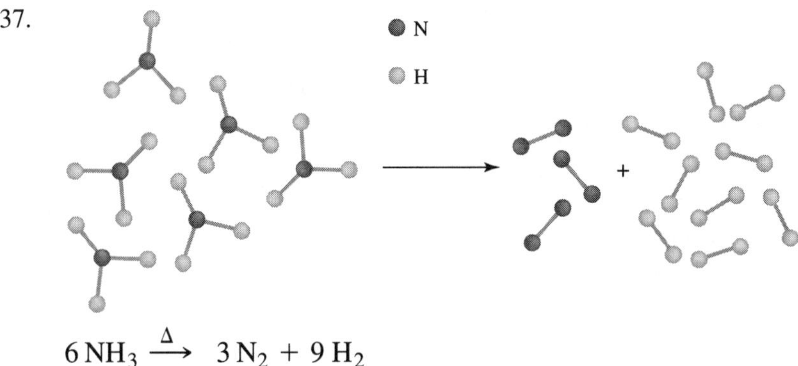

$$6\,NH_3 \xrightarrow{\Delta} 3\,N_2 + 9\,H_2$$

38. Zn metal is below Mg on the activity series.

39. $Ti + Ni(NO_3)_2 \longrightarrow$ Reaction occurs
    $Ti + Pb(NO_3)_2 \longrightarrow$ Reaction occurs
    $Ti + Mg(NO_3)_2 \longrightarrow$ no reaction

    Ti is above Ni and Pb in the activity series since both react. Ti is below Mg in the series since it will not replace Mg. From the printed activity series in the chapter Ni lies above Pb so the order is:

    Mg
    Ti
    Ni
    Pb

40. (a) $4\,K + O_2 \longrightarrow 2\,K_2O$
    (b) $2\,Al + 3\,Cl_2 \longrightarrow 2\,AlCl_3$
    (c) $CO_2 + H_2O \longrightarrow H_2CO_3$
    (d) $CaO + H_2O \longrightarrow Ca(OH)_2$

41. (a) $2\,HgO \xrightarrow{\Delta} 2\,Hg + O_2$
    (b) $2\,NaClO_3 \xrightarrow{\Delta} 2\,NaCl + 3\,O_2$
    (c) $MgCO_3 \xrightarrow{\Delta} MgO + CO_2$
    (d) $2\,PbO_2 \xrightarrow{\Delta} 2\,PbO + O_2$

42. (a) $Zn + H_2SO_4 \longrightarrow H_2 + ZnSO_4$
    (b) $2\,AlI_3 + 3\,Cl_2 \longrightarrow 2\,AlCl_3 + 3\,I_2$
    (c) $Mg + 2\,AgNO_3 \longrightarrow Mg(NO_3)_2 + 2\,Ag$
    (d) $2\,Al + 3\,CoSO_4 \longrightarrow Al_2(SO_4)_3 + 3\,Co$

43. (a) $ZnCl_2 + 2\,KOH \longrightarrow Zn(OH)_2 + 2\,KCl$
    (b) $CuSO_4 + H_2S \longrightarrow H_2SO_4 + CuS$
    (c) $3\,Ca(OH)_2 + 2\,H_3PO_4 \longrightarrow 6\,H_2O + Ca_3(PO_4)_2$
    (d) $2\,(NH_4)_3PO_4 + 3\,Ni(NO_3)_2 \longrightarrow 6\,NH_4NO_3 + Ni_3(PO_4)_2$
    (e) $Ba(OH)_2 + 2\,HNO_3 \longrightarrow 2\,H_2O + Ba(NO_3)_2$
    (f) $(NH_4)_2S + 2\,HCl \longrightarrow H_2S + 2\,NH_4Cl$

44. (a) $AgNO_3(aq) + KCl(aq) \longrightarrow AgCl(s) + KNO_3(aq)$
    (b) $Ba(NO_3)_2(aq) + MgSO_4(aq) \longrightarrow Mg(NO_3)_2(aq) + BaSO_4(s)$
    (c) $H_2SO_4(aq) + Mg(OH)_2(aq) \longrightarrow 2\,H_2O(l) + MgSO_4(aq)$
    (d) $MgO(s) + H_2SO_4(aq) \longrightarrow H_2O(l) + MgSO_4(aq)$
    (e) $Na_2CO_3(aq) + NH_4Cl(aq) \longrightarrow$ no reaction

45. (a) $2\,C_2H_6 + 7\,O_2 \longrightarrow 4\,CO_2 + 6\,H_2O$
    (b) $2\,C_6H_6 + 15\,O_2 \longrightarrow 12\,CO_2 + 6\,H_2O$
    (c) $C_7H_{16} + 11\,O_2 \longrightarrow 7\,CO_2 + 8\,H_2O$

46. 1. combustion of fossil fuels
    2. destruction of the rain forests by burning
    3. increased population

47. Carbon dioxide, methane, and water are all considered to be greenhouse gases. They each act to trap the heat near the surface of the earth in the same manner in which a greenhouse is warmed.

48. The effects of global warming can be reduced by:
    1. developing new energy sources (not dependant on fossil fuels)
    2. conservation of energy resources
    3. recycling
    4. decreased destruction of the rain forests and other forests

49. About half the carbon dioxide released into the atmosphere remains in the air. The rest is absorbed by plants and used in photosynthesis or is dissolved in the oceans.

Challenge Exercises

50.

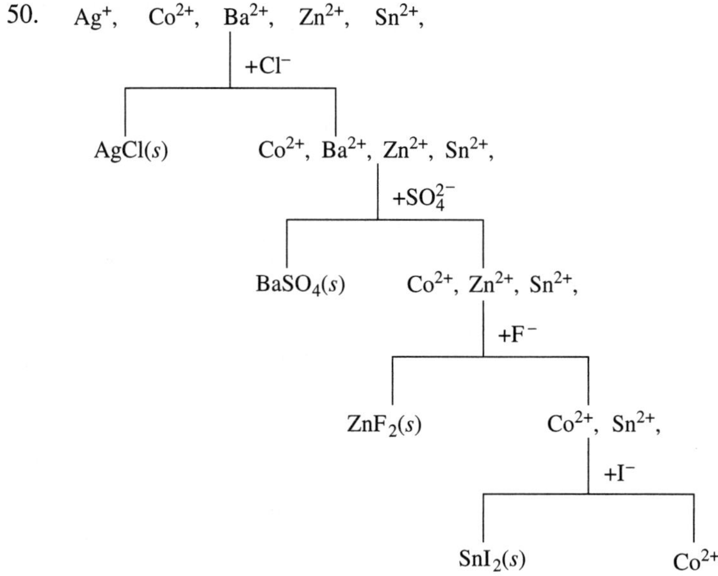

CHAPTER 9

CALCULATIONS FROM CHEMICAL EQUATIONS

1. The theoretical yield of a chemical reaction is the maximum amount of product that can be produced based on a balanced equation. The actual yield of a reaction is the actual amount of product obtained.

2. You can calculate the percent yield of a chemical reaction by dividing the actual yield by the theoretical yield and multiplying by one hundred.

3. A mole ratio is the ratio between the mole amounts of two atoms and/or molecules involved in a chemical reaction.

4. In order to convert grams to moles the molar mass of the compound under consideration needs to be determined.

5. The balanced equation is
   $Ca_3P_2 + 6\,H_2O \longrightarrow 3\,Ca(OH)_2 + 2\,PH_3$

   (a) Correct: $(1\text{ mol }Ca_3P_2)\left(\dfrac{2\text{ mol }PH_3}{1\text{ mol }Ca_3P_2}\right) = 2\text{ mol }PH_3$

   (b) Incorrect: 1 g Ca_3P_2 would produce 0.4 g PH_3

   $$(1\text{ g }Ca_3P_2)\left(\frac{1\text{ mol}}{182.2\text{ g}}\right)\left(\frac{2\text{ mol }PH_3}{1\text{ mol }Ca_3P_2}\right)\left(\frac{33.99\text{ g}}{\text{mol}}\right) = 0.4\text{ g }PH_3$$

   (c) Correct: see equation

   (d) Correct: see equation

   (e) Incorrect: 2 mol Ca_3P_2 requires 12 mol H_2O to produce 4.0 mol PH_3.

   $$(2\text{ mol }Ca_3P_2)\left(\frac{6\text{ mol }H_2O}{1\text{ mol }Ca_3P_2}\right) = 12\text{ mol }H_2O$$

   (f) Correct: 2 mol Ca_3P_2 will react with 12 mol H_2O (3 mol H_2O are present in excess) and 6 mol $Ca(OH)_2$ will be formed.

   $$(2\text{ mol }Ca_3P_2)\left(\frac{3\text{ mol }Ca(OH)_2}{1\text{ mol }Ca_3P_2}\right) = 6\text{ mol }Ca(OH)_2$$

   (g) Incorrect: $(200.\text{ g }Ca_3P_2)\left(\dfrac{1\text{ mol}}{182.2\text{ g}}\right)\left(\dfrac{6\text{ mol }H_2O}{1\text{ mol }Ca_3P_2}\right)\left(\dfrac{18.02\text{ g}}{\text{mol}}\right) = 119\text{ g }H_2O$

The amount of water present (100. g) is less than needed to react with 200. g Ca_3P_2. H_2O is the limiting reactant.

(h) Incorrect: water is the limiting reactant.

$$(100.\ \text{g H}_2\text{O})\left(\frac{1\ \text{mol}}{18.02\ \text{g}}\right)\left(\frac{2\ \text{mol PH}_3}{6\ \text{mol H}_2\text{O}}\right)\left(\frac{33.99\ \text{g}}{\text{mol}}\right) = 62.9\ \text{g PH}_3$$

6. The balanced equation is

$$2\,CH_4 + 3\,O_2 + 2\,NH_3 \longrightarrow 2\,HCN + 6\,H_2O$$

(a) Correct

(b) Incorrect: $(16\ \text{mol O}_2)\left(\frac{2\ \text{mol HCN}}{3\ \text{mol O}_2}\right) = 10.7\ \text{mol HCN}$ (not 12 mol HCN)

(c) Correct

(d) Incorrect: $(12\ \text{mol HCN})\left(\frac{6\ \text{mol H}_2\text{O}}{2\ \text{mol HCN}}\right) = 36\ \text{mol H}_2\text{O}$ (not 4 mol H_2O)

(e) Correct

(f) Incorrect: O_2 is the limiting reactant

$$(3\ \text{mol O}_2)\left(\frac{2\ \text{mol HCN}}{3\ \text{mol O}_2}\right) = 2\ \text{mol HCN (not 3 mol HCN)}$$

7. (a) $(25.0\ \text{g KNO}_3)\left(\frac{1\ \text{mol}}{101.1\ \text{g}}\right) = 0.247\ \text{mol KNO}_3$

(b) $(56\ \text{mmol NaOH})\left(\frac{1\ \text{mol}}{1000\ \text{mmol}}\right) = 0.056\ \text{mol NaOH}$

(c) $(5.4 \times 10^2\ \text{g (NH}_4)_2\text{C}_2\text{O}_4)\left(\frac{1\ \text{mol}}{124.1\ \text{g}}\right) = 4.4\ \text{mol (NH}_4)_2\text{C}_2\text{O}_4$

(d) The conversion is: mL sol $\longrightarrow$ g sol $\longrightarrow$ g H_2SO_4 $\longrightarrow$ mol H_2SO_4

$$(16.8\ \text{mL solution})\left(\frac{1.727\ \text{g}}{\text{mL}}\right)\left(\frac{0.800\ \text{g H}_2\text{SO}_4}{\text{g solution}}\right)\left(\frac{1\ \text{mol}}{98.09\ \text{g}}\right) = 0.237\ \text{mol H}_2\text{SO}_4$$

8. (a) $(2.10\ \text{kg NaHCO}_3)\left(\frac{1000\ \text{g}}{\text{kg}}\right)\left(\frac{1\ \text{mol}}{84.01\ \text{g}}\right) = 25.0\ \text{mol NaHCO}_3$

(b) $(525\ \text{mg ZnCl}_2)\left(\frac{1\ \text{g}}{1000\ \text{mg}}\right)\left(\frac{1\ \text{mol}}{136.3\ \text{g}}\right) = 3.85 \times 10^{-3}\ \text{mol ZnCl}_2$

(c) $(9.8 \times 10^{24}\ \text{molecules CO}_2)\left(\frac{1\ \text{mol}}{6.022 \times 10^{23}\ \text{molecules}}\right) = 16\ \text{mol CO}_2$

(d) $(250\ \text{mL } C_2H_5OH)\left(\frac{0.789\ \text{g}}{\text{mL}}\right)\left(\frac{1\ \text{mol}}{46.07\ \text{g}}\right) = 4.3\ \text{mol } C_2H_5OH$

9. (a) $(2.55\ \text{mol Fe(OH)}_3)\left(\frac{106.9\ \text{g}}{\text{mol}}\right) = 273\ \text{g Fe(OH)}_3$

(b) $(125\ \text{kg } CaCO_3)\left(\frac{1000\ \text{g}}{\text{kg}}\right) = 1.25 \times 10^5\ \text{g } CaCO_3$

(c) $(10.5\ \text{mol } NH_3)\left(\frac{17.03\ \text{g}}{\text{mol}}\right) = 179\ \text{g } NH_3$

(d) $(72\ \text{mmol HCl})\left(\frac{1\ \text{mol}}{1000\ \text{mmol}}\right)\left(\frac{36.46\ \text{g}}{\text{mol}}\right) = 2.6\ \text{g HCl}$

(e) $(500.0\ \text{mL } Br_2)\left(\frac{3.119\ \text{g}}{\text{mL}}\right) = 1559.5\ \text{g } Br_2 = 1.560 \times 10^3\ \text{g } Br_2$

10. (a) $(0.00844\ \text{mol } NiSO_4)\left(\frac{154.8\ \text{g}}{\text{mol}}\right) = 1.31\ \text{g } NiSO_4$

(b) $(0.0600\ \text{mol } HC_2H_3O_2)\left(\frac{60.05\ \text{g}}{\text{mol}}\right) = 3.60\ \text{g } HC_2H_3O_2$

(c) $(0.725\ \text{mol } Bi_2S_3)\left(\frac{514.2\ \text{g}}{\text{mol}}\right) = 373\ \text{g } Bi_2S_3$

(d) $(4.50 \times 10^{21}\ \text{molecules } C_6H_{12}O_6)\left(\frac{1\ \text{mol}}{6.022 \times 10^{23}\ \text{molecules}}\right)\left(\frac{180.2\ \text{g}}{\text{mol}}\right)$

$= 1.35\text{g } C_6H_{12}O_6$

(e) $(75\ \text{mL solution})\left(\frac{1.175\ \text{g}}{\text{mL}}\right)\left(\frac{0.200\ \text{g } K_2CrO_4}{\text{g solution}}\right) = 18\ \text{g } K_2CrO_4$

11. 10.0 g H_2O or 10.0 g H_2O_2

Water has a lower molar mass than hydrogen peroxide. 10.0 grams of water has a lower molar mass, contains more moles, and therefore more molecules than 10.0 g of H_2O_2.

12. Larger number of molecules: 25.0 g HCl or 85 g $C_6H_{12}O_6$

$(25.0\ \text{g HCl})\left(\frac{1\ \text{mol}}{36.46\ \text{g}}\right)\left(\frac{6.022 \times 10^{23}\ \text{molecules}}{\text{mol}}\right) = 4.13 \times 10^{23}\ \text{molecules HCl}$

$$(85.0\text{ g }C_6H_{12}O_6)\left(\frac{1\text{ mol}}{180.2\text{ g}}\right)\left(\frac{6.022\times 10^{23}\text{ molecules}}{\text{mol}}\right)$$

$$= 2.84\times 10^{23}\text{ molecules }C_6H_{12}O_6$$

HCl contains more molecules

13. Mole ratios

$2\,C_3H_7OH + 9\,O_2 \longrightarrow 6\,CO_2 + 8\,H_2O$

(a) $\dfrac{6\text{ mol }CO_2}{2\text{ mol }C_3H_7OH}$

(b) $\dfrac{2\text{ mol }C_3H_7OH}{9\text{ mol }O_2}$

(c) $\dfrac{9\text{ mol }O_2}{6\text{ mol }CO_2}$

(d) $\dfrac{8\text{ mol }H_2O}{2\text{ mol }C_3H_7OH}$

(e) $\dfrac{6\text{ mol }CO_2}{8\text{ mol }H_2O}$

(f) $\dfrac{8\text{ mol }H_2O}{9\text{ mol }O_2}$

14. Mole ratios

$3\,CaCl_2 + 2\,H_3PO_4 \longrightarrow Ca_3(PO_4)_2 + 6\,HCl$

(a) $\dfrac{3\text{ mol }CaCl_2}{1\text{ mol }Ca_3(PO_4)_2}$

(b) $\dfrac{6\text{ mol HCl}}{2\text{ mol }H_3PO_4}$

(c) $\dfrac{3\text{ mol }CaCl_2}{2\text{ mol }H_3PO_4}$

(d) $\dfrac{1\text{ mol }Ca_3(PO_4)_2}{2\text{ mol }H_3PO_4}$

(e) $\dfrac{6\text{ mol HCl}}{1\text{ mol }Ca_3(PO_4)_2}$

(f) $\dfrac{2\text{ mol }H_3PO_4}{6\text{ mol HCl}}$

15. $C_2H_5OH + 3\,O_2 \longrightarrow 2\,CO_2 + 3\,H_2O$

$$(7.75\text{ mol }C_2H_5OH)\left(\frac{2\text{ mol }CO_2}{1\text{ mol }C_2H_5OH}\right) = 15.5\text{ mol }CO_2$$

16. Moles of Cl_2

$4\,HCl + O_2 \longrightarrow 2\,Cl_2 + 2\,H_2O$

$$(5.60\text{ mol HCl})\left(\frac{2\text{ mol }Cl_2}{4\text{ mol HCl}}\right) = 2.80\text{ mol }Cl_2$$

17. $MnO_2(s) + 4\,HCl(aq) \longrightarrow Cl_2(g) + MnCl_2(aq) + 2\,H_2O(l)$

$$(1.05 \text{ mol } MnO_2)\left(\frac{4 \text{ mol HCl}}{1 \text{ mol } MnO_2}\right) = 4.20 \text{ mol HCl}$$

18. $Al_4C_3 + 12\,H_2O \longrightarrow 4\,Al(OH)_3 + 3\,CH_4$

(a) $$(100.\text{ g } Al_4C_3)\left(\frac{1 \text{ mol}}{144.0 \text{ g}}\right)\left(\frac{12 \text{ mol } H_2O}{1 \text{ mol } Al_4C_3}\right) = 8.33 \text{ mol } H_2O$$

(b) $$(0.600 \text{ mol } CH_4)\left(\frac{4 \text{ mol } Al(OH)_3}{3 \text{ mol } CH_4}\right) = 0.800 \text{ mol } Al(OH)_3$$

19. Grams of NaOH

$Ca(OH)_2 + Na_2CO_3 \longrightarrow 2\,NaOH + CaCO_3$

The conversion is: g $Ca(OH)_2 \longrightarrow$ mol $Ca(OH)_2 \longrightarrow$ mol NaOH $\longrightarrow$ g NaOH

$$(500 \text{ g } Ca(OH)_2)\left(\frac{1 \text{ mol}}{74.10 \text{ g}}\right)\left(\frac{2 \text{ mol NaOH}}{1 \text{ mol } Ca(OH)_2}\right)\left(\frac{40.00 \text{ g}}{\text{mol}}\right) = 5 \times 10^2 \text{ g NaOH}$$

20. Grams of $Zn_3(PO_4)_2$

$3\,Zn + 2\,H_3PO_4 \longrightarrow Zn_3(PO_4)_2 + 3\,H_2$

The conversion is: g Zn $\longrightarrow$ mol Zn $\longrightarrow$ mol $Zn_3(PO_4)_2 \longrightarrow$ g $Zn_3(PO_4)_2$

$$(10.0 \text{ g Zn})\left(\frac{1 \text{ mol}}{65.39 \text{ g}}\right)\left(\frac{1 \text{ mol } Zn_3(PO_4)_2}{3 \text{ mol Zn}}\right)\left(\frac{386.1 \text{ g}}{\text{mol}}\right) = 19.7 \text{ g } Zn_3(PO_4)_2$$

21. The balanced equation is $Fe_2O_3 + 3\,C \longrightarrow 2\,Fe + 3\,CO$

The conversion is: kg $Fe_2O_3 \longrightarrow$ kmol $Fe_2O_3 \longrightarrow$ kmol Fe $\longrightarrow$ kg Fe

$$(125 \text{ kg } Fe_2O_3)\left(\frac{1 \text{ kmol}}{159.7 \text{ kg}}\right)\left(\frac{2 \text{ kmol Fe}}{1 \text{ kmol } Fe_2O_3}\right)\left(\frac{55.85 \text{ kg}}{\text{kmol}}\right) = 87.4 \text{ kg Fe}$$

22. The balanced equation is $3\,Fe + 4\,H_2O \longrightarrow Fe_3O_4 + 4\,H_2$

Calculate the grams of both H_2O and Fe to produce 375 g Fe_3O_4

$$(375 \text{ g } Fe_3O_4)\left(\frac{1 \text{ mol}}{231.6 \text{ g}}\right)\left(\frac{4 \text{ mol } H_2O}{1 \text{ mol } Fe_3O_4}\right)\left(\frac{18.02 \text{ g}}{\text{mol}}\right) = 117 \text{ g } H_2O$$

$$(375 \text{ g } Fe_3O_4)\left(\frac{1 \text{ mol}}{231.6 \text{ g}}\right)\left(\frac{3 \text{ mol Fe}}{1 \text{ mol } Fe_3O_4}\right)\left(\frac{55.85 \text{ g}}{\text{mol}}\right) = 271 \text{ g Fe}$$

23. The balanced equation is $2\,C_2H_6 + 7\,O_2 \longrightarrow 4\,CO_2 + 6\,H_2O$

(a) $$(15.0\text{ mol } C_2H_6)\left(\frac{7\text{ mol } O_2}{2\text{ mol } C_2H_6}\right) = 52.5\text{ mol } O_2$$

(b) $$(8.00\text{ g } H_2O)\left(\frac{1\text{ mol}}{18.02\text{ g}}\right)\left(\frac{4\text{ mol } CO_2}{6\text{ mol } H_2O}\right)\left(\frac{44.01\text{ g}}{\text{mol}}\right) = 13.0\text{ g } CO_2$$

(c) $$(75.0\text{ g } C_2H_6)\left(\frac{1\text{ mol}}{30.07\text{ g}}\right)\left(\frac{4\text{ mol } CO_2}{2\text{ mol } C_2H_6}\right)\left(\frac{44.01\text{ g}}{\text{mol}}\right) = 2.20 \times 10^2\text{ g } CO_2$$

24. $4\,FeS_2 + 11\,O_2 \longrightarrow 2\,Fe_2O_3 + 8\,SO_2$

(a) $$(1.00\text{ mol } FeS_2)\left(\frac{2\text{ mol } Fe_2O_3}{4\text{ mol } FeS_2}\right) = 0.500\text{ mol } Fe_2O_3$$

(b) $$(4.50\text{ mol } FeS_2)\left(\frac{11\text{ mol } O_2}{4\text{ mol } FeS_2}\right) = 12.4\text{ mol } O_2$$

(c) $$(1.55\text{ mol } Fe_2O_3)\left(\frac{8\text{ mol } SO_2}{2\text{ mol } Fe_2O_3}\right) = 6.20\text{ mol } SO_2$$

(d) $$(0.512\text{ mol } FeS_2)\left(\frac{8\text{ mol } SO_2}{4\text{ mol } FeS_2}\right)\left(\frac{64.07\text{ g}}{\text{mol}}\right) = 65.6\text{ g } SO_2$$

(e) $$(40.6\text{ g } SO_2)\left(\frac{1\text{ mol}}{64.07\text{ g}}\right)\left(\frac{11\text{ mol } O_2}{8\text{ mol } SO_2}\right) = 0.871\text{ mol } O_2$$

(f) $$(221\text{ g } Fe_2O_3)\left(\frac{1\text{ mol}}{159.7\text{ g}}\right)\left(\frac{4\text{ mol } FeS_2}{2\text{ mol } Fe_2O_3}\right)\left(\frac{120.0\text{ g}}{\text{mol}}\right) = 332\text{ g } FeS_2$$

25. (a) Hydrogen Oxygen

Hydrogen is the limiting reactant.

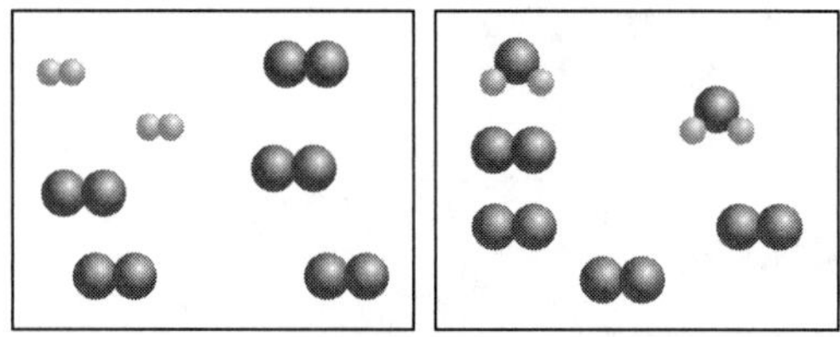

(b) Hydrogen Bromine

No limiting reactant.

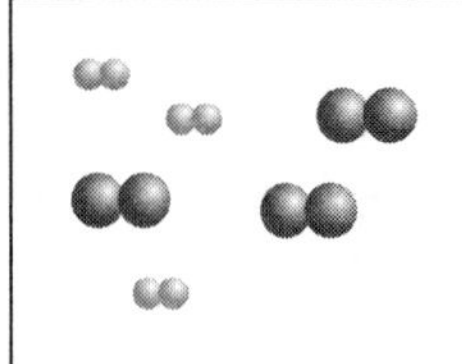 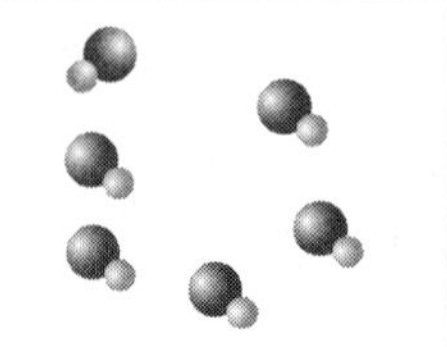

(c) Potassium Chlorine

Potassium is the limiting reactant.

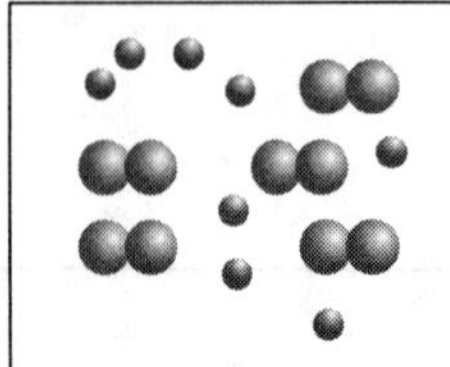 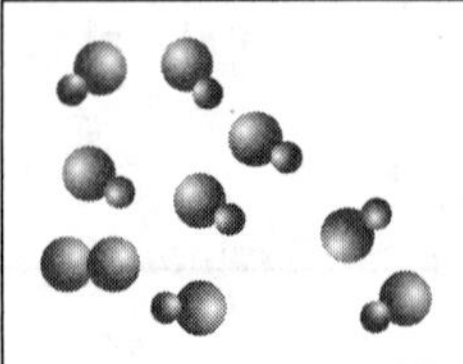

(d) Aluminum Oxygen

Oxygen is the limiting reactant.

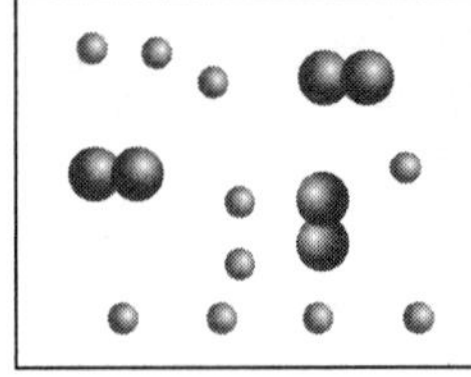 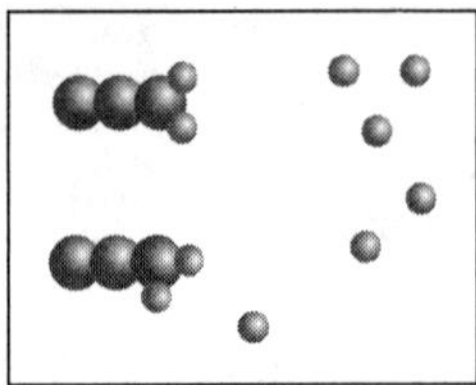

26. (a) Lithium Iodine

No limiting reactant.

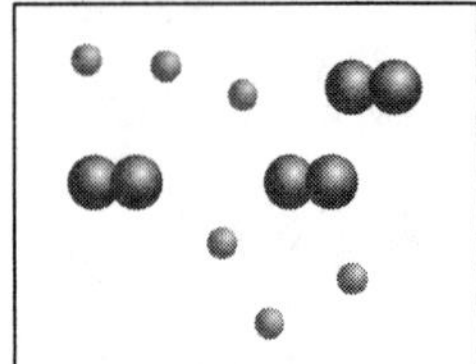 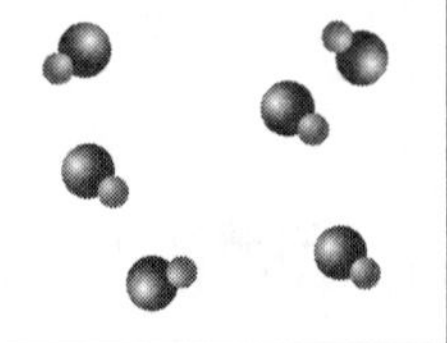

(b) Silver Chlorine

Silver is the limiting reactant.

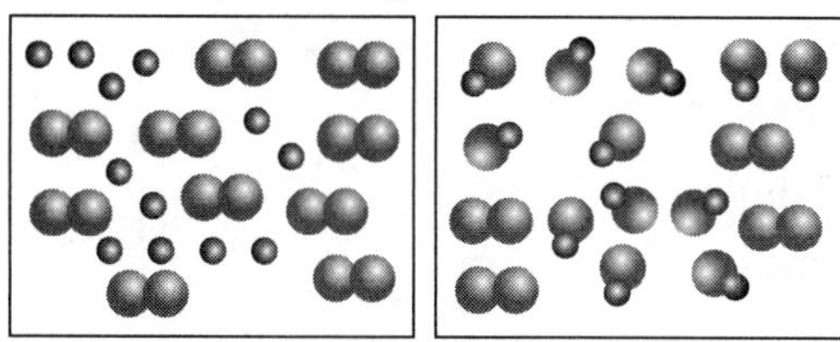

(c) Nitrogen Oxygen

Oxygen is the limiting reactant.

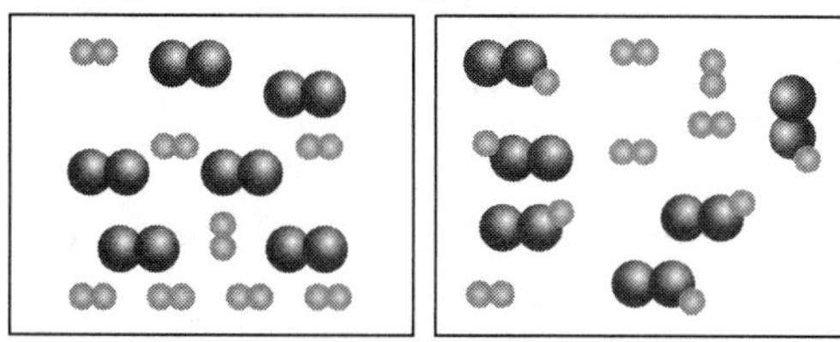

(d) Iron Hydrogen Oxygen

Water is the limiting reactant.

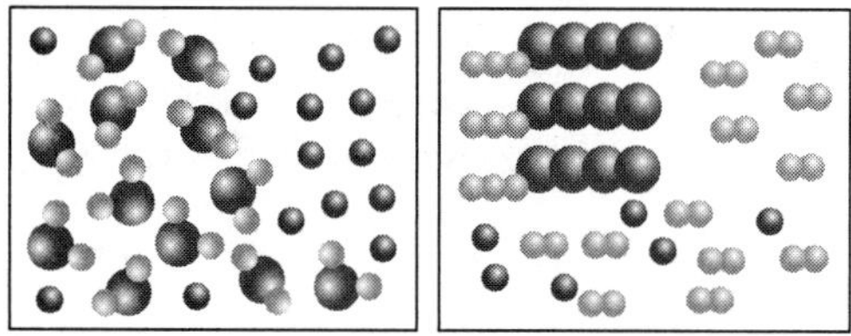

27. (a) KOH + HNO_3 $\longrightarrow$ KNO_3 + H_2O

16.0 g   12.0 g

Choose one of the products and calculate its mass that would be produced from each given reactant. Using KNO_3 as the product:

$$(16.0 \text{ g KOH})\left(\frac{1 \text{ mol}}{56.10 \text{ g}}\right)\left(\frac{1 \text{ mol KNO}_3}{1 \text{ mol KOH}}\right)\left(\frac{101.1 \text{ g}}{\text{mol}}\right) = 28.8 \text{ g KNO}_3$$

$$(12.0 \text{ g HNO}_3)\left(\frac{1 \text{ mol}}{63.02 \text{ g}}\right)\left(\frac{1 \text{ mol KNO}_3}{1 \text{ mol KOH}}\right)\left(\frac{101.1 \text{ g}}{\text{mol}}\right) = 19.3 \text{ g KNO}_3$$

Since HNO_3 produces less KNO_3, it is the limiting reactant and KOH is in excess.

(b) $2\,NaOH \quad + \quad H_2SO_4 \quad \longrightarrow \quad Na_2SO_4 \quad + \quad 2\,H_2O$

10.0 g   10.0 g

Choose one of the products and calculate its mass that would be produced from each given reactant. Using H_2O as the product:

$$(10.0\text{ g NaOH})\left(\frac{1\text{ mol}}{40.00\text{ g}}\right)\left(\frac{2\text{ mol H}_2\text{O}}{2\text{ mol NaOH}}\right)\left(\frac{18.02\text{ g}}{\text{mol}}\right) = 4.51\text{ g H}_2\text{O}$$

$$(10.0\text{ g H}_2\text{SO}_4)\left(\frac{1\text{ mol}}{98.09\text{ g}}\right)\left(\frac{2\text{ mol H}_2\text{O}}{1\text{ mol H}_2\text{SO}_4}\right)\left(\frac{18.02\text{ g}}{\text{mol}}\right) = 3.67\text{ g H}_2\text{O}$$

Since H_2SO_4 produces less H_2O, it is the limiting reactant and NaOH is in excess.

28. (a) $2\,Bi(NO_3)_3 \quad + \quad 3\,H_2S \quad \longrightarrow \quad Bi_2S_3 \quad + \quad 6\,HNO_3$

50.0 g   6.00 g

Choose one of the products and calculate its mass that would be produced from each given reactant. Using Bi_2S_3 as the product:

$$(50.0\text{ g Bi(NO}_3)_3)\left(\frac{1\text{ mol}}{395.0\text{ g}}\right)\left(\left(\frac{1\text{ mol Bi}_2\text{S}_3}{2\text{ mol Bi(NO}_3)_3}\right)\left(\frac{514.2\text{ g}}{\text{mol}}\right) = 32.5\text{ g Bi}_2\text{S}_3$$

$$(6.00\text{ g H}_2\text{S})\left(\frac{1\text{ mol}}{34.09\text{ g}}\right)\left(\frac{1\text{ mol Bi}_2\text{S}_3}{3\text{ mol H}_2\text{S}}\right)\left(\frac{514.2\text{ g}}{\text{mol}}\right) = 30.2\text{ g Bi}_2\text{S}_3$$

Since H_2S produces less Bi_2S_3, it is the limiting reactant and $Bi(NO_3)_3$ is in excess.

(b) $3\,Fe \quad + \quad 4\,H_2O \quad \longrightarrow \quad Fe_3O_4 \quad + \quad 4\,H_2$

40.0 g   16.0 g

Choose one of the products and calculate its mass that would be produced from each given reactant. Using H_2 as the product:

$$(40.0\text{ g Fe})\left(\frac{1\text{ mol}}{55.85\text{ g}}\right)\left(\frac{4\text{ mol H}_2}{3\text{ mol Fe}}\right)\left(\frac{2.016\text{ g}}{\text{mol}}\right) = 1.93\text{ g H}_2$$

$$(16.0\text{ g H}_2\text{O})\left(\frac{1\text{ mol}}{18.02\text{ g}}\right)\left(\frac{4\text{ mol H}_2}{4\text{ mol H}_2\text{O}}\right)\left(\frac{2.016\text{ g}}{\text{mol}}\right) = 1.79\text{ g H}_2$$

Since H_2O produces less H_2, it is the limiting reactant and Fe is in excess.

29. Limiting reactant calculations
$C_3H_8 + 5\,O_2 \longrightarrow 3\,CO_2 + 4\,H_2O$

(a) Reaction between 20.0 g C_3H_8 and 20.0 g O_2
Convert each amount to grams of CO_2

$$(20.0 \text{ g } C_3H_8)\left(\frac{1 \text{ mol}}{44.09 \text{ g}}\right)\left(\frac{3 \text{ mol } CO_2}{1 \text{ mol } C_3H_8}\right)\left(\frac{44.01 \text{ g}}{\text{mol}}\right) = 59.9 \text{ g } CO_2$$

$$(20.0 \text{ g } O_2)\left(\frac{1 \text{ mol}}{32.00 \text{ g}}\right)\left(\frac{3 \text{ mol } CO_2}{5 \text{ mol } O_2}\right)\left(\frac{44.01 \text{ g}}{\text{mol}}\right) = 16.5 \text{ g } CO_2$$

O_2 is the limiting reactant. The yield is 16.5 g CO_2.

(b) Reaction between 20.0 g C_3H_8 and 80.0 g O_2
Convert each amount to grams of CO_2

$$(20.0 \text{ g } C_3H_8)\left(\frac{1 \text{ mol}}{44.09 \text{ g}}\right)\left(\frac{3 \text{ mol } CO_2}{1 \text{ mol } C_3H_8}\right)\left(\frac{44.01 \text{ g}}{\text{mol}}\right) = 59.9 \text{ g } CO_2$$

$$(80.0 \text{ g } O_2)\left(\frac{1 \text{ mol}}{32.00 \text{ g}}\right)\left(\frac{3 \text{ mol } CO_2}{5 \text{ mol } O_2}\right)\left(\frac{44.01 \text{ g}}{\text{mol}}\right) = 66.0 \text{ g } CO_2$$

C_3H_8 is the limiting reactant. The yield is 59.9 g CO_2.

(c) Reaction between 2.0 mol C_3H_8 and 14.0 mol O_2
According to the equation, 2 mol C_3H_8 will react with 10 mol O_2. Therefore, C_3H_8 is the limiting reactant and 4.0 mol O_2 will remain unreacted.

$$(2.0 \text{ mol } C_3H_8)\left(\frac{3 \text{ mol } CO_2}{1 \text{ mol } C_3H_8}\right) = 6.0 \text{ mol } CO_2 \text{ produced}$$

$$(2.0 \text{ mol } C_3H_8)\left(\frac{4 \text{ mol } H_2O}{1 \text{ mol } C_3H_8}\right) = 8.0 \text{ mol } H_2O \text{ produced}$$

When the reaction is completed, 6.0 mol CO_2, 8.0 H_2O, and 4.0 mol O_2 will be in the container.

30. Limiting reactant calculations
$C_3H_8 + 5\,O_2 \longrightarrow 3\,CO_2 + 4\,H_2O$

(a) Reaction between 5.0 mol C_3H_8 and 5 mol O_2

$$(5.0 \text{ mol } C_3H_8)\left(\frac{3 \text{ mol } CO_2}{1 \text{ mol } C_3H_8}\right) = 15.0 \text{ mol } CO_2$$

$$(5.0 \text{ mol } O_2)\left(\frac{3 \text{ mol } CO_2}{5 \text{ mol } O_2}\right) = 3.0 \text{ mol } CO_2$$

The O_2 is the limiting reactant; 3.0 mol CO_2 produced.

(b) Reaction between 3.0 mol C_3H_8 and 20.0 mol O_2

$$(3.0\text{ mol } C_3H_8)\left(\frac{3\text{ mol } CO_2}{1\text{ mol } C_3H_8}\right) = 9.0\text{ mol } CO_2$$

$$(20.0\text{ mol } O_2)\left(\frac{3\text{ mol } CO_2}{5\text{ mol } O_2}\right) = 12.0\text{ mol } CO_2$$

The C_3H_8 is the limiting reactant; 9.0 mol CO_2 produced.

(c) Reaction between 20.0 mol C_3H_8 and 3.0 mol O_2
According to the equation, 1 mol C_3H_8 will react with 5 mol O_2, O_2 is clearly the limiting reactant.

$$(3.0\text{ mol } O_2)\left(\frac{3\text{ mol } CO_2}{5\text{ mol } O_2}\right) = 1.8\text{ mol } CO_2\text{ produced}$$

31. $X_8 + 12\,O_2 \longrightarrow 8\,XO_3$
The conversion is: g O_2 $\longrightarrow$ mol O_2 $\longrightarrow$ mol X_8

$$(120.0\text{ g } O_2)\left(\frac{1\text{ mol}}{32.00\text{ g}}\right)\left(\frac{1\text{ mol } X_8}{12\text{ mol } O_2}\right) = 0.3125\text{ mol } X_8 \qquad 80.0\text{ g } X_8 = 0.3125\text{ mol } X_8$$

$$\frac{80.0\text{ g}}{0.3125\text{ mol}} = 256\text{ g/mol } X_8$$

$$\text{molar mass } X = \frac{256\frac{\text{g}}{\text{mol}}}{8} = 32.0\frac{\text{g}}{\text{mol}}$$

Using the periodic table we find that the element with 32.0 g/mol is sulfur.

32. $X + 2\,HCl \longrightarrow XCl_2 + H_2$
The conversion is: g H_2 $\longrightarrow$ mol H_2 $\longrightarrow$ mol X

$$(2.42\text{ g } H_2)\left(\frac{1\text{ mol}}{2.016\text{ g}}\right)\left(\frac{1\text{ mol } X}{1\text{ mol } H_2}\right) = 1.20\text{ mol } X \qquad 78.5\text{ g } X = 1.20\text{ mol} X$$

$$\frac{78.5\text{ g}}{1.20\text{ mol}} = 65.4\text{ g/mol}$$

Using the periodic table we find that the element with atomic mass 65.4 is zinc.

33. Limiting reactant calculation and percentage yield
$2\,Al + 3\,Br_2 \longrightarrow 2\,AlBr_3$
Reaction between 25.0 g Al and 100. g Br_2
Calculate the grams of $AlBr_3$ from each reactant.

$$(25.0 \text{ g Al})\left(\frac{1 \text{ mol}}{26.98 \text{ g}}\right)\left(\frac{2 \text{ mol AlBr}_3}{2 \text{ mol Al}}\right)\left(\frac{266.7 \text{ g}}{\text{mol}}\right) = 247 \text{ g AlBr}_3$$

$$(100. \text{ g Br}_2)\left(\frac{1 \text{ mol}}{159.8 \text{ g}}\right)\left(\frac{2 \text{ mol AlBr}_3}{3 \text{ mol Br}_2}\right)\left(\frac{266.7 \text{ g}}{\text{mol}}\right) = 111 \text{ g AlBr}_3$$

Br_2 is limiting; 111 g $AlBr_3$ is the theoretical yield of product.

$$\text{Percent yield} = \left(\frac{\text{actual yield}}{\text{theoretical yield}}\right)(100) = \left(\frac{64.2 \text{ g}}{111 \text{ g}}\right)(100) = 57.8\%$$

34. Percent yield calculation

$Fe(s) + CuSO_4(aq) \longrightarrow Cu(s) + FeSO_4(aq)$

$$(400. \text{ g CuSO}_4)\left(\frac{1 \text{ mol}}{159.6 \text{ g}}\right)\left(\frac{1 \text{ mol Cu}}{1 \text{ mol CuSO}_4}\right)\left(\frac{63.55 \text{ g}}{\text{mol}}\right) = 159 \text{ g Cu (theoretical yield)}$$

$$\% \text{ yield} = \left(\frac{\text{actual yield}}{\text{theoretical yield}}\right)(100) = \left(\frac{151 \text{ g}}{159 \text{ g}}\right)(100) = 95.0\% \text{ yield of Cu}$$

35. The balanced equation is $3\,C + 2\,SO_2 \longrightarrow CS_2 + 2\,CO_2$

Calculate the g C needed to produce 950 g CS_2 taking into account that the yield of CS_2 is 86.0%. First calculate the theoretical yield of CS_2.

$$\frac{950 \text{ g CS}_2}{0.860} = 1.1 \times 10^3 \text{ g CS}_2 \text{ (theoretical yield)}$$

Now calculate the grams of coke needed to produce 1.1×10^3 g CS_2.

$$(1.1 \times 10^3 \text{ g CS}_2)\left(\frac{1 \text{ mol}}{76.15 \text{ g}}\right)\left(\frac{3 \text{ mol C}}{1 \text{ mol CS}_2}\right)\left(\frac{12.01 \text{ g}}{\text{mol}}\right) = 5.2 \times 10^2 \text{ g C}$$

36. The balanced equation is $CaC_2 + 2\,H_2O \longrightarrow C_2H_2 + Ca(OH)_2$

First calculate the grams of pure CaC_2 in the sample from the amount of C_2H_2 produced.

$$(0.540 \text{ mol C}_2\text{H}_2)\left(\frac{1 \text{ mol CaC}_2}{1 \text{ mol C}_2\text{H}_2}\right)\left(\frac{64.10 \text{ g}}{\text{mol}}\right) = 34.6 \text{ g of pure CaC}_2 \text{ in the sample}$$

Now calculate the percent CaC_2 in the impure sample.

$$\left(\frac{34.6 \text{ g CaC}_2}{44.5 \text{ g sample}}\right)(100) = 77.8\% \text{ CaC}_2 \text{ in the impure sample}$$

37. No. There are not enough screwdrivers, wrenches or pliers. 2400 screwdrivers, 3600 wrenches and 1200 pliers are needed for 600 tool sets.

38. A subscript is used to indicate the number of atoms in a formula. It cannot be changed without changing the identity of the substance. Coefficients are used only to balance atoms in chemical equations. They may be changed as needed to achieve a balanced equation.

39. Consider the reaction A $\longrightarrow$ 2B and assume that you have 1 gram of A. This does not guarantee that you will produce 1 gram of B because A and B have different molar masses. One gram of A does not contain the same number of molecules as 1 gram of B. However, 1 mole of A does have the same number of molecules as one mole of B. (Remember, 1 mole $= 6.022 \times 10^{23}$ molecules always.) If you determine the number of moles in one gram of A and multiply by 2 to get the number of moles of B . . . then from that you can determine the grams of B using its molar mass. Equations are written in terms of moles not grams.

40. $4\,KO_2 + 2\,H_2O + 4\,CO_2 \longrightarrow 4\,KHCO_3 + 3\,O_2$

(a) $$\left(\frac{0.85\text{ g }CO_2}{\text{min}}\right)\left(\frac{1\text{ mol}}{44.01\text{ g}}\right)\left(\frac{4\text{ mol }KO_2}{4\text{ mol }CO_2}\right) = \frac{0.019\text{ mol }KO_2}{\text{min}}$$

$$\left(\frac{0.019\text{ mol }KO_2}{\text{min}}\right)(10.0\text{ min}) = 0.19\text{ mol }KO_2$$

(b) The conversion is: $\frac{\text{g }CO_2}{\text{min}} \longrightarrow \frac{\text{mol }CO_2}{\text{min}} \longrightarrow \frac{\text{mol }O_2}{\text{min}} \longrightarrow \frac{\text{g }O_2}{\text{min}} \longrightarrow \frac{\text{g }O_2}{\text{hr}}$

$$\left(\frac{0.85\text{ g }CO_2}{\text{min}}\right)\left(\frac{1\text{ mol}}{44.01\text{ g}}\right)\left(\frac{3\text{ mol }O_2}{4\text{ mol }CO_2}\right)\left(\frac{32.00\text{ g}}{\text{mol}}\right)\left(\frac{60.0\text{ min}}{1.0\text{ hr}}\right) = \frac{28\text{ g }O_2}{\text{hr}}$$

41. (a) $$(750\text{ g }C_6H_{12}O_6)\left(\frac{1\text{ mol}}{180.2\text{ g}}\right)\left(\frac{2\text{ mol }C_2H_5OH}{1\text{ mol }C_6H_{12}O_6}\right)\left(\frac{46.07\text{ g}}{\text{mol}}\right) = 380\text{ g }C_2H_5OH$$

$$(750\text{ g }C_6H_{12}O_6)\left(\frac{1\text{ mol}}{180.2\text{ g}}\right)\left(\frac{2\text{ mol }CO_2}{1\text{ mol }C_6H_{12}O_6}\right)\left(\frac{44.01\text{ g}}{\text{mol}}\right) = 370\text{ g }CO_2$$

(b) $$(380\text{ g }C_2H_5OH)\left(\frac{1\text{ mL}}{0.79\text{ g}}\right) = 480\text{ mL }C_2H_5OH$$

42. $4\,P + 5\,O_2 \longrightarrow P_4O_{10}$

$P_4O_{10} + 6\,H_2O \longrightarrow 4\,H_3PO_4$

In the first reaction:

$$(20.0\text{gP})\left(\frac{1\text{ mol}}{30.97\text{ g}}\right) = 0.646\text{ mol P}$$

$$(30.0 \text{ g } O_2)\left(\frac{1 \text{ mol}}{32.00 \text{ g}}\right) = 0.938 \text{ mol } O_2$$

This is a ratio of $\dfrac{0.646 \text{ mol P}}{0.938 \text{ mol } O_2} = \dfrac{3.44 \text{ mol P}}{5.00 \text{ mol } O_2}$

Therefore, P is the limiting reactant and the P_4O_{10} produced is:

$$(0.646 \text{ mol P})\left(\frac{1 \text{ mol } P_4O_{10}}{4 \text{ mol P}}\right) = 0.162 \text{ mol } P_4O_{10}$$

In the second reaction:

$$(15.0 \text{ g } H_2O)\left(\frac{1 \text{ mol}}{18.02 \text{ g}}\right) = 0.832 \text{ mol } H_2O$$

and we have 0.162 mol P_4O_{10}. The ratio of $\dfrac{H_2O}{P_4O_{10}}$ is $\dfrac{0.832 \text{ mol}}{0.162 \text{ mol}} = \dfrac{5.14 \text{ mol}}{1.00 \text{ mol}}$

Therefore, H_2O is the limiting reactant and the H_3PO_4 produced is:

$$(0.832 \text{ mol } H_2O)\left(\frac{4 \text{ mol } H_3PO_4}{6 \text{ mol } H_2O}\right)\left(\frac{97.99 \text{ g}}{\text{mol}}\right) = 54.4 \text{ g } H_3PO_4$$

43. $2\,CH_3OH + 3\,O_2 \longrightarrow 2\,CO_2 + 4\,H_2O$

The conversion is:

mL $CH_3OH \longrightarrow$ g $CH_3OH \longrightarrow$ mol $CH_3OH \longrightarrow$ mol $O_2 \longrightarrow$ g O_2

$$(60.0 \text{ mL } CH_3OH)\left(\frac{0.72 \text{ g}}{\text{mL}}\right)\left(\frac{1 \text{ mol}}{32.04 \text{ g}}\right)\left(\frac{3 \text{ mol } O_2}{2 \text{ mol } CH_3OH}\right)\left(\frac{32.00 \text{ g}}{\text{mol}}\right) = 65 \text{ g } O_2$$

44. $7\,H_2O_2 + N_2H_4 \longrightarrow 2\,HNO_3 + 8\,H_2O$

(a) $(0.33 \text{ mol } N_2H_4)\left(\dfrac{2 \text{ mol } HNO_3}{1 \text{ mol } N_2H_4}\right) = 0.66 \text{ mol } HNO_3$

(b) $(2.75 \text{ mol } H_2O)\left(\dfrac{7 \text{ mol } H_2O_2}{8 \text{ mol } H_2O}\right) = 2.41 \text{ mol } H_2O_2$

(c) $(8.72 \text{ mol } HNO_3)\left(\dfrac{8 \text{ mol } H_2O}{2 \text{ mol } HNO_3}\right) = 34.9 \text{ mol } H_2O$

(d) $(120 \text{ g } N_2H_4)\left(\dfrac{1 \text{ mol}}{32.05 \text{ g}}\right)\left(\dfrac{7 \text{ mol } H_2O_2}{1 \text{ mol } N_2H_4}\right)\left(\dfrac{34.02 \text{ g}}{\text{mol}}\right) = 8.9 \times 10^2 \text{ g } H_2O_2$

45. $4\,Ag + 2\,H_2S + O_2 \longrightarrow 2\,Ag_2S + 2\,H_2O$

Determine the limiting reactant.

$$(1.1\text{ g Ag})\left(\frac{1\text{ mol}}{107.9\text{ g}}\right)\left(\frac{2\text{ mol }Ag_2S}{4\text{ mol Ag}}\right)\left(\frac{247.9\text{ g}}{\text{mol}}\right) = 1.3\text{ g }Ag_2S$$

$$(0.14\text{ g }H_2S)\left(\frac{1\text{ mol}}{34.09\text{ g}}\right)\left(\frac{2\text{ mol }Ag_2S}{2\text{ mol }H_2S}\right)\left(\frac{247.9\text{ g}}{\text{mol}}\right) = 1.0\text{ g }Ag_2S$$

$$(0.080\text{ g }O_2)\left(\frac{1\text{ mol}}{32.00\text{ g}}\right)\left(\frac{2\text{ mol }Ag_2S}{1\text{ mol }O_2}\right)\left(\frac{247.9\text{ g}}{\text{mol}}\right) = 1.2\text{ g }Ag_2S$$

The H_2S is the limiting reactant so 1.0 g Ag_2S is formed.

46. Mass of the beaker

26.500 g	beaker + $Ca(OH)_2$
26.095 g	beaker + CaO
0.405 g	H_2O absorbed

$$(0.405\text{ g }H_2O)\left(\frac{1\text{ mol}}{18.02\text{ g}}\right) = 2.25 \times 10^{-2}\text{ mol }H_2O\text{ absorbed}$$

Since the reaction is a 1:1 mole, the amount of CaO in the beaker is 2.25×10^{-2} mol. Convert to grams.

$$(2.25 \times 10^{-2}\text{ mol CaO})\left(\frac{56.08\text{ g CaO}}{\text{mol}}\right) = 1.26\text{ g CaO in the beaker.}$$

26.095 g	beaker + CaO
−1.26 g	CaO
24.835 g	mass of the beaker

47. $Pb(NO_3)_2(aq) + 2\,KI(aq) \longrightarrow PbI_2(s) + 2\,KNO_3(aq)$

(a) The solid is lead (II) iodide, PbI_2.

(b) Double displacement reaction.

(c) Calculate the moles of each reactant.

$$(15\text{ g }Pb(NO_3)_2) + \left(\frac{1\text{ mol}}{331.2\text{ g}}\right) = 0.045\text{ mol }Pb(NO_3)_2$$

$$(15\text{ g KI})\left(\frac{1\text{ mol}}{166.0\text{ g}}\right) = 0.090\text{ mol KI}$$

Stoichiometric quantities of reactants are used.

Theoretical yield of PbI_2 is 0.045 mol.

Actual yield: $(6.68\text{ g PbI}_2)\left(\dfrac{1\text{ mol}}{461.1\text{ g}}\right) = 0.0145\text{ mol PbI}_2$

Percent yield: $\left(\dfrac{0.0145\text{ mol}}{0.045\text{ mol}}\right)(100) = 32\%\text{ yield}$

48. Composition of a mixture of KNO_3 and KCl.
In the mixture only KCl reacts with $AgNO_3$.

$KCl(aq) + AgNO_3(aq) \longrightarrow AgCl(s) + KNO_3$

$(4.33\text{ g AgCl})\left(\dfrac{74.55\text{ g KCl}}{143.7\text{ g AgCl}}\right) = 2.25\text{ g KCl in the mixture}$

$10.00\text{ g mixture} - 2.25\text{ g KCl} = 7.75\text{ g KNO}_3$

$\left(\dfrac{2.25\text{ g KCl}}{10.00\text{ g mixture}}\right)(100) = 22.5\%\text{ KCl}$

$\left(\dfrac{7.75\text{ g KNO}_3}{10.00\text{ g mixture}}\right)(100) = 77.5\%\text{ KNO}_3$

49. The balanced equation is $Zn + 2\,HCl \longrightarrow ZnCl_2 + H_2$

$180.0\text{ g Zn} - 35\text{ g Zn} = 145\text{ g Zn reacted with HCl}$

(a) $(145\text{ g Zn})\left(\dfrac{1\text{ mol}}{65.39\text{ g}}\right)\left(\dfrac{1\text{ mol H}_2}{1\text{ mol Zn}}\right) = 2.22\text{ mol H}_2\text{ produced}$

(b) $(145\text{ g Zn})\left(\dfrac{1\text{ mol}}{65.39\text{ g}}\right)\left(\dfrac{2\text{ mol HCl}}{1\text{ mol Zn}}\right)\left(\dfrac{36.46\text{ g}}{\text{mol}}\right) = 162\text{ g HCl reacted}$

50. $Fe(s) \quad + \quad CuSO_4(aq) \quad \longrightarrow \quad Cu(s) \quad + \quad FeSO_4(aq)$
2.0 mol   3.0 mol

(a) 2.0 mol Fe react with 2.0 mol $CuSO_4$ to yield 2.0 mol Cu and 2.0 mol $FeSO_4$. 1.0 mol $CuSO_4$ is unreacted. At the completion of the reaction, there will be 2.0 mol Cu, 2.0 mol $FeSO_4$, and 1.0 mol $CuSO_4$.

(b) Determine which reactant is limiting and then calculate the g $FeSO_4$ produced from that reactant.

$(20.0\text{ g Fe})\left(\dfrac{1\text{ mol}}{55.85\text{ g}}\right)\left(\dfrac{1\text{ mol Cu}}{1\text{ mol Fe}}\right)\left(\dfrac{63.55\text{ g}}{\text{mol}}\right) = 22.8\text{ g Cu}$

$(40.0\text{ g CuSO}_4)\left(\dfrac{1\text{ mol}}{159.6\text{ g}}\right)\left(\dfrac{1\text{ mol Cu}}{1\text{ mol CuSO}_4}\right)\left(\dfrac{63.55\text{ g}}{\text{mol}}\right) = 15.9\text{ g Cu}$

Since $CuSO_4$ produces less Cu, it is the limiting reactant. Determine the mass of $FeSO_4$ produced from 40.0 g $CuSO_4$.

$$(40.0\text{ g CuSO}_4)\left(\frac{1\text{ mol}}{159.6\text{ g}}\right)\left(\frac{1\text{ mol FeSO}_4}{1\text{ mol CuSO}_4}\right)\left(\frac{151.9\text{ g}}{\text{mol}}\right) = 38.1\text{ g FeSO}_4\text{ produced}$$

Calculate the mass of unreacted Fe.

$$(40.0\text{ g CuSO}_4)\left(\frac{1\text{ mol}}{159.6\text{ g}}\right)\left(\frac{1\text{ mol Fe}}{1\text{ mol CuSO}_4}\right)\left(\frac{55.85\text{ g}}{\text{mol}}\right) = 14.0\text{ g Fe will react}$$

Unreacted Fe $= 20.0\text{ g} - 14.0\text{ g} = 6.0\text{ g}$. Therefore, at the completion of the reaction, 15.9 g Cu, 38.1 g $FeSO_4$, 6.0 g Fe, and no $CuSO_4$ remain.

51. Limiting reactant calculation

$CO(g) + 2\,H_2(g) \longrightarrow CH_3OH(l)$

Reaction between 40.0 g CO and 10.0 g H_2: determine the limiting reactant by calculating the amount of CH_3OH that would be formed from each reactant.

$$(40.0\text{ g CO})\left(\frac{1\text{ mol}}{28.01\text{ g}}\right)\left(\frac{1\text{ mol CH}_3\text{OH}}{1\text{ mol CO}}\right)\left(\frac{32.04\text{ g}}{\text{mol}}\right) = 45.8\text{ g CH}_3\text{OH}$$

$$(10.0\text{ g H}_2)\left(\frac{1\text{ mol}}{2.016\text{ g}}\right)\left(\frac{1\text{ mol CH}_3\text{OH}}{2\text{ mol H}_2}\right)\left(\frac{32.04\text{ g}}{\text{mol}}\right) = 79.5\text{ g CH}_3\text{OH}$$

CO is limiting; H_2 is in excess; 45.8 g CH_3OH will be produced.
Calculate the mass of unreacted H_2:

$$(40.0\text{ g CO})\left(\frac{1\text{ mol}}{28.01\text{ g}}\right)\left(\frac{2\text{ mol H}_2}{1\text{ mol CO}}\right)\left(\frac{2.016\text{ g}}{\text{mol}}\right) = 5.76\text{ g H}_2\text{ react}$$

$10.0\text{ g H}_2 - 5.76\text{ g H}_2 = 4.2\text{ g H}_2$ remain unreacted

52. The balanced equation is $C_6H_{12}O_6 \longrightarrow 2\,C_2H_5OH + 2\,CO_2$

(a) First calculate the theoretical yield.

$$(750\text{ g C}_6\text{H}_{12}\text{O}_6)\left(\frac{1\text{ mol}}{180.2\text{ g}}\right)\left(\frac{2\text{ mol C}_2\text{H}_5\text{OH}}{1\text{ mol C}_6\text{H}_{12}\text{O}_6}\right)\left(\frac{46.07\text{ g}}{\text{mol}}\right)$$

$$= 3.8 \times 10^2\text{g C}_2\text{H}_5\text{OH (theoretical yield)}$$

Then take 84.6% of the theoretical yield to obtain the actual yield.

$$\text{actual yield} = \frac{(\text{theoretical yield})(84.6)}{100} = \frac{(3.8 \times 10^2\text{ g C}_2\text{H}_5\text{OH})(84.6)}{100}$$

$$= 3.2 \times 10^2\text{ g C}_2\text{H}_5\text{OH}$$

(b) 475 g C_2H_5OH represents 84.6% of the theoretical yield. Calculate the theoretical yield.

$$\text{theoretical yield} = \frac{475\text{ g}}{0.846} = 561\text{ g } C_2H_5OH$$

Now calculate the g $C_6H_{12}O_6$ needed to produce 561 g C_2H_5OH.

$$(561\text{ g } C_2H_5OH)\left(\frac{1\text{ mol}}{46.07\text{ g}}\right)\left(\frac{1\text{ mol } C_6H_{12}O_6}{2\text{ mol } C_2H_5OH}\right)\left(\frac{180.2\text{ g}}{\text{mol}}\right) = 1.10 \times 10^3\text{ g } C_6H_{12}O_6$$

53. The balanced equations are:

$CaCl_2(aq) + 2\,AgNO_3(aq) \longrightarrow Ca(NO_3)_2(aq) + 2\,AgCl(s)$

$MgCl_2(aq) + 2\,AgNO_3(aq) \longrightarrow Mg(NO_3)_2(aq) + 2\,AgCl(s)$

1 mol of each salt will produce the same amount (2 mol) of AgCl. $MgCl_2$ has a higher percentage of Cl than $CaCl_2$ because Mg has a lower atomic mass than Ca. Therefore, on an equal mass basis, $MgCl_2$ will produce more AgCl than will $CaCl_2$.

Calculations show that 1.00 g $MgCl_2$ produces 3.01 g AgCl, and 1.00 g $CaCl_2$ produces 2.56 g AgCl.

54. The balanced equation is $Li_2O + H_2O \longrightarrow 2\,LiOH$

The conversion is: g $H_2O \longrightarrow$ mol $H_2O \longrightarrow$ mol $Li_2O \longrightarrow$ g $Li_2O \longrightarrow$ kg Li_2O

$$\left(\frac{2500\text{ g } H_2O}{\text{astronaut day}}\right)\left(\frac{1\text{ mol}}{18.02\text{ g}}\right)\left(\frac{1\text{ mol } Li_2O}{1\text{ mol } H_2O}\right)\left(\frac{29.88\text{ g}}{\text{mol}}\right)\left(\frac{1\text{ kg}}{1000\text{ g}}\right) = \frac{4.1\text{ kg } Li_2O}{\text{astronaut day}}$$

$$\left(\frac{4.1\text{ kg } Li_2O}{\text{astronaut day}}\right)(30\text{ days})(3\text{ astronauts}) = 3.7 \times 10^2\text{ kg } Li_2O$$

55. The balanced equation is

$H_2SO_4 + 2\,NaCl \longrightarrow Na_2SO_4 + 2\,HCl$

First calculate the g HCl to be produced

$$(20.0\text{ L HCl solution})\left(\frac{1000\text{ mL}}{1\text{ L}}\right)\left(\frac{1.20\text{ g}}{1.00\text{ mL}}\right)(0.420) = 1.01 \times 10^4\text{ g HCl}$$

Then calculate the g H_2SO_4 required to produce the HCl

$$(1.01 \times 10^4\text{ g HCl})\left(\frac{1\text{ mol}}{36.46\text{ g}}\right)\left(\frac{1\text{ mol } H_2SO_4}{2\text{ mol HCl}}\right)\left(\frac{98.09\text{ g}}{1\text{ mol}}\right) = 1.36 \times 10^4\text{ g } H_2SO_4$$

Finally, calculate the kg H_2SO_4 (96%)

$$(1.36 \times 10^4\text{ g } H_2SO_4)\left(\frac{1.00\text{ g } H_2SO_4\text{ solution}}{0.96\text{ g } H_2SO_4}\right)\left(\frac{1\text{ kg}}{1000\text{ g}}\right) = 14\text{ kg concentrated } H_2SO_4$$

56. Percent yield of H_2SO_4

$$(100.0\text{ g S})\left(\frac{1\text{ mol}}{32.07\text{ g}}\right) = 3.118\text{ mol S to start with}$$

$$3.118\text{ mol S} \longrightarrow \begin{array}{r} 3.118 \\ -0.3118 \\ \hline 2.8062 \end{array}\text{ mol }SO_2 - 10\% = 2.806\text{ mol }SO_2$$

$$2.806\text{ mol }SO_2 \longrightarrow \begin{array}{r} 2.806 \\ -0.2806 \\ \hline 2.5254 \end{array}\text{ mol }SO_3 - 10\% = 2.525\text{ mol }SO_3$$

$$2.525\text{ mol }SO_3 \longrightarrow \begin{array}{r} 2.525 \\ -0.2525 \\ \hline 2.2725 \end{array}\text{ mol }H_2SO_4 - 10\% = 2.273\text{ mol }H_2SO_4$$

$$(2.273\text{ mol }H_2SO_4)\left(\frac{98.09\text{ g}}{\text{mol}}\right) = 223.0\text{ g }H_2SO_4\text{ formed}$$

$$(3.118\text{ mol S})\left(\frac{1\text{ mol }H_2SO_4}{1\text{ mol S}}\right) = 3.118\text{ mol }H_2SO_4\text{ (theoretical yield)}$$

$$\left(\frac{2.273\text{ mol }H_2SO_4}{3.118\text{ mol }H_2SO_4}\right)(100) = 72.90\%\text{ yield}$$

Alternate Solution:

Calculation of yield. There are three chemical steps to the formation of H_2SO_4. Each step has a 10% loss of yield.

Step 1: 100% yield − 10% = 90.00% yield
Step 2: 90.00% yield − 10% = 81.00% yield
Step 3: 81.00% yield − 10% = 72.90% yield

Now calculate the grams of product. One mole of sulfur will yield a maximum of 1 mol H_2SO_4. Therefore 3.118 mol S will give a maximum of 3.118 mol H_2SO_4.

$$(3.118\text{ mol S})\left(\frac{1\text{ mol }H_2SO_4}{1\text{ mol S}}\right)\left(\frac{98.09\text{ g}}{\text{mol}}\right)(0.7290) = 223.0\text{ g }H_2SO_4\text{ yield}$$

57. According to the equations, the moles of CO_2 come from both reactions and the moles H_2O come from only the first reaction.
So the mol $NaHCO_3$ = 2 × mol H_2O = 2 × 0.0357 mol = 0.0714 mol $NaHCO_3$

$$(0.0714\text{ mol }NaHCO_3)\left(\frac{84.01\text{ g}}{\text{mol}}\right) = 6.00\text{ g }NaHCO_3\text{ in the sample}$$

$$(10.00\text{ g }NaHCO_3 + Na_2CO_3) - 6.00\text{ g }NaHCO_3 = 4.00\text{ g }Na_2CO_3\text{ in the sample}$$

$$\left(\frac{6.00 \text{ g NaHCO}_3}{10.00 \text{ g}}\right)(100) = 60.0\% \text{ NaHCO}_3$$

$$\left(\frac{4.00 \text{ g Na}_2\text{CO}_3}{10.00 \text{ g}}\right)(100) = 40.0\% \text{ Na}_2\text{CO}_3$$

58. The balanced equation is $2\ KClO_3 \longrightarrow 2\ KCl + 3\ O_2$

12.82 g mixture − 9.45 g residue = 3.37 g O_2 lost by heating

Because the O_2 lost came only from $KClO_3$, we can use it to calculate the amount of $KClO_3$ in the mixture.

The conversion is: g $O_2 \longrightarrow$ mol $O_2 \longrightarrow$ mol $KClO_3 \longrightarrow$ g $KClO_3$

$$(3.37 \text{ g O}_2)\left(\frac{1 \text{ mol}}{32.00 \text{ g}}\right)\left(\frac{2 \text{ mol KClO}_3}{3 \text{ mol O}_2}\right)\left(\frac{122.6 \text{ g}}{\text{mol}}\right) = 8.61 \text{ g KClO}_3 \text{ in the mixture}$$

$$\left(\frac{8.61 \text{ g KClO}_3}{12.82 \text{ g sample}}\right)(100) = 67.2\% \text{ KClO}_3$$

59. The balanced equation is

$Al(OH)_3(s) + 3\ HCl(aq) \longrightarrow AlCl_3(aq) + 3\ H_2O(l)$

The conversion is: L HCl $\longrightarrow$ g HCl $\longrightarrow$ mol HCl $\longrightarrow$ mol $Al(OH)_3 \rightarrow$ g $Al(OH)_3$

$$\left(\frac{2.5 \text{ L}}{\text{day}}\right)\left(\frac{3.0 \text{ g HCl}}{\text{L}}\right)\left(\frac{1 \text{ mol}}{36.46 \text{ g}}\right)\left(\frac{1 \text{ mol Al(OH)}_3}{3 \text{ mol HCl}}\right)\left(\frac{78.00 \text{ g}}{\text{mol}}\right) = 5.3 \text{ g Al(OH)}_3/\text{day}$$

Now calculate the number of 400. mg tablets that can be made from 5.3 g $Al(OH)_3$

$$\left(\frac{5.3 \text{ g Al(OH)}_3}{\text{day}}\right)\left(\frac{1000 \text{ mg}}{\text{g}}\right)\left(\frac{1 \text{ tablet}}{400. \text{ mg}}\right) = 13 \text{ tablets/day}$$

CHAPTER 10

MODERN ATOMIC THEORY

1. An electron orbital is a region in space around the nucleus of an atom where an electron is most probably found.

2. A second electron may enter an orbital already occupied by an electron if its spin is opposite that of the electron already in the orbital and all other orbitals of the same sublevel contain an electron.

3. The valence shell is the outermost energy level of an atom.

4. Valence electrons are the electrons located in the outermost energy level of an atom. Valence electrons are involved in bonding.

5. All the electrons in the atom are located in the orbitals closest to the nucleus.

6. Both 1s and 2s orbitals are spherical in shape and located symmetrically around the nucleus. The sizes of the spheres are different—the radius of the 2s orbital is larger than the 1s. The electrons in 2s orbitals are located further from the nucleus.

7. The letters used to designate the energy sublevels are s, p, d, and f.

8. 1s, 2s, 2p, 3s, 3p, 4s, 3d, 4p

9. s–2 electrons per shell
   p–6 electrons per shell after the first energy level
   d–10 electrons per shell after the second energy level

10. The main difference is that the Bohr orbit has an electron traveling a specific path around the nucleus while an orbital is a region in space where the electron is most probably found.

11. Bohr's model was inadequate since it could not account for atoms more complex than hydrogen. It was modified by Schrödinger into the modern concept of the atom in which electrons exhibit wave and particle properties. The motion of electrons is determined only by probability functions as a region in space, or a cloud surrounding the nucleus.

12. *s* orbital

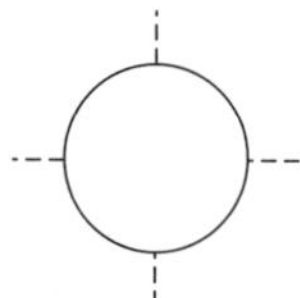

p orbitals

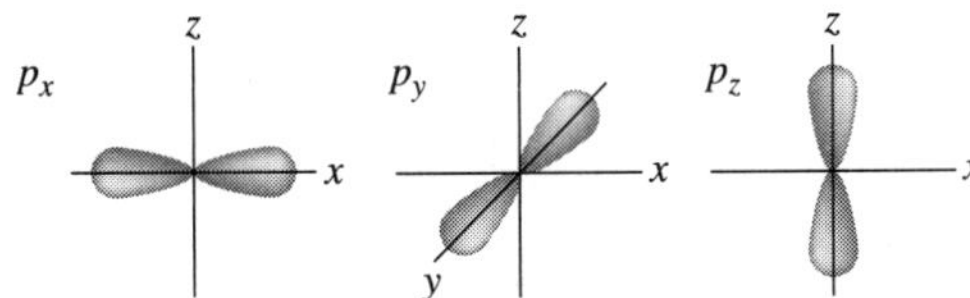

13. 3 is the third energy level
    d indicates an energy sublevel
    7 indicates the number of electrons in the d sublevel

14. Transition elements are found in the center of the periodic table. The last electrons for these elements are found in the d or f orbitals.
    Representative elements are located on either side of the periodic table (Group IA–VIIA). The valence electrons for these elements are found in s and/or p orbitals.

15. Elements in the s-block all have one or two electrons in their outermost energy level. These valence electrons are located in an s-orbital.

16.

Atomic #	Symbol
8	O
16	S
34	Se
52	Te
84	Po

All of these elements have an outermost electron structure of s^2p^4

17. F, Cl, Br, I, At (Halogens)

18. The greatest number of elements in any period is 32. The 6th period has this number of elements.

19. The elements in Group A always have their last electrons in the outermost energy level, while the last electrons in Group B lie in an inner level.

20. Pairs of elements which are out of sequence with respect to atomic masses are: Ar and K; Co and Ni; Te and I; Th and Pa; U and Np; Lr and Rf; Pu and Am.

21.

(a)	H	1 proton	(c)	Sc	21 protons
(b)	B	5 protons	(d)	U	92 protons

22.

(a)	F	9 protons	(c)	Br	35 protons
(b)	Ag	47 protons	(d)	Sb	51 protons

23. (a) B $1s^22s^22p^1$
    (b) Ti $1s^22s^22p^63s^23p^64s^23d^2$
    (c) Zn $1s^22s^22p^63s^23p^64s^23d^{10}$
    (d) Sr $1s^22s^22p^63s^23p^64s^23d^{10}4p^65s^2$

24. (a) Cl $1s^22s^22p^63s^23p^5$
    (b) Ag $1s^22s^22p^63s^23p^64s^23d^{10}4p^65s^14d^{10}$
    (c) Li $1s^22s^1$
    (d) Fe $1s^22s^22p^63s^23p^64s^23d^6$
    (e) I $1s^22s^22p^63s^23p^64s^23d^{10}4p^65s^24d^{10}5p^5$

25. The spectral lines of hydrogen are produced by energy emitted when the electron from on hydrogen atom, which has absorbed energy, falls from a higher energy level to a lower energy level (closer to the nucleus).

26. Bohr said that a number of orbits were available for electrons, each corresponding to an energy level. When an electron falls from a higher energy orbit to a lower energy orbit, energy is given off as a specific wavelength of light. Only those energies in the visible range are seen in the hydrogen spectrum. Each line corresponds to a change from one orbit to another.

27. 9 orbitals in the third energy level: 3s, $3p_x$, $3p_y$, $3p_z$ plus five d orbitals

28. 32 electrons in the fourth energy level

29.
    (a) (7p 7n) $2e^-5e^-$ $^{14}_{7}N$
    (b) (17p 18n) $2e^-8e^-7e^-$ $^{35}_{17}Cl$
    (c) (30p 35n) $2e^-8e^-18e^-2e^-$ $^{65}_{30}Zn$
    (d) (40p 51n) $2e^-8e^-18e^-10e^-2e^-$ $^{91}_{40}Zr$
    (e) (53p 74n) $2e^-8e^-18e^-18e^-7e^-$ $^{127}_{53}I$

30.

(a) 14p, 14n $2e^-8e^-4e^-$ $^{28}_{14}Si$

(b) 16p, 16n $2e^-8e^-6e^-$ $^{32}_{16}S$

(c) 18p, 22n $2e^-8e^-8e^-$ $^{40}_{18}Ar$

(d) 23p, 28n $2e^-8e^-11e^-2e^-$ $^{51}_{23}V$

(e) 15p, 16n $2e^-8e^-5e^-$ $^{31}_{15}P$

31. (a) O $1s^22s^22p^4$

(b) Ca $1s^22s^22p^63s^23p^64s^2$

(c) Ar $1s^22s^22p^63s^23p^6$

(d) Br $1s^22s^22p^63s^23p^64s^23d^{10}4p^5$

(e) Fe $1s^22s^22p^63s^23p^64s^23d^6$

32. (a) Li $1s^22s^1$

(b) P $1s^22s^22p^63s^23p^3$

(c) Zn $1s^22s^22p^63s^23p^64s^23d^{10}$

(d) Na $1s^22s^22p^63s^1$

(e) K $1s^22s^22p^63s^23p^64s^1$

33. (a) Mg (c) Ni

(b) Al (d) Mn

34. (a) Sc (c) Sn

(b) Zn (d) Cs

35. atomic No. — electron structure

	atomic No.	electron structure
(a)	8	$1s^22s^22p^4$
(b)	11	$1s^22s^22p^63s^1$
(c)	17	$1s^22s^22p^63s^23p^5$
(d)	23	$1s^22s^22p^63s^23p^64s^23d^3$
(e)	28	$1s^22s^22p^63s^23p^64s^23d^8$
(f)	34	$1s^22s^22p^63s^23p^64s^23d^{10}4p^4$

36. atomic No. — electron structure

	atomic No.	electron structure
(a)	9	$[He]2s^22p^5$
(b)	26	$[Ar]4s^23d^6$
(c)	31	$[Ar]4s^23d^{10}4p^1$
(d)	39	$[Kr]5s^24d^1$
(e)	52	$[Kr]5s^24d^{10}5p^4$
(f)	10	$[He]2s^22p^6$

37. (a) F [↑↓] [↑↓] [↑↓ | ↑↓ | ↑]

(b) S [↑↓] [↑↓] [↑↓ | ↑↓ | ↑↓] [↑↓] [↑↓ | ↑ | ↑]

(c) Co [↑↓] [↑↓] [↑↓ | ↑↓ | ↑↓] [↑↓] [↑↓ | ↑↓ | ↑↓] [↑↓]
[↑↓ | ↑↓ | ↑ | ↑ | ↑]

(d) Kr [↑↓] [↑↓] [↑↓ | ↑↓ | ↑↓] [↑↓] [↑↓ | ↑↓ | ↑↓] [↑↓]
[↑↓ | ↑↓ | ↑↓ | ↑↓ | ↑↓] [↑↓ | ↑↓ | ↑↓]

(e) Ru [↑↓] [↑↓] [↑↓ | ↑↓ | ↑↓] [↑↓] [↑↓ | ↑↓ | ↑↓] [↑↓]
[↑↓ | ↑↓ | ↑↓ | ↑↓ | ↑↓] [↑↓ | ↑↓ | ↑↓] [↑↓] [↑↓ | ↑ | ↑ | ↑ | ↑]

38. (a) Cl [↑↓] [↑↓] [↑↓ | ↑↓ | ↑↓] [↑↓] [↑↓ | ↑↓ | ↑]

(b) Mg [↑↓] [↑↓] [↑↓ | ↑↓ | ↑↓] [↑↓]

(c) Ni [↑↓] [↑↓] [↑↓ | ↑↓ | ↑↓] [↑↓] [↑↓ | ↑↓ | ↑↓] [↑↓]
[↑↓ | ↑↓ | ↑↓ | ↑ | ↑]

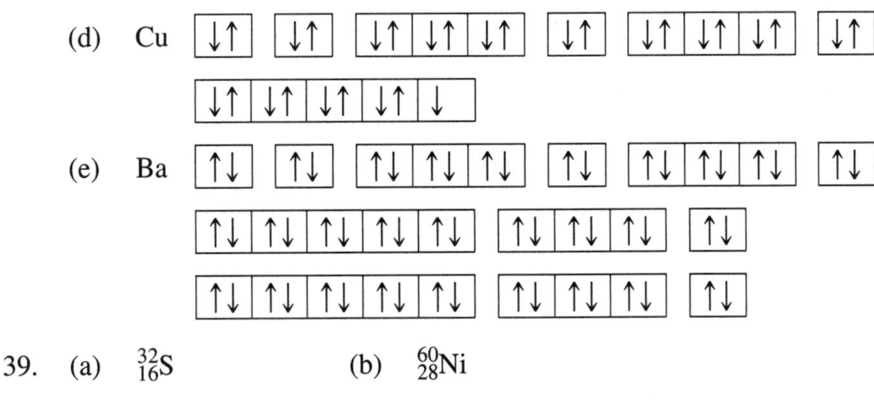

39. (a) $^{32}_{16}S$ (b) $^{60}_{28}Ni$

40.
(a) 13p 14n $2e^- 8e^- 3e^-$ $^{27}_{13}Al$

(b) 22p 26n $2e^- 8e^- 10e^- 2e^-$ $^{48}_{22}Ti$

41. The eleventh electron of sodium is located in the third energy level because the first and second levels are filled. Also the properties of sodium are similar to the other elements in Group IA.

42. The last electron in potassium is located in the fourth energy level because the 4s orbital is at a lower energy level than the 3d orbital. Also the properties of potassium are similar to the other elements in Group IA.

43. Noble gases all have filled s and p orbitals in the outermost energy level.

44. Noble gases each have filled s and p orbitals in the outermost energy level.

45. Moving from left to right in any period of elements, the atomic number increases by one from one element to the next and the atomic radius generally decreases. Each period (except period 1) begins with an alkali metal and ends with a noble gas. There is a trend in properties of the elements changing from metallic to nonmetallic from the beginning to the end of the period.

46. The elements in a group have the same number of outer energy level electrons. They are located vertically on the periodic table.

47. (a) 4 (b) 6 (c) 1 (d) 7 (e) 3

48. (a) 5 (b) 5 (c) 6 (d) 2 (e) 3

49. The outermost energy level contains one electron in an s orbital.

50. All of these elements have a s^2d^{10} electron configuration in their outermost energy levels.

51. (a) and (g)
(b) and (d)

52. (a) and (f)
(d) and (h)

53. 12, 38 since they are in the same periodic group

54. 7, 33 since they are in the same periodic group

55. (a) K, metal
(b) Pu, metal
(c) S, nonmetal
(d) Sb, metalloid

56. (a) I, nonmetal
(b) W, metal
(c) Mo, metal
(d) Ge, metalloid

57. Period 6, lanthanide series, contains the first element with an electron in an f orbital.

58. Period 4 Group IIIB contains the first element with an electron in a d orbital.

59. Group VIIA contain 7 valence electrons.
Group VIIB contain 2 electrons in the outermost level and 5 electrons in an inner d orbital.
Group A elements are representative while Group B elements are transition elements.

60. Group IIIA contain 3 valence electrons.
Group IIIB contain 2 electrons in the outermost level and one electron in an inner d orbital.
Group A elements are representative while Group B elements are transition elements.

61. The valence energy level of an atom can be determined by looking at what period the element is in. Period 1 corresponds to valence energy level 1, period 2 to valence energy level 2 and so on. The number of valence electrons for element's 1–18 can be determined by looking at the Roman numeral for the element's group. For example, boron is under Group IIIA, therefore it has three valence shell electrons.

62. (a) valence energy level 2, 1 valence electron
(b) valence energy level 3, 7 valence electrons
(c) valence energy level 3, 4 valence electrons
(d) valence energy level 3, 6 valence electrons
(e) valence energy level 2, 2 valence electrons

63.

$1s^3$	Li	Atomic No. 3
$1s^3 2s^3 2p^9$	P	Atomic No. 15
$1s^3 2s^3 2p^9 3s^3 3p^9$	Co	Atomic No. 27

64. Nitrogen has more valence electrons on more energy levels. More varied electron transitions are possible.

65. (a) all 100
(b) all but H, He, Li, Be (96)
(c) 80 (all but those from H to Ca)
(d) the 44 elements beginning after Ba

66. (a) $\frac{2}{2} \times 100 = 100\%$

(b) $\frac{4}{4} \times 100 = 100\%$

(c) $\frac{10}{54} \times 100 = 19\%$

(d) $\frac{8}{34} \times 100 = 24\%$

(e) $\frac{11}{55} \times 100 = 20\%$

67. (a) $:\dot{\underset{\cdot}{\text{O}}}:$ 2 pairs of valence electrons

(b) $\cdot\ddot{\underset{\cdot}{\text{P}}}\cdot$ 1 pair of valence electrons

(c) $:\ddot{\underset{\cdot}{\text{I}}}:$ 3 pairs of valence electrons

(d) $:\ddot{\underset{\cdot\cdot}{\text{Xe}}}:$ 4 pairs of valence electrons

(e) $\text{Rb}\cdot$ 0 pairs of valence electrons

68. $\dfrac{1.5}{1.0 \times 10^{-8}} = 150{,}000{,}000 = \dfrac{1.5 \times 10^{8}}{1}$

69. (a) Ne (b) Ge (c) F (d) N

70. The outermost electron structure for both sulfur and oxygen is s^2p^4.

71. Transition elements are found in Groups IB–VIIB, VIII, lanthanides and actinides.

72. In transition elements the last electron added is in a d or f orbital. The last electron added in a representative element is in an s or p orbital.

73. Elements number 8, 16, 34, 52, 84 all have 6 electrons in their outer shell.

74. (a) scandium (b) zirconium (c) mercury (d) cesium

75. (a) alkali metals (b) alkaline earth metals (c) halogens

76. (a) 3p sublevel (b) 4d sublevel (c) 4f sublevel

77.

(a)	Na	representative element	metal
(b)	N	representative element	nonmetal
(c)	Mo	transition element	metal
(d)	Ra	representative element	metal
(e)	As	representative element	metalloid
(f)	Ne	noble gas	nonmetal

78. The outermost energy level is the 7th. Element 87 contains one electron in the 7s orbital.

79. If 36 is a noble gas, 35 would be in periodic Group VIIA and 37 would be in periodic Group IA.

80. Answers will vary but should at least include a statement about: (1) Numbering of the elements and their relationship to atomic structure; (2) division of the elements into periods and groups; (3) division of the elements into metals, nonmetals, and metalloids; (4) identification and location of the representative and transition elements.

81. (a) $[Rn]7s^25f^{14}6d^{10}7p^5$
(b) 7 valence electrons, $7s^27p^5$
(c) F, Cl, Br, I, At
(d) halogen family, Period 7

82. (a) The two elements are isotopes.
    (b) The two elements are adjacent to each other in the same period.

83. Most gases are located in the upper right part of the periodic table (H is an exception). They are nonmetals. Liquids show no pattern. Neither do solids, except the vast majority of solids are metals.

84. excited sulfur atom:
electron configuration: $1s^22s^22p^63s^13p^5$
orbital diagram:

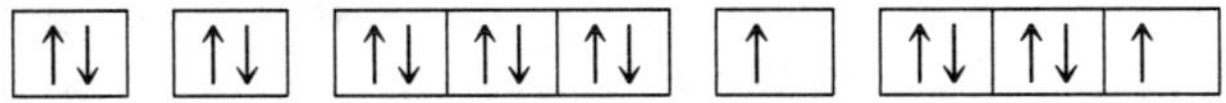

CHAPTER 11

CHEMICAL BONDS: THE FORMATION OF COMPOUNDS FROM ATOMS

1. smallest Cl, Mg, Na, K, Rb largest.

2. More energy is required for neon because it has a very stable outer electron structure consisting of an octet of electrons in filled orbitals (noble gas electron structure). Sodium, an alkali metal, has a relatively unstable outer electron structure with a single electron in an unfilled orbital. The sodium electron is also farther away from the nucleus and is shielded by more inner electrons than are neon outer electrons.

3. When a third electron is removed from beryllium, it must come from a very stable electron structure corresponding to that of the noble gas, helium. In addition, the third electron must be removed from a +2 beryllium ion, which increases the difficulty of removing it.

4. The first ionization energy decreases from top to bottom because in the successive alkali metals, the outermost electron is farther away from the nucleus and is more shielded from the positive nucleus by additional electron energy levels.

5. The first ionization energy decreases from top to bottom because the outermost electrons in the successive noble gases are farther away from the nucleus and are more shielded by additional inner electron energy levels.

6. Barium and beryllium are in the same family. The electron to be removed from barium is, however, located in an energy level farther away from the nucleus than is the energy level holding the electron in beryllium. Hence, it requires less energy to remove the electron from barium than to remove the electron from beryllium. Barium, therefore, has a lower ionization energy than beryllium.

7. The first electron removed from a sodium atom is the one valence electron, which is shielded from most of its nuclear charge by the electrons of lower levels. To remove a second electron from the sodium ion requires breaking into the noble gas structure. This requires much more energy than that necessary to remove the first electron, because the Na^+ is already positive.

8. (a) K > Na
   (b) Na > Mg
   (c) O > F
   (d) I > Br
   (e) Zr > Ti

9. The first element in each group has the smallest radius.

10. Atomic size increases down a column since each successive element has an additional energy level which contains electrons located farther from the nucleus.

11. A Lewis structure is a representation of the bonding in a compound. Valence electrons are responsible for bonding, therefore they are the only electrons that need to be shown in a Lewis structure.

12. Metals are less electronegative than nonmetals. Therefore, metals lose electrons more easily than nonmetals. So, metals will transfer electrons to nonmetals leaving the metals with a positive charge and the nonmetals with a negative charge.

13. The noble gases are the most stable of all the elements because they have a complete octet (8 electrons) in their valence level. When the elements in Groups IA, IIA, VIA and VIIA form ions they want to establish a stable electron structure. Therefore, the elements in Groups IA and IIA lose electrons (obtain a positive charge) to achieve a noble gas electron configuration and the elements in Groups VIA and VIIA gain electrons (obtain a negative charge) to achieve a noble gas electron configuration.

14. A polar bond is a bond between two atoms with very different electronegativities. A polar molecule is a dipole due to unequal electrical charge between bonded atoms resulting from unequal sharing of electrons.

15. The electron pair arrangement is the arrangement of both bonding and nonbonding electrons in a Lewis structure. The molecular shape of a molecule is the three dimensional arrangement of its' atoms in space.

16. A Lewis structure is a visual representation of the arrangement of atoms and electrons in a molecule or an ion. It shows how the atoms in a molecule are bonded together.

17. The more electronegative atom in the bond between two atoms will more strongly attract electrons so it will have a partial negative charge (δ−). The less electronegative atom will have a partial positive charge (δ+) because the bonding pair of electrons has been pulled away by the more electronegative atom.

18. The dots in a Lewis structure represent nonbonding pairs of electrons. The lines represent bonding pairs of electrons.

19.

Group	IA	IIA	IIIA	IVA	VA	VIA	VIIA
	E·	E:	Ẹ:	·Ẹ:	·Ė̩:	·Ė̤:	·Ë̤:

20. Lewis structures:

    Cs·   Ba: Ṫl: ·Ṗb: ·Ṗo: ·Ät: :R̈n:

    Each of these is a representative element and has the same number of electrons in its outer energy level as its periodic group.

21. (a) Elements with the highest electronegativities are found in the upper right hand corner of the periodic table.
    (b) Elements with the lowest electronegativities are found in the lower left of the periodic table.

22. Valence electrons are the electrons found in the outermost energy level of an atom.

23. By losing one electron, a potassium atom acquires a noble gas structure and becomes a K^+ ion. To become a K^{2+} ion requires the loss of a second electron and breaking into the noble gas structure of the K^+ ion. This requires too much energy to generally occur.

24. An aluminum ion has a +3 charge because it has lost 3 electrons in acquiring a noble gas electron structure.

25. Magnesium atom is larger because it has electrons in the 3rd shell, while a magnesium ion does not. Also, the Mg^{2+} ion has 12 protons and 10 electrons, creating a charge imbalance and drawing the electrons in towards the nucleus more closely.

26. A bromine atom is smaller because it has one less electron than the bromine ion in its outer shell. Also, the ion has 35 protons and 36 electrons, creating a charge imbalance, resulting in a lessening of the attraction of the electrons towards the nucleus.

27.

	+	−		+	−
(a)	H	O	(d)	Pb	S
(b)	Na	F	(e)	N	O
(c)	H	N	(f)	H	C

28.

	+	−		+	−
(a)	H	Cl	(d)	I	Br
(b)	Li	H	(e)	Mg	H
(c)	C	Cl	(f)	O	F

29. (a) ionic (b) covalent (c) covalent (d) ionic

30. (a) covalent (b) ionic (c) covalent (d) ionic

31. Magnesium has an electron structure $1s^22s^22p^63s^2$, while the structure for chlorine is $1s^22s^22p^63s^23p^5$. When these two elements react with each other, each magnesium atom loses its $3s^2$ electrons, one to each of two chlorine atoms. The resulting structures for both magnesium and chlorine are noble gas configurations.

32. (a) $F + 1\,e^- \longrightarrow F^-$ (b) $Ca \longrightarrow Ca^{2+} + 2\,e^-$

33. (a) $Mg{:} + \cdot\ddot{\underset{\cdot\cdot}{F}}{:} + \cdot\ddot{\underset{\cdot\cdot}{F}}{:} \rightarrow MgF_2$

(b) $K\cdot + \cdot K + \cdot\ddot{\underset{\cdot}{O}}{:} \rightarrow K_2O$

34. (a) $Ca{:} + \cdot\ddot{\underset{\cdot}{O}}{:} \rightarrow CaO$

(b) $Na\cdot + {:}\ddot{\underset{\cdot\cdot}{Br}}{:} \rightarrow NaBr$

35. Valence electrons: H(1) K(1) Mg(2) He(2) Al(3)

36. Valence electrons: Si(4) N(5) P(5) O(6) Cl(7)

37. Noble gas structures:
(a) Calcium atom, loses $2\,e^-$
(b) Sulfur atom, gains $2\,e^-$
(c) Helium, none

38. (a) Chloride ion, none
(b) Nitrogen atom, gain $3\,e^-$ or lose $5\,e^-$
(c) Potassium atom, lose $1\,e^-$

39. (a) A magnesium ion, Mg^{2+}, is larger than an aluminum ion, Al^{3+}. Both ions have the same number of electrons (isoelectronic), with identical distribution of these electrons in their energy levels. The increased nuclear charge of the Al^{3+} ion pulls the electrons in the Al^{3+} ion closer to its nucleus, making it smaller than the Mg^{2+} ion.

(b) The Fe^{2+} ion is larger than the Fe^{3+} ion because the Fe^{2+} ion has one more electron than the Fe^{3+} ion. Both ions have the same nuclear charge.

40. (a) A potassium atom is larger than a potassium ion because the atom has one more energy level containing an electron than the ion.

(b) A bromide ion is larger than a bromine atom because it has, within the same principal energy level, one more electron than a bromine atom.

41. (a) ionic, NaCl (b) molecular, CH_4 (c) ionic, $MgBr_2$ (d) molecular, Br_2 (e) molecular, CO_2

42. (a) covalent, O_2 (b) polar covalent, HBr (c) polar covalent, H_2O (d) covalent, I_2 (e) covalent, CCl_4

43. (a) NaH, Na_2O (b) CaH_2, CaO (c) AlH_3, Al_2O_3 (d) SnH_4, SnO_2

44. (a) SbH_3, Sb_2O_3 (b) H_2Se, SeO_3 (c) HCl, Cl_2O_7 (d) CH_4, CO_2

45.

Li_2SO_4	lithium sulfate	K_2SO_4	potassium sulfate
Rb_2SO_4	rubidium sulfate	Cs_2SO_4	cesium sulfate
Fr_2SO_4	francium sulfate		

46.

$BeBr_2$	beryllium bromide	$BaBr_2$	barium bromide
$MgBr_2$	magnesium bromide	$RaBr_2$	radium bromide
$SrBr_2$	strontium bromide		

47. Lewis structures:

(a) Na· (b) $[:\ddot{\underset{\cdot\cdot}{Br}}:]^-$ (c) $[:\ddot{\underset{\cdot\cdot}{O}}:]^{2-}$

48. (a) $\underset{\cdot}{Ga}:$ (b) Ga^{3+} (c) Ca^{2+}

49. (a) covalent (b) ionic (c) ionic (d) covalent

50. (a) covalent (b) ionic (c) covalent (d) covalent

51. (a) ionic (b) covalent (c) covalent

52. (a) covalent (b) covalent (c) covalent

53. (a) H:H (b) :N:::N: (c) $:\ddot{\underset{\cdot\cdot}{Cl}}:\ddot{\underset{\cdot\cdot}{Cl}}:$

54. (a) $:\ddot{O}::\ddot{O}:$ (b) $:\ddot{\underset{\cdot\cdot}{Br}}:\ddot{\underset{\cdot\cdot}{Br}}:$ (c) $:\ddot{\underset{\cdot\cdot}{I}}:\ddot{\underset{\cdot\cdot}{I}}:$

55.

(a)
```
  ··  ··  ··
 :Cl :N :Cl:
  ··  ··  ··
     :Cl:
      ··
```

(c)
```
   H H
   ·· ··
 H:C:C:H
   ·· ··
   H H
```

(b)
```
    ··   ··
 H:O:C::O:
    ·· ··
      :O:
       ··
       H
```

(d)
```
            ··    ··     ·· ]-
 [Na]+   [ :O:N::O:        
            ··  ··    ··
              :O:
               ··    ]
```

56.

(a)
```
  ··
 :S:H
  ··
  H
```

(c)
```
    ··
 H:N:H
    ··
    H
```

(b)
```
  ··      ··
 :S::C::S:
```

(d)
```
 [    H   ]+
 [    ··  ]      ··   -
 [ H:N:H  ]   [ :Cl: ]
 [    ··  ]      ··
 [    H   ]
```

57.

(a) $[Ba]^{2+}$

(d) $[:C:::N:]^{-}$

(b) $[Al]^{3+}$

(e)
```
 [  ··   ··  ··  ]-
 [ :O::C:O:      ]
 [     ··  ··    ]
 [    :O:        ]
 [     ··        ]
 [     H         ]
```

(c)
```
 [  ··  ··  ··  ]2-
 [ :O:S:O:      ]
 [  ··  ··  ··  ]
 [    :O:       ]
 [     ··       ]
```

58.

(a)
```
    ··   -
 [ :I: ]
    ··
```

(d)
```
 [  ··  ··  ··  ]-
 [ :O:Cl:O:     ]
 [  ··  ··  ··  ]
 [    :O:       ]
 [     ··       ]
```

(b)
```
    ··   2-
 [ :S: ]
    ··
```

(e)
```
 [  ··       ··  ]-
 [ :O:N::O:      ]
 [  ··  ··   ··  ]
 [    :O:        ]
 [     ··        ]
```

(c)
```
 [  ··      ··  ]2-
 [ :O:C::O      ]
 [  ··  ··  ··  ]
 [    :O:       ]
 [     ··       ]
```

59. (a) H_2O, polar (b) HBr, polar (c) CF_4, nonpolar

60. (a) F_2, nonpolar (b) CO_2, nonpolar (c) NH_3, polar

61. (a) 4 electron pairs, tetrahedral
    (b) 4 electron pairs, tetrahedral
    (c) 3 electron pairs, trigonal planar

62. (a) 3 electron pairs, trigonal planar
    (b) 4 electron pairs, tetrahedral
    (c) 4 electron pairs, tetrahedral

63. (a) tetrahedral (b) pyramidal (c) tetrahedral

64. (a) tetrahedral (b) pyramidal (c) tetrahedral

65. (a) tetrahedral (b) pyramidal (c) bent

66. (a) tetrahedral (b) bent (c) bent

67. Oxygen

68. Potassium

69. N_2H_4 14 e^-

```
H H
| |
:N:N:
| |
H H
```

70. (a) NO_2^- 18 e^- $[:\ddot{O}-\ddot{N}=\ddot{O}:]^-$ linear

(b) SO_4^{2-} 32 e^- tetrahedral

```
[    :Ö:    ]2-
[     |     ]
[ :Ö—S—Ö:   ]
[     |     ]
[    :Ö:    ]
```

(c) $SOCl_2$ 26 e^- pyramidal

```
:Cl—S—Cl:
    |
   :O:
```

(d) Cl_2O 20 e^- bent

```
     O
   /   \
 Cl     Cl
```

71. (a) C_2H_6 14 e⁻

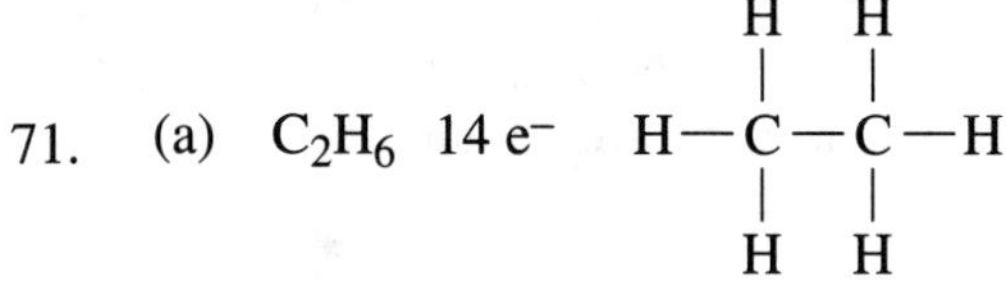

(b) C_2H_4 12 e⁻ $H-C{=}C-H$ (each C also bonded to one H)

(c) C_2H_2 10 e⁻ $H-C{\equiv}C-H$

72. (a) Hg (b) Be (c) N (d) Fr (e) Au

73. (a) Zn (b) Be (c) N

74. (1) Fluorine is missing one electron from its 2p level while neon has a full energy level.
(2) Fluorine's valence electrons are closer to the nucleus. The attraction between the electrons and the positive nucleus is greater.

75. Lithium has a +1 charge after the first electron is removed. It takes more energy to overcome that charge and to remove another electron than to remove a single electron from an uncharged He atom.

76. Yes. Each of these elements have an ns^1 electron and they could lose that electron in the same way elements in Group IA do. They would then form +1 ions and ionic compounds such as CuCl, AgCl, and AuCl.

77. $SnBr_2$, $GeBr_2$

78. The bond between sodium and chlorine is ionic. An electron has been transferred from a sodium atom to a chlorine atom. The substance is composed of ions not molecules. Use of the word molecule implies covalent bonding.

79. A covalent bond results from the sharing of a pair of electrons between two atoms, while an ionic bond involves the transfer of one or more electrons from one atom to another.

80. This structure shown is incorrect since the bond is ionic. It should be represented as:

$$[Na]^{+}[:\ddot{\underset{..}{O}}:]^{2-}[Na]^{+}$$

81. The four most electronegative elements are F, O, N, Cl

82. highest F, O, S, H, Mg, Cs lowest

83. It is possible for a molecule to be nonpolar even though it contains polar bonds. If the molecule is symmetrical, the polarities of the bonds will cancel (in a manner similar to a positive and negative number of the same size) resulting in a nonpolar molecule. An example is CO_2 which is linear and nonpolar.

84. Both molecules contain polar bonds. CO_2 is symmetrical about the C atom, so the polarities cancel. In CO, there is only one polar bond, therefore the molecule is polar.

85. (a) 105° (b) 107° (c) 109.5° (d) 109.5°

86. (a) Both use the p orbitals for bonding. B uses one s and two p orbitals while N uses three p orbitals for bonding.
(b) BF_3 is trigonal planar while NF_3 is pyramidal.
(c) BF_3 no lone pairs
NF_3 one lone pair
(d) BF_3 has 3 very polar covalent bonds. NF_3 has 3 polar covalent bonds

87. Fluorine's electronegativity is greater than any other element. Ionic bonds form between atoms of widely different electronegativities. Therefore, Fr–F, Cs–F, Rb–F or K–F would be ionic substances with the greatest electronegativity difference.

88. Each element in a particular column has the same number of valence electrons and therefore the same Lewis structure.

89. S $\quad \dfrac{1.40\text{ g}}{32.07\dfrac{\text{g}}{\text{mol}}} = 0.0437\text{ mol} \qquad \dfrac{0.0437}{0.0437} = 1.00$

O $\quad \dfrac{2.10\text{ g}}{16.00\dfrac{\text{g}}{\text{mol}}} = 0.131\text{ mol} \qquad \dfrac{0.131}{0.0437} = 3.00$

Empirical formula is SO_3

```
     ..
    :O:
     |
 ..      ..
:O—S=O:
 ..
```

90. We need to know the molecular formula before we can draw the Lewis structure. From the data, determine the empirical and then the molecular formula.

$$\text{C} \quad \frac{14.5\ \text{g}}{12.01\,\frac{\text{g}}{\text{mol}}} = 1.21\ \text{mol} \qquad \frac{1.21}{1.21} = 1.00$$

$$\text{Cl} \quad \frac{85.5\ \text{g}}{35.45\,\frac{\text{g}}{\text{mol}}} = 2.41\ \text{mol} \qquad \frac{2.41}{1.21} = 2.01$$

CCl_2 is the empirical formula

empirical mass $= 1(12.01) + 2(35.45) = 82.91$

$$\frac{166}{82.91} = 2.00$$

Therefore, the molecular formula is $(CCl_2)_2$ or C_2Cl_4

$$\begin{array}{ccc} :\ddot{\text{Cl}} & & :\ddot{\text{Cl}}: \\ & \text{C}=\text{C} & \\ \ddot{\text{Cl}} & & \ddot{\text{Cl}} \end{array}$$

91. (a) incorrect, correct structure is:

$$:\ddot{\text{O}}=\text{C}=\ddot{\text{O}}:$$

(b) correct

(c) correct

(d) incorrect, correct structure is:

$$\left[\begin{array}{c} :\ddot{\underset{\cdot\cdot}{\text{O}}}-\text{N}-\ddot{\underset{\cdot\cdot}{\text{O}}}: \\ \| \\ :\text{O}: \end{array}\right]^-$$

(e) incorrect, correct structure is:

$$\text{H}-\text{C}\equiv\text{N}:$$

(f) incorrect, correct structure is:

$$\left[\begin{array}{ccc} & :\ddot{\text{O}}: & \\ & | & \\ :\ddot{\underset{\cdot\cdot}{\text{O}}}- & \text{S} & -\ddot{\underset{\cdot\cdot}{\text{O}}}: \\ & | & \\ & :\underset{\cdot\cdot}{\text{O}}: & \end{array}\right]^{2-}$$

CHAPTER 12

THE GASEOUS STATE OF MATTER

1. In Figure 12.1, color is the evidence of diffusion; bromine is colored and air is colorless. If hydrogen and oxygen had been the two gases, this would not work because both gases are colorless. Two ways could be used to show the diffusion. The change of density would be one method. Before diffusion the gas in the flask containing hydrogen would be much less dense. After diffusion, the gas densities in both flasks would be equal. A second method would require the introduction of spark gaps into both flasks. Before diffusion, neither gas would show a reaction when sparked. After diffusion, the gases in both flasks would explode because of the mixture of hydrogen and oxygen.

2. The pressure of a gas is the force that gas particles exert on the walls of a container. It depends on the temperature, the number of molecules of the gas and the volume of the container.

3. The air pressure inside the balloon is greater than the air pressure outside the balloon. The pressure inside must equal the sum of the outside air pressure plus the pressure exerted by the stretched rubber of the balloon.

4. The major components of dry air are nitrogen and oxygen.

5. 1 torr $=$ 1 mm Hg

6. The molecules of H_2 at 100°C are moving faster. Temperature is a measure of average kinetic energy. At higher temperatures, the molecules will have more kinetic energy.

7. 1 atm corresponds to 4 L.

8. The pressure times the volume at any point on the curve is equal to the same value. This is an inverse relationship as is Boyle's law. ($PV = k$)

9. If $T_2 < T_1$, the volume of the cylinder would decrease (the piston would move downward).

10. The pressure inside the bottle is less than atmospheric pressure. We come to this conclusion because the water inside the bottle is higher than the water in the trough (outside the bottle).

11. The density of air is 1.29 g/L. Any gas listed below air in Table 12.3 has a density greater than air. For example: O_2, H_2S, HCl, F_2, CO_2.

12. Basic assumptions of Kinetic Molecular Theory include:

    (a) Gases consist of tiny particles.
    (b) The distance between particles is great compared to the size of the particles.
    (c) Gas particles move in straight lines. They collide with one another and with the walls of the container with no loss of energy.
    (d) Gas particles have no attraction for each other.
    (e) The average kinetic energy of all gases is the same at any given temperature. It varies directly with temperature.

13. The order of increasing molecular velocities is the order of decreasing molar masses.

    increasing molecular velocity

    ⟶

    Rn, F_2, N_2CH_4, He, H_2

    ⟵

    decreasing molar mass

    At the same temperature the kinetic energies of the gases are the same and equal to $\frac{1}{2}mv^2$. For the kinetic energies to be the same, the velocities must increase as the molar masses decrease.

14. Average kinetic energies of all these gases are the same, since the gases are all at the same temperature.

15. Gases are described by the following parameters:

    (a) pressure
    (b) volume
    (c) temperature
    (d) number of moles

16. An ideal gas is one which follows the described gas laws at all P, V and T and whose behavior is described exactly by the Kinetic Molecular Theory.

17. Boyle's law: $P_1V_1 = P_2V_2$, ideal gas equation: $PV = nRT$
    If you have an equal number of moles of two gases at the same temperature the right side of the ideal gas equation will be the same for both gases. You can then set PV for the first gas equal to PV for the second gas (Boyle's law) because the right side of both equations will cancel.

18. Charles' law: $V_1/T_1 = V_2/T_2$, ideal gas equation: $PV = nRT$
    Rearrange the ideal gas equation to: $V/T = nR/P$
    If you have an equal number of moles of two gases at the same pressure the right side of the rearranged ideal gas equation will be the same for both. You can set V/T for the first gas equal to V/T for the second gas (Charles' law) because the right side of both equations will cancel.

19. A gas is least likely to behave ideally at low temperatures. Under this condition, the velocities of the molecules decrease and attractive forces between the molecules begin to play a significant role.

20. A gas is least likely to behave ideally at high pressures. Under this condition, the molecules are forced close enough to each other so that their volume is no longer small compared to the volume of the container. Attractive forces may also occur here and sooner or later, the gas will liquefy.

21. Equal volumes of H_2 and O_2 at the same T and P:

   (a) have equal number of molecules (Avogadro's law)

   (b) mass O_2 = 16 times mass of H_2

   (c) moles O_2 = moles H_2

   (d) average kinetic energies are the same (T same)

   (e) rate H_2 = 4 times the rate of O_2 (Graham's Law of Effusion)

   (f) density O_2 = 16 times the density of H_2

$$\text{density } O_2 = \left(\frac{\text{mass } O_2}{\text{volume } O_2}\right) \qquad \text{density } H_2 = \left(\frac{\text{mass } H_2}{\text{volume } H_2}\right)$$

$$\text{volume } O_2 = \text{volume } H_2$$

$$\left(\frac{\text{mass } O_2}{\text{density } O_2}\right) = \left(\frac{\text{mass } H_2}{\text{density } H_2}\right) \qquad \text{density } O_2 = \left(\frac{\text{mass } O_2}{\text{mass } H_2}\right)(\text{density } H_2)$$

$$\text{density } O_2 = \left(\frac{32}{2}\right)(\text{density } H_2) = 16(\text{density } H_2)$$

22. Behavior of gases as described by the Kinetic Molecular Theory.

   (a) Boyle's law. Boyle's law states that the volume of a fixed mass of gas is inversely proportional to the pressure, at constant temperature. The Kinetic Molecular Theory assumes the volume occupied by gases is mostly empty space. Decreasing the volume of a gas by compressing it, increases the concentration of gas molecules, resulting in more collisions of the molecules and thus increased pressure upon the walls of the container.

   (b) Charles' law. Charles' law states that the volume of a fixed mass of gas is directly proportional to the absolute temperature, at constant pressure. According to Kinetic Molecular Theory, the kinetic energies of gas molecules are proportional to the absolute temperature. Increasing the temperature of a gas causes the molecules to move faster, and in order for the pressure not to increase, the volume of the gas must increase.

(c) Dalton's law. Dalton's law states that the pressure of a mixture of gases is the sum of the pressures exerted by the individual gases. According to the Kinetic Molecular Theory, there are no attractive forces between gas molecules; therefore, in a mixture of gases, each gas acts independently and the total pressure exerted will be the sum of the pressures exerted by the individual gases.

23. $N_2(g) + O_2(g) \longrightarrow 2\,NO(g)$

$1 \text{ vol} + 1 \text{ vol} \longrightarrow 2 \text{ vol}$

According to Avogadro's Law, equal volumes of nitrogen and oxygen at the same temperature and pressure contain the same number of molecules. In the reaction, nitrogen and oxygen molecules react in a 1:1 ratio. Since two volumes of nitrogen monoxide are produced, one molecule of nitrogen and one molecule of oxygen must produce two molecules of nitrogen monoxide. Therefore each nitrogen and oxygen molecule must be made up on two atoms (diatomic).

24. We refer gases to STP because some reference point is needed to relate volume to moles. A temperature and pressure must be specified to determine the moles of gas in a given volume, and 0°C and 760 torr are convenient reference points.

25. Conversion of oxygen to ozone is an endothermic reaction. Evidence for this statement is that energy ($286\text{ kJ}/3\text{ mol }O_2$) is required to convert O_2 to O_3.

26. Chlorofluorocarbons, (Freons, CCl_3F, and CCl_2F_2), are responsible for damaging the ozone layer. When these compounds are carried up to the stratosphere (the outer part of the atmosphere) they absorb ultraviolet radiation and produce chlorine free radicals that react with ozone (O_3) and destroy it.

27. Heating a mole of N_2 gas at constant pressure has the following effects:

(a) Density will decrease. Heating the gas at constant pressure will increase its volume. The mass does not change, so the increased volume results in a lower density.

(b) Mass does not change. Heating a substance does not change its mass.

(c) Average kinetic energy of the molecules increases. This is a basic assumption of the Kinetic Molecular Theory.

(d) Average velocity of the molecules will increase. Increasing the temperature increases the average kinetic energies of the molecules; hence, the average velocity of the molecules will increase also.

(e) Number of N_2 molecules remains unchanged. Heating does not alter the number of molecules present, except if extremely high temperatures were attained. Then, the N_2 molecules might dissociate into N atoms resulting in fewer N_2 molecules.

28. Oxygen atom = O   Oxygen molecule = O_2 Ozone molecule = O_3
An oxygen molecule contains 16 electrons.

29. (a) $(715\text{ mm Hg})\left(\dfrac{1\text{ atm}}{760\text{ mm Hg}}\right) = 0.941\text{ atm}$

(b) $(715\text{ mm Hg})\left(\dfrac{1\text{ in. Hg}}{25.4\text{ mm Hg}}\right) = 28.1\text{ in. Hg}$

(c) $(715\text{ mm Hg})\left(\dfrac{14.7\text{ lb/in.}^2}{760\text{ mm Hg}}\right) = 13.8\text{ lb/in.}^2$

30. (a) $(715\text{ mm Hg})\left(\dfrac{1\text{ torr}}{1\text{ mm Hg}}\right) = 715\text{ torr}$

(b) $(715\text{ mm Hg})\left(\dfrac{1013\text{ mbar}}{760\text{ mm Hg}}\right) = 953\text{ mbar}$

(c) $(715\text{ mm Hg})\left(\dfrac{101.325\text{ kPa}}{760\text{ mm Hg}}\right) = 95.3\text{ kPa}$

31. (a) $(28\text{ mm Hg})\left(\dfrac{1\text{ atm}}{760\text{ mm Hg}}\right) = 0.037\text{ atm}$

(b) $(6000.\text{ cm Hg})\left(\dfrac{1\text{ atm}}{76\text{ cm Hg}}\right) = 78.95\text{ atm}$

(c) $(795\text{ torr})\left(\dfrac{1\text{ atm}}{760\text{ torr}}\right) = 1.05\text{ atm}$

(d) $(5.00\text{ kPa})\left(\dfrac{1\text{ atm}}{101.325\text{ kPa}}\right) = 0.0493\text{ atm}$

32. (a) $(62\text{ mm Hg})\left(\dfrac{1\text{ atm}}{760\text{ mm Hg}}\right) = 0.082\text{ atm}$

(b) $(4250.\text{ cm Hg})\left(\dfrac{1\text{ atm}}{76\text{ cm Hg}}\right) = 55.92\text{ atm}$

(c) $(225\text{ torr})\left(\dfrac{1\text{ atm}}{760\text{ torr}}\right) = 0.296\text{ atm}$

(d) $(0.67\text{ kPa})\left(\dfrac{1\text{ atm}}{101.325\text{ kPa}}\right) = 0.0066\text{ atm}$

33. $P_1V_1 = P_2V_2$ or $V_2 = \dfrac{P_1V_1}{P_2}$.

(a) $\dfrac{(500.\text{ mm Hg})(400.\text{ mL})}{760\text{ mm Hg}} = 2.6 \times 10^2\text{ mL}$

(b) $\dfrac{(500.\text{ torr})(400.\text{ mL})}{250\text{ torr}} = 8.0 \times 10^2\text{ mL}$

34. $P_1V_1 = P_2V_2$ or $V_2 = \dfrac{P_1V_1}{P_2}$. Change 500. mm Hg to atmospheres and torrs.

(a) $\dfrac{(500.\text{ mm Hg})(1\text{ atm}/760\text{ mm Hg})(400.\text{ mL})}{2.00\text{ atm}} = 132\text{ mL}$

(b) $\dfrac{(500.\text{ mm Hg})(1\text{ torr}/1\text{ mm Hg})(400.\text{ mL})}{325\text{ torr}} = 615\text{ mL}$

35. $P_2 = \dfrac{P_1V_1}{V_2}$ $\quad \dfrac{(640.\text{ mm Hg})(500.\text{ mL})}{855\text{ mL}} = 374\text{ mm Hg}$

36. $P_2 = \dfrac{P_1V_1}{V_2}$ $\quad \dfrac{(640.\text{ mm Hg})(500.\text{ mL})}{450.\text{ mL}} = 711\text{ mm Hg}$

37. $\dfrac{V_1}{T_1} = \dfrac{V_2}{T_2}$ or $V_2 = \dfrac{V_1T_2}{T_1}$; Temperatures must be in Kelvin (°C + 273)

(a) $\dfrac{(6.00\text{ L})(273\text{ K})}{248\text{ K}} = 6.60\text{ L}$

(b) $\dfrac{(6.00\text{ L})(100.\text{ K})}{248\text{ K}} = 2.42\text{ L}$

38. $\dfrac{V_1}{T_1} = \dfrac{V_2}{T_2}$ or $V_2 = \dfrac{V_1T_2}{T_1}$; Temperatures must be in Kelvin (°C + 273)

0.0°F = −18°C

(a) $\dfrac{(6.00\text{ L})(255\text{ K})}{248\text{ K}} = 6.17\text{ L}$

(b) $\dfrac{(6.00\text{ L})(345\text{ K})}{248\text{ K}} = 8.35\text{ L}$

39. Use the combined gas law $\frac{P_1V_1}{T_1} = \frac{P_2V_2}{T_2}$ or $V_2 = \frac{P_1V_1T_2}{P_2T_1}$

$$V_2 = \frac{(740 \text{ mm Hg})(410 \text{ mL})(273 \text{ K})}{(760 \text{ mm Hg})(300. \text{ K})} = 3.6 \times 10^2 \text{ mL}$$

40. Use the combined gas law $\frac{P_1V_1}{T_1} = \frac{P_2V_2}{T_2}$ or $V_2 = \frac{P_1V_1T_2}{P_2T_1}$

$$V_2 = \frac{(740 \text{ mm Hg})(410 \text{ mL})(523 \text{ K})}{(680 \text{ mm Hg})(300. \text{ K})} = 7.8 \times 10^2 \text{ mL}$$

41. Use the combined gas law $\frac{P_1V_1}{T_1} = \frac{P_2V_2}{T_2}$ or $V_2 = \frac{P_1V_1T_2}{P_2T_1}$

$$V_2 = \frac{(0.950 \text{ atm})(1400. \text{ mL})(275 \text{ K})}{(4.0 \text{ torr})(1 \text{ atm}/760 \text{ torr})(291 \text{ K})} = 2.4 \times 10^5 \text{ L}$$

42. Use the combined gas law $\frac{P_1V_1}{T_1} = \frac{P_2V_2}{T_2}$ or $V_2 = \frac{P_1V_1T_2}{P_2T_1}$

$$V_2 = \frac{(2.50 \text{ atm})(22.4 \text{ L})(268 \text{ K})}{(1.50 \text{ atm})(300. \text{ K})} = 33.4 \text{ L}$$

43. $P_{total} = P_{N_2} + P_{H_2O \text{ vapor}} = 720. \text{ torr}$

$P_{H_2O \text{ vapor}} = 17.5 \text{ torr}$

$P_{N_2} = 720. \text{ torr} - 17.5 \text{ torr} = 703 \text{ torr}$

44. $P_{total} = P_{N_2} + P_{H_2O \text{ vapor}} = 705 \text{ torr}$

$P_{H_2O \text{ vapor}} = 23.8 \text{ torr}$

$P_{N_2} = 705 \text{ torr} - 23.8 \text{ torr} = 681 \text{ torr}$

45. $P_{total} = P_{N_2} + P_{H_2} + P_{O_2}$

$= 200. \text{ torr} + 600. \text{ torr} + 300. \text{ torr} = 1100. \text{ torr} = 1.100 \times 10^3 \text{ torr}$

46. $P_{total} = P_{H_2} + P_{N_2} + P_{O_2}$

$= 325 \text{ torr} + 475 \text{ torr} + 650. \text{ torr} = 1450. \text{ torr} = 1.450 \times 10^3 \text{ torr}$

47. $P_{total} = P_{CH_4} + P_{H_2O \text{ vapor}}$

$P_{H_2O \text{ vapor}} = 23.8 \text{ torr}$

$P_{CH_4} = 720.\text{ torr} - 23.8\text{ torr} = 696\text{ torr}$

To calculate the volume of dry methane, note that the temperature is constant, so $P_1V_1 = P_2V_2$ can be used.

$$V_2 = \frac{P_1V_1}{P_2} = \frac{(696\text{ torr})(2.50\text{ L})}{(760.\text{ torr})} = 2.29\text{ L}$$

48. $P_{total} = P_{C_3H_8} + P_{H_2O\text{ vapor}}$ C_3H_8 in propane

$P_{H_2O\text{ vapor}} = 20.5\text{ torr}$

$P_{C_3H_8} = 745\text{ torr} - 20.5\text{ torr} = 725\text{ torr}$

To calculate the volume of dry propane, note that the temperature is constant, so $P_1V_1 = P_2V_2$ can be used.

$$V_2 = \frac{P_1V_1}{P_2} = \frac{(725\text{ torr})(1.25\text{ L})}{(760.\text{ torr})} = 1.19\text{ L } C_3H_8$$

49. 1 mol of a gas occupies 22.4 L at STP

$$(2.5\text{ mol})\left(\frac{22.4\text{ L}}{\text{mol}}\right) = 56\text{ L } Cl_2$$

50. $(1.25\text{ mol})\left(\frac{22.4\text{ L}}{\text{mol}}\right) = 28.0\text{ L } N_2$

51. $(2500\text{ mL})\left(\frac{1\text{ L}}{1000\text{ mL}}\right)\left(\frac{1\text{ mol}}{22.4\text{ L}}\right)\left(\frac{44.01\text{ g } CO_2}{\text{mol}}\right) = 4.9\text{ g } CO_2$

52. $(1.75\text{ L})\left(\frac{1\text{ mol}}{22.4\text{ L}}\right)\left(\frac{17.03\text{ g } NH_3}{\text{mol}}\right) = 1.33\text{ g } NH_3$

53. (a) $(1.0\text{ mol } NO_2)\left(\frac{22.4\text{ L}}{\text{mol}}\right) = 22.4\text{ L } NO_2$

(b) $(17.05\text{ g } NO_2)\left(\frac{1\text{ mol}}{46.01\text{ g}}\right)\left(\frac{22.4\text{ L}}{\text{mol}}\right) = 8.30\text{ L } NO_2$

(c) $(1.20 \times 10^{24}\text{ molecules } NO_2)\left(\frac{1\text{ mol}}{6.022 \times 10^{23}\text{ molecules}}\right)\left(\frac{22.4\text{ L}}{\text{mol}}\right) = 44.6\text{ L } NO_2$

54. (a) $(0.50\text{ mol } H_2S)\left(\frac{22.4\text{ L}}{\text{mol}}\right) = 11\text{ L } H_2S$

(b) $(22.41 \text{ g H}_2\text{S})\left(\dfrac{1 \text{ mol}}{34.09 \text{ g}}\right)\left(\dfrac{22.4 \text{ L}}{\text{mol}}\right) = 14.73 \text{ L H}_2\text{S}$

(c) $(8.55 \times 10^{23} \text{ molecules H}_2\text{S})\left(\dfrac{1 \text{ mol}}{6.022 \times 10^{23} \text{ molecules}}\right)\left(\dfrac{22.4 \text{ L}}{\text{mol}}\right) = 31.8 \text{ L H}_2\text{S}$

55. $(1.00 \text{ L NH}_3)\left(\dfrac{1 \text{ mol}}{22.4 \text{ L}}\right)\left(\dfrac{6.022 \times 10^{23} \text{ molecules}}{\text{mol}}\right) = 2.69 \times 10^{22} \text{ molecules NH}_3$

56. $(1.00 \text{ L CH}_4)\left(\dfrac{1 \text{ mol}}{22.4 \text{ L}}\right)\left(\dfrac{6.022 \times 10^{23} \text{ molecules}}{\text{mol}}\right) = 2.69 \times 10^{22} \text{ molecules CH}_4$

57. density of Cl_2 gas $= 3.17$ g/L (from table 12.3)
$(10.0 \text{ g})/(3.17 \text{ g/L}) = 3.15$ liters

58. density of CH_4 gas $= 0.716$ g/L (from table 12.3)
$(3.0 \text{ L})(0.716 \text{ g/L}) = 2.1 \text{ g CH}_4$

59. (a) $d = \left(\dfrac{83.80 \text{ g Kr}}{\text{mol}}\right)\left(\dfrac{1 \text{ mol}}{22.4 \text{ L}}\right) = 3.741 \text{ g/L Kr}$

(b) $d = \left(\dfrac{80.07 \text{ g SO}_3}{\text{mol}}\right)\left(\dfrac{1 \text{ mol}}{22.4 \text{ L}}\right) = 3.575 \text{ g/L SO}_3$

60. (a) $d = \left(\dfrac{4.003 \text{ g He}}{\text{mol}}\right)\left(\dfrac{1 \text{ mol}}{22.4 \text{ L}}\right) = 0.1788 \text{ g/L He}$

(b) $d = \left(\dfrac{56.10 \text{ g C}_4\text{H}_8}{\text{mol}}\right)\left(\dfrac{1 \text{ mol}}{22.4 \text{ L}}\right) = 2.504 \text{ g/L C}_4\text{H}_8$

61. (a) $d = \left(\dfrac{38.00 \text{ g F}_2}{\text{mol}}\right)\left(\dfrac{1 \text{ mol}}{22.4 \text{ L}}\right) = 1.696 \text{ g/L F}_2$

(b) Assume 1.00 mol of F_2 and determine the volume using the ideal gas equation, PV = nRT.

$$V = \frac{nRT}{P} = \frac{(1.00 \text{ mol})(0.0821 \text{ L atm/mol K})(300. \text{ K})}{1.00 \text{ atm}}$$

$= 24.6$ L at 27°C and 1.00 atm

$$d = \frac{38.00 \text{ g}}{24.6 \text{ L}} = 1.54 \text{ g/L F}_2$$

62. (a) $d = \left(\frac{70.90\text{ g Cl}_2}{\text{mol}}\right)\left(\frac{1\text{ mol}}{22.4\text{ L}}\right) = 3.165\text{ g/L Cl}_2$

(b) Assume 1.00 mol of Cl_2 and determine the volume using the ideal gas equation, PV = nRT.

$$V = \frac{nRT}{P} = \frac{(1.00\text{ mol})(0.0821\text{ L atm/mol K})(295\text{ K})}{0.500\text{ atm}}$$

$$= 48.4\text{ L at }22°\text{C and }0.500\text{ atm}$$

$$d = \frac{70.90\text{ g}}{48.4\text{ L}} = 1.46\text{ g/L Cl}_2$$

63. PV = nRT $\quad V = \frac{nRT}{P} \quad V = \frac{(2.3\text{ mol})(0.0821\text{ L atm/mol K})(300.\text{ K})}{\dfrac{750\text{ torr}}{760\frac{\text{torr}}{\text{atm}}}} = 57\text{ L Ne}$

64. PV = nRT $\quad V = \frac{nRT}{P} \quad V = \frac{(0.75\text{ mol})(0.0821\text{ L atm/mol K})(298\text{ K})}{\dfrac{725\text{ torr}}{760\frac{\text{torr}}{\text{atm}}}} = 19\text{ L Kr}$

65. When working with gases, the identity of the gas does not affect the volume, as long as the number of moles are known.

Total moles = mol H_2 + mol CO_2 = 5.00 mol + 0.500 mol = 5.50 mol

$$V = (5.50\text{ mol})\left(\frac{22.4\text{ L}}{\text{mol}}\right) = 123\text{ L}$$

66. When working with gases, the identity of the gas does not affect the volume, as long as the number of moles are known.

Total moles = mol N_2 + mol HCl = 2.50 mol + 0.750 mol = 3.25 mol

$$V = (3.25\text{ mol})\left(\frac{22.4\text{ L}}{\text{mol}}\right) = 72.8\text{ L}$$

67. (a) $4\,NH_3(g) + 5\,O_2(g) \longrightarrow 4\,NO(g) + 6\,H_2O(g)$

$$(5.5\text{ mol NO})\left(\frac{4\text{ mol NH}_3}{4\text{ mol NO}}\right) = 5.5\text{ mol NH}_3$$

(b) Limiting reactant problem. Remember, volume-volume relationships are the same as mole-mole relationships when dealing with gases at constant T and P.

$$(12\ \text{L O}_2)\left(\frac{4\ \text{mol NO}}{5\ \text{mol O}_2}\right) = 9.6\ \text{L NO (from O}_2)$$

$$(10.\ \text{L NH}_3)\left(\frac{4\ \text{mol NO}}{4\ \text{mol NH}_3}\right) = 10.\ \text{L NO (from NH}_3)$$

Oxygen is the limiting reactant, 9.6 L NO is formed.

(c) Limiting reactant problem. Remember, volume-volume relationships are the same as mole-mole relationships when dealing with gases at constant T and P.

$$(3.0\ \text{L NH}_3)\left(\frac{4\ \text{mol NO}}{4\ \text{mol NH}_3}\right) = 3.0\ \text{L NO (from NH}_3)$$

$$(3.0\ \text{L O}_2)\left(\frac{4\ \text{mol NO}}{5\ \text{mol O}_2}\right) = 2.4\ \text{L NO (from O}_2)$$

Oxygen is the limiting reactant, 2.4 L NO is formed.

68. $4\,NH_3(g) + 5\,O_2(g) \longrightarrow 4\,NO(g) + 6\,H_2O(g)$

(a) $$(7.0\ \text{mol O}_2)\left(\frac{4\ \text{mol NH}_3}{5\ \text{mol O}_2}\right) = 5.6\ \text{mol NH}_3$$

(b) $$(800.\ \text{mL O}_2)\left(\frac{4\ \text{mol NO}}{5\ \text{mol O}_2}\right) = 640.\ \text{mL NO} = 0.640\ \text{L NO}$$

(c) $$(60.\ \text{L NO})\left(\frac{1\ \text{mol}}{22.4\ \text{L}}\right)\left(\frac{5\ \text{mol O}_2}{4\ \text{mol NO}}\right)\left(\frac{32.00\ \text{g}}{\text{mol}}\right) = 1.1 \times 10^2\ \text{g O}_2$$

69. The balanced equation is

$4\,FeS + 7\,O_2 \xrightarrow{\Delta} 2\,Fe_2O_3 + 4\,SO_2$ $\quad 0.600\ \text{kg} = 600.\ \text{g}$

$$(600.\ \text{g FeS})\left(\frac{1\ \text{mol}}{87.92\ \text{g}}\right)\left(\frac{7\ \text{mol O}_2}{4\ \text{mol FeS}}\right)\left(\frac{22.4\ \text{L}}{\text{mol}}\right) = 268\ \text{L O}_2$$

70. $4\,FeS + 7\,O_2 \xrightarrow{\Delta} 2\,Fe_2O_3 + 4\,SO_2$ $\quad 0.600\ \text{kg} = 600.\ \text{g}$

$$(600.\ \text{g FeS})\left(\frac{1\ \text{mol}}{87.92\ \text{g}}\right)\left(\frac{4\ \text{mol SO}_2}{4\ \text{mol FeS}}\right)\left(\frac{22.4\ \text{L}}{\text{mol}}\right) = 153\ \text{L SO}_2$$

Additional Exercises

71. Like any other gas, water in the gaseous state occupies a much larger volume than in the liquid state.

72. During the winter the air in a car's tires is colder, the molecules move slower and the pressure decreases. In order to keep the pressure at the manufacturer's recommended psi air needs to be added to the tire. The opposite is true during the summer.

73. (a) the pressure will be cut in half

    (b) the pressure will double

    (c) the pressure will be cut in half

    (d) the pressure will increase to 3.7 atm or 2836 torr

$$PV = nRT \qquad P = \frac{nRT}{V}$$

$$P = \frac{(1.5\ \text{mol})(0.0821\ \text{L atm/mol K})(303\ \text{K})}{10.\ \text{L}} = 3.7\ \text{atm}$$

$$P = 3.7\ \text{atm}\left(\frac{760\ \text{torr}}{1\ \text{atm}}\right) = 2.8 \times 10^3\ \text{torr}$$

74. (a) (c)

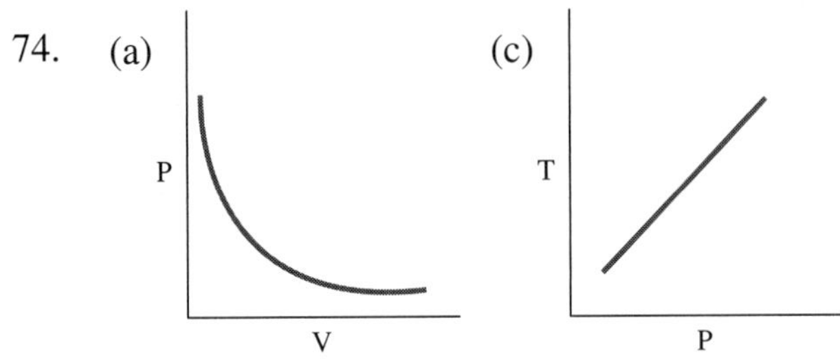

(b) (d)

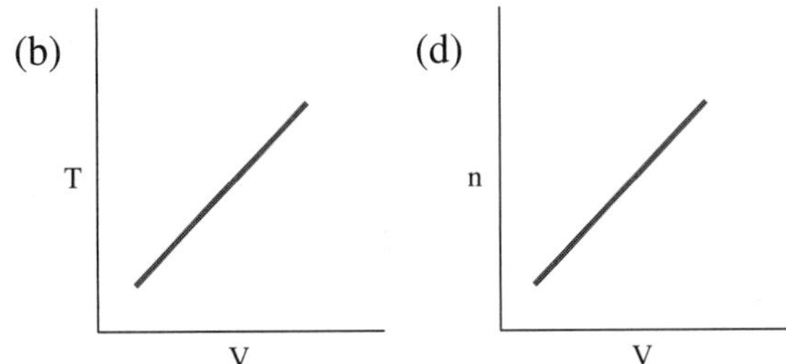

75. The can is a sealed unit and very likely still contains some of the aerosol. As the can is heated, pressure builds up in it eventually causing the can to explode and rupture with possible harm from flying debris.

76. One mole of an ideal gas occupies 22.4 liters at standard conditions. (0°C and 1 atm pressure)

$$PV = nRT$$

$$(1.00\ \text{atm})(V) = (1.00\ \text{mol})(0.0821\ \text{L atm/mol K})(273\ \text{K})$$

$$V = 22.4\ \text{L}$$

77. Solve for volume using $PV = nRT$

(a) $$V = \frac{(0.2\ \text{mol}\ Cl_2)(0.0821\ \text{L atm/mol K})(321\ \text{K})}{(80\ \text{cm}/76\ \text{cm})\ \text{atm}} = 5\ \text{L}\ Cl_2$$

(b) $$V = \frac{(4.2\ \text{g}\ NH_3)\left(\frac{1\ \text{mol}}{17.03\ \text{g}}\right)\left(\frac{0.0821\ \text{L atm}}{\text{mol K}}\right)(161\ \text{K})}{0.65\ \text{atm}} = 5.0\ \text{L}\ NH_3$$

(c) Assume 25°C for room temperature

$$V = \frac{(21\ \text{g}\ SO_3)\left(\frac{1\ \text{mol}}{80.07\ \text{g}}\right)\left(\frac{0.0821\ \text{L atm}}{\text{mol K}}\right)(298\ \text{K})}{\dfrac{110\ \text{kPa}}{101.3\dfrac{\text{kPa}}{\text{atm}}}} = 5.9\ \text{L}\ SO_3$$

21 g SO_3 has the greatest volume

78. Assume 1 mol of each gas

(a) $SF_6 = 146.1\ \text{g/mol}$

$$d = \left(\frac{146.1\ \text{g}}{\text{mol}}\right)\left(\frac{1\ \text{mol}}{22.4\ \text{L}}\right) = 6.52\ \text{g/L}\ SF_6$$

(b) Assume 25°C and 1 atm pressure

$$V(\text{at } 25°\text{C}) = (22.4\ \text{L})\left(\frac{298\ \text{K}}{273\ \text{K}}\right) = 24.5\ \text{L}$$

$$C_2H_6 = 30.07\ \text{g/mol}$$

$$d = \left(\frac{30.07\ \text{g}}{\text{mol}}\right)\left(\frac{1\ \text{mol}}{24.5\ \text{L}}\right) = 1.23\ \text{g/L}\ C_2H_6$$

(c) He at $-80°C$ and 2.15 atm

$$V = \frac{(1\ \text{mol})(0.0821\ \text{L atm/mol K})(193\ \text{K})}{2.15\ \text{atm}} = 7.37\ \text{L}$$

$$d = \left(\frac{4.003\ \text{g}}{\text{mol}}\right)\left(\frac{1\ \text{mol}}{7.37\ \text{L}}\right) = 0.543\ \text{g/L He}$$

SF_6 has the greatest density

79. (a) Empirical formula. Assume 100 g starting material

$$\frac{80.0\ \text{g C}}{12.01\ \text{g/mol}} = 6.66\ \text{mol C} \qquad \frac{6.66}{6.66} = 1$$

$$\frac{20.0\ \text{g H}}{1.008\ \text{g/mol}} = 19.8\ \text{mol H} \qquad \frac{19.8}{6.66} = 2.97$$

Empirical formula $= CH_3$

Empirical mass $= 12.01\ \text{g} + 3.024\ \text{g} = 15.03\ \text{g/mol}$

(b) Molecular formula. $\left(\frac{2.01\ \text{g}}{1.5\ \text{L}}\right)\left(\frac{22.4\ \text{L}}{\text{mol}}\right) = 30.\ \text{g/mol}$ (molar mass)

$\frac{30.\ \text{g/mol}}{15.03\ \text{g/mol}} = 2$; Molecular formula is C_2H_6

(c) Valence electrons $= 2(4) + 6 = 14$

```
   H   H
   |   |
H— C — C —H
   |   |
   H   H
```

Lewis structure

80. $PV = nRT$

(a) $\left(\frac{(790\ \text{torr})(1\ \text{atm})}{760\ \text{torr}}\right)(2.0\ \text{L}) = (n)(0.0821\ \text{L atm/mol K})(298\ \text{K})$

$n = 0.085$ mol (total moles)

(b) mol N_2 = total moles − mol O_2 − mol CO_2

$$= 0.085\ \text{mol} - \frac{0.65\ \text{g}\ O_2}{32.00\ \text{g/mol}} - \frac{0.58\ \text{g}\ CO_2}{44.01\ \text{g/mol}}$$

mol $N_2 = 0.085\ \text{mol} - 0.020\ \text{mol}\ O_2 - 0.013\ \text{mol}\ CO_2 = 0.052\ \text{mol}$

$$(0.052\ \text{mol}\ N_2)\left(\frac{28.02\ \text{g}\ N_2}{\text{mol}}\right) = 1.5\ \text{g}\ N_2$$

(c) $P_{O_2} = (790 \text{ torr})\left(\frac{0.020 \text{ mol } O_2}{0.085 \text{ mol}}\right) = 1.9 \times 10^2 \text{ torr}$

$P_{CO_2} = (790 \text{ torr})\left(\frac{0.013 \text{ mol } CO_2}{0.085 \text{ mol}}\right) = 1.2 \times 10^2 \text{ torr}$

$P_{N_2} = (790 \text{ torr})\left(\frac{0.051 \text{ mol } N_2}{0.085 \text{ mol}}\right) = 4.7 \times 10^2 \text{ torr}$

81. $2\,CO + O_2 \longrightarrow 2\,CO_2$
Calculate the moles of O_2 and CO to find the limiting reactant.
$PV = nRT$

O_2: $(1.8 \text{ atm})(0.500 \text{ L } O_2) = (n)(0.0821 \text{ L atm/mol K})(288 \text{ K})$
mol O_2 = 0.038 mol

CO: $\left(\frac{800 \text{ mm Hg} \times 1 \text{ atm}}{760 \text{ mm Hg}}\right)(0.500 \text{ L}) = (n)(0.0821 \text{ L atm/mol K})(333 \text{ K})$

mol CO = 0.019 mol   Limiting reactant is CO

0.038 mol O_2 will react with 0.076 mol CO.

$(0.019 \text{ mol CO})\left(\frac{2 \text{ mol } CO_2}{2 \text{ mol CO}}\right)\left(\frac{22.4 \text{ L}}{\text{mol}}\right) = 0.43 \text{ L } CO_2 = 430 \text{ mL } CO_2$

82. $PV = nRT$ or $PV = \left(\frac{g}{\text{molar mass}}\right)RT$

$\left(\frac{1.4 \text{ g}}{\text{cm}^3}\right)\left(\frac{1000 \text{ cm}^3}{\text{L}}\right) = 1.4 \times 10^3 \text{ g/L}$

$(1.3 \times 10^9 \text{ atm})(1.0 \text{ L}) = \left(\frac{1.4 \times 10^3 \text{ g}}{2.0 \text{ g/mol}}\right)(0.0821 \text{ L atm/mol K})(T)$

$T = \frac{(1.3 \times 10^9 \text{ atm})(1.0 \text{ L})(2.0 \text{ g/mol})}{(0.0821 \text{ L atm/mol K})(1.4 \times 10^3 \text{ g})} = 2.3 \times 10^7 \text{ K}$

83. (a) Assume atmospheric pressure of 14.7 lb/in.^{2} to begin with.

Total pressure in the ball = 14.7 lb/in.^{2} + 13 lb/in.^{2} = 28 lb/in.^{2}
$PV = nRT$

$(28 \text{ lb/in.}^2)\left(\frac{1 \text{ atm}}{14.7 \text{ lb/in.}^2}\right)(2.24 \text{ L}) = (n)(0.0821 \text{ L atm/mol K})(293 \text{ K})$

n = 0.18 mol air

(b) mass of air in the ball

$$m = (0.18\ \text{mol})\left(\frac{29\ \text{g}}{\text{mol}}\right) = 5.2\ \text{g air}$$

(c) Actually the pressure changes when the temperature changes. Since pressure is directly proportional to moles we can calculate the change in moles required to keep the pressure the same at 30°C as it was at 20°C.

$PV = nRT$

$$(28\ \text{lb/in.}^2)\left(\frac{1\ \text{atm}}{14.7\ \text{lb/in.}^2}\right)(2.24\ \text{L}) = (n)(0.0821\ \text{L atm/mol K})(303\ \text{K})$$

$n = 0.17$ mol of air required to keep the pressure the same at 30°C.

0.01 mol air (0.18 − 0.17) must be allowed to escape from the ball.

$$(0.01\ \text{mol air})\left(\frac{29\ \text{g}}{\text{mol}}\right) = 0.29\ \text{g or } 0.3\ \text{g air must be allowed to escape.}$$

84. Use the combined gas laws to calculate the bursting temperature (T_2).

$$\frac{P_1V_1}{T_1} = \frac{P_2V_2}{T_2}$$

$P_1 = 65$ cm $\qquad P_2 = 1.00$ atm (76 cm)

$V_1 = 1.75$ L $\qquad V_2 = 2.00$ L

$T_1 = 20°\text{C}\ (293\ \text{K})$ $\qquad T_2 = T_2$

$$T_2 = \frac{P_2V_2T_1}{P_1V_1} = \frac{(76\ \text{cm})(2.00\ \text{L})(293\ \text{K})}{(65\ \text{cm})(1.75\ \text{L})} = 392\ \text{K}\ (119°\text{C})$$

85. $P_1V_1 = P_2V_2$ or $P_2 = \dfrac{P_1V_1}{V_2}$

$$P_2 = \frac{(1.0\ \text{atm})(2500\ \text{L})}{25\ \text{L}} = 1.0 \times 10^2\ \text{atm}$$

86. To double the volume of a gas, at constant pressure, the temperature (K) must be doubled.

$$\frac{V_1}{T_1} = \frac{V_2}{T_2} \qquad V_2 = 2\,V_1$$

$$\frac{V_1}{T_1} = \frac{2\,V_1}{T_2} \qquad T_2 = \frac{2\,V_1T_1}{V_1} \qquad T_2 = 2\,T_1$$

$T_2 = 2(300.\ \text{K}) = 600.\ \text{K} = 327°\text{C}$

87. V = volume at 22°C and 740 torr

$2\,V$ = volume after change in temperature (P constant)

V = volume after change in pressure (T constant)

Since temperature is constant, $P_1V_1 = P_2V_2$ or $P_2 = \frac{P_1V_1}{V_2}$

$$P_2 = (740 \text{ torr})\left(\frac{2\text{ V}}{\text{V}}\right) = 1.5 \times 10^3 \text{ torr (pressure to change 2 V to V)}$$

88. Volume is constant, so $\frac{P_1}{T_1} = \frac{P_2}{T_2}$ or $T_2 = \frac{T_1P_2}{P_1}$;

$$T_2 = \frac{(500.\text{ torr})(295\text{ K})}{700.\text{ torr}} = 211\text{ K} = -62°\text{C}$$

89. The volume of the cylinder remains constant, so

$-196°\text{C} + 273 = 77\text{ K}$

$\frac{P_1}{T_1} = \frac{P_2}{T_2}$ or $P_2 = \frac{P_1T_2}{T_1}$;

$$P_2 = \frac{(252\text{ atm})(77\text{ K})}{298\text{ K}} = 65\text{ atm}$$

90. The volume of the tires remains constant (until they burst), so

$\frac{P_1}{T_1} = \frac{P_2}{T_2}$ or $T_2 = \frac{T_1P_2}{P_1}$;

$71.0°\text{F} = 21.7°\text{C} = 295\text{ K}$

$$T_2 = \frac{(44\text{ psi})(295\text{ K})}{30.\text{ psi}} = 433\text{ K} = 160°\text{C} = 320°\text{F}$$

91. Use the combined gas laws.

$\frac{P_1V_1}{T_1} = \frac{P_2V_2}{T_2}$ or $V_2 = \frac{P_1V_1T_2}{P_2T_1}$

P_1 and V_1 are at STP

$$V_2 = \frac{(760\text{ torr})(5.30\text{ L})(343\text{ K})}{(830\text{ torr})(273\text{ K})} = 6.1\text{ L}$$

92. Use the combined gas laws.

$\frac{P_1V_1}{T_1} = \frac{P_2V_2}{T_2}$ or $P_2 = \frac{P_1V_1T_2}{V_2T_1}$ P_1 and T_1 are at STP

$$P_2 = \frac{(1.00\text{ atm})(800.\text{ mL})(303\text{ K})}{(250.\text{ mL})(273\text{ K})} = 3.55\text{ atm}$$

93. Use the combined gas law $\frac{P_1V_1}{T_1} = \frac{P_2V_2}{T_2}$ or $V_2 = \frac{P_1V_1T_2}{P_2T_1}$

First calculate the volume at STP.

$$V_2 = \frac{(400.\ \text{torr})(600.\ \text{mL})(273\ \text{K})}{(760.\ \text{torr})(313\ \text{K})} = 275\ \text{mL} = 0.275\ \text{L}$$

At STP, a mole of any gas has a volume of 22.4 L

$$(0.275\ \text{L})\left(\frac{1\ \text{mol}}{22.4\ \text{L}}\right)\left(\frac{6.022 \times 10^{23}\ \text{molecules}}{1\ \text{mol}}\right) = 7.39 \times 10^{21}\ \text{molecules}$$

Each molecule of N_2O contains 3 atoms, so:

$$(7.39 \times 10^{21}\ \text{molecules})\left(\frac{3\ \text{atoms}}{1\ \text{molecule}}\right) = 2.22 \times 10^{22}\ \text{atoms}$$

94. Each gas acts independently, so calculate the pressure of each gas in the 10.0 L container and add them together.

$$CO_2 \quad \frac{(5.00\ \text{L})(500.\ \text{torr})}{10.0\ \text{L}} = 250.\ \text{torr}$$

$$CH_4 \quad \frac{(3.00\ \text{L})(400.\ \text{torr})}{10.0\ \text{L}} = 120.\ \text{torr}$$

$$P_{total} = P_{CO_2} + P_{CH_4} = 250.\ \text{torr} + 120.\ \text{torr} = 370.\ \text{torr}$$

95. Pressure varies directly with absolute temperature.

$$\frac{P_1}{T_1} = \frac{P_2}{T_2} \qquad P_2 = \frac{P_1T_2}{T_1} \qquad T_1 = 25°\text{C} + 273 = 298\ \text{K}$$

$$T_2 = 212°\text{F} = 100°\text{C} = 373\ \text{K}$$

$$P_2 = \frac{(32\ \text{lb/in.}^2)(373\ \text{K})}{298\ \text{K}} = 40.\ \text{lb/in.}^2$$

At 212°F the tire pressure is 40. lb/in.^2

The tire will not burst.

96. Pressure varies directly with temperature.

$$P_2 = \frac{P_1T_2}{T_1} \qquad T_1 = 25°\text{C} + 273 = 298\ \text{K}$$

$$T_2 = 91°\text{C} + 273 = 364\ \text{K}$$

$$P_2 = \frac{(225\ \text{lb/in.}^2)(364\ \text{K})}{298\ \text{K}} = 275\ \text{lb/in.}^2$$

97. Use the ideal gas equation

$PV = nRT \qquad V = \frac{nRT}{P} \qquad\qquad P = \frac{(400.0\ \text{torr})}{}\left(\frac{1\ \text{atm}}{760\ \text{torr}}\right) = 0.5263\ \text{atm}$

$$V = (1100.0\ \text{g}\ CO_2)\left(\frac{1\text{mol}}{44.01\ \text{g}}\right)\left(\frac{0.0821\ \text{L}\cdot\text{atm}}{\text{mol}\cdot\text{K}}\right)\left(\frac{268\ \text{K}}{0.5263\ \text{atm}}\right) = 1045\ \text{L}$$

$$(1100.0\ \text{g}\ CO_2)\left(\frac{1\ \text{mol}}{44.01\ \text{g}}\right) = 24.99\ \text{mol}\ CO_2$$

24.99 mol He will have the some number of atoms as 24.99 mol CO_2

$$(24.99\ \text{mol He})\left(\frac{4.003\ \text{g}}{\text{mol}}\right) = 100.0\ \text{g He}$$

98. A column of mercury at 1 atm pressure is 760 mm Hg high. The density of mercury is 13.6 times that of water, so a column of water at 1 atm pressure should be 13.6 times as high as that for mercury.

$(760\ \text{mm})(13.6) = 1.03 \times 10^4\ \text{mm}\ (33.8\ \text{ft})$

99. Use the ideal gas equation

$PV = nRT \qquad P = \frac{nRt}{V}$

$$P = \frac{(2.00\ \text{mol})(0.0821\ \text{L}\cdot\text{atm})(300\ \text{K})}{(2.00\ \text{atm})(\text{mol}\cdot\text{K})} = 24.6\ \text{atm}$$

100. Use the ideal gas equation

$PV = nRT \qquad n = \frac{RT}{PV}$

Change 2.20×10^3 lb/in.^{2} to atmosphere

$$(2.20 \times 10^3\ \text{lb/in.}^2)\left(\frac{1\ \text{atm}}{14.7\ \text{lb/in.}^2}\right) = 150.\ \text{atm}$$

$$n = \frac{(150.\ \text{atm})(55\ \text{L})}{\left(\frac{0.0821\ \text{L}\cdot\text{atm}}{\text{mol}\cdot\text{K}}\right)(300.\ \text{K})} = 3.3 \times 10^2\ \text{mol}\ O_2$$

101. The number of moles of gas is proportional to pressure. (T and V constant)

$$\frac{P_1}{n_1} = \frac{P_2}{n_2} \quad \text{or} \quad n_2 = \frac{n_1P_2}{P_1}$$

(a) $(60.0 \text{ mol } H_2)\left(\frac{850 \text{ lb/in.}^2}{1500 \text{ lb/in.}^2}\right) = 34 \text{ mol } H_2$

(b) $(60.0 \text{ mol } H_2)\left(\frac{2.016 \text{ g}}{\text{mol}}\right) = 121 \text{ g } H_2$

102. The conversion is: $m^3 \longrightarrow cm^3 \longrightarrow mL \longrightarrow L \longrightarrow mol$

$$(1.00 \text{ m}^3)\left(\frac{100 \text{ cm}}{1 \text{ m}}\right)^3\left(\frac{1 \text{ mL}}{1 \text{ cm}^3}\right)\left(\frac{1 \text{ L}}{1000 \text{ mL}}\right)\left(\frac{1 \text{ mol}}{22.4 \text{ L}}\right) = 44.6 \text{ mol } Cl_2$$

103. First calculate the moles of gas and then convert moles to molar mass.

$$(0.560 \text{ L})\left(\frac{1 \text{ mol}}{22.4 \text{ L}}\right) = 0.0250 \text{ mol}$$

$$\frac{1.08 \text{ g}}{0.0250 \text{ mol}} = 43.2 \text{ g/mol (molar mass)}$$

104. At STP 22.4 L of CH_4 has a mass of 16.04 g. $\frac{22.4 \text{ L}}{16.04 \text{ g}} = \frac{1.40 \text{ L}}{1.00 \text{ g}}$

Using 1.00 g as the mass of the sample:

$P_1 = 1$ atm	$P_2 = 1$ atm
$V_1 = 1.40$ L	$V_2 = 1.0$ L
$T_1 = 273$ K	$T_2 = T_2$

Since the pressure is constant, $\frac{V_1}{T_1} = \frac{V_2}{T_2}$

$$T_2 = \frac{V_2T_1}{V_1} = \frac{(1.00 \text{ L})(273 \text{ K})}{1.40 \text{ L}} = 195 \text{ K} = -78°\text{C}$$

105. The conversion is: $g/L \longrightarrow g/mol$

$$\left(\frac{1.78 \text{ g}}{\text{L}}\right)\left(\frac{22.4 \text{ L}}{\text{mol}}\right) = 39.9 \text{ g/mol (molar mass)}$$

106. $PV = nRT$

(a) $V = \frac{nRT}{P} = \frac{(0.510 \text{ mol})(0.0821 \text{ L atm/mol K})(320. \text{ K})}{1.6 \text{ atm}} = 8.4 \text{ L } H_2$

(b) $n = \frac{PV}{RT} = \frac{(0.789 \text{ atm})(16.0 \text{ L})}{(0.0821 \text{ L atm/mol K})(300. \text{ K})} = 0.513 \text{ mol } CH_4$

The molar mass for CH_4 is 16.04 g/mol
$(16.04\ \text{g/mol})(0.513\ \text{mol}) = 8.23\ \text{g}\ CH_4$

(c) $PV = nRT$, but $n = \frac{g}{M}$ where M is the molar mass and g is the grams of the gas.

Thus, $PV = \frac{gRT}{M}$. To determine density, $d = g/V$.

Solving $PV = \frac{gRT}{M}$ for $\frac{g}{V}$ produces $\frac{g}{V} = \frac{PM}{RT}$.

$$d = \frac{g}{V} = \frac{(4.00\ \text{atm})(44.01\ \text{g/mol})}{(0.0821\ \text{L atm/mol K})(253\ \text{K})} = 8.48\ \text{g/L}\ CO_2$$

(d) Since $d = \frac{g}{V} = \frac{PM}{RT}$ from part (c), solve for M (molar mass)

$$M = \frac{dRT}{P} = \frac{(2.58\ \text{g/L})(0.0821\ \text{L atm/mol K})(300.\ \text{K})}{1.00\ \text{atm}} = 63.5\ \text{g/mol (molar mass)}$$

107. $PV = nRT$

$$n = \frac{PV}{RT} = \frac{(0.813\ \text{atm})(0.215\ \text{L})}{(0.0821\ \text{L atm/mol K})(303\ \text{K})} = 7.03 \times 10^{-3}\ \text{mol}$$

$$\text{molar mass} = \left(\frac{1.15\ \text{g}}{7.03 \times 10^{-3}\ \text{mol}}\right) = 164\ \text{g/mol}$$

108. $PV = nRT$

$$T = \frac{PV}{nR} = \frac{(4.15\ \text{atm})(0.250\ \text{L})}{(4.50\ \text{mol})(0.0821\ \text{L atm/mol K})} = 2.81\ \text{K}$$

109. $PV = nRT$

$$n = \frac{PV}{RT} = \frac{(0.500\ \text{atm})(5.20\ \text{L})}{(0.0821\ \text{L atm/mol K})(250\ \text{K})} = 0.13\ \text{mol}\ N_2$$

110. $C_2H_2(g) + 2HF(g) \longrightarrow C_2H_4F_2(g)$
$1.0\ \text{mol}\ C_2H_2 \longrightarrow 1.0\ \text{mol}\ C_2H_4F_2$

$$(5.0\ \text{mol HF})\left(\frac{1\ \text{mol}\ C_2H_4F_2}{2\ \text{mol HF}}\right) = 2.5\ \text{mol}\ C_2H_4F_2$$

C_2H_2 is the limiting reactant. 1.0 mol $C_2H_4F_2$ forms, no moles C_2H_2 remain.
According to the equation, 2.0 mol HF yields 1.0 mol $C_2H_4F_2$. Therefore,
5.0 mol HF − 2.0 mol HF = 3.0 mol HF unreacted

The flask contains 1.0 mol $C_2H_4F_2$ and 3.0 mol HF when the reaction is complete.
The flask contains 4.0 mol of gas.

$$P = \frac{nRT}{V} = \frac{(4.0\ \text{mol})(0.0821\ \text{L atm/mol K})(273\ \text{K})}{10.0\ \text{L}} = 9.0\ \text{atm}$$

111. $(8.30\ \text{mol Al})\left(\frac{3\ \text{mol}\ H_2}{2\ \text{mol Al}}\right)\left(\frac{22.4\ \text{L}}{\text{mol}}\right) = 279\ \text{L}\ H_2$ at STP

112. According to Graham's Law of Effusion, the rates of effusion are inversely proportional to the molar mass.

$$\frac{\text{rate He}}{\text{rate } N_2} = \sqrt{\frac{\text{molar mass } N_2}{\text{molar mass He}}} = \sqrt{\frac{28.02}{4.003}} = \sqrt{7.000} = 2.646$$

Helium effuses 2.646 times faster than nitrogen.

113. (a) According to Graham's Law of Effusion, the rates of effusion are inversely proportional to the molar mass.

$$\frac{\text{rate He}}{\text{rate } CH_4} = \sqrt{\frac{16.04}{4.003}} = \sqrt{4.007} = 2.002$$

Helium effuses twice as fast as CH_4.

(b) x = distance He travels

$100 - x$ = distance CH_4 travels

$D_{He} = 2\,D_{CH_4}$ D = distance traveled

$x = 2(100 - x)$

$3x = 200$

$x = 66.7$ cm

The gases meet 66.7 cm from the helium end.

114. Assume 100. g of material to start with. Calculate the empirical formula.

C $(85.7\ \text{g})\left(\frac{1\ \text{mol}}{12.01\ \text{g}}\right) = 7.14\ \text{mol}$ $\frac{7.14}{7.14} = 1.00\ \text{mol}$

H $(14.3\ \text{g})\left(\frac{1\ \text{mol}}{1.008\ \text{g}}\right) = 14.2\ \text{mol}$ $\frac{14.2}{7.14} = 1.99\ \text{mol}$

The empirical formula is CH_2. To determine the molecular formula, the molar mass must be known.

$$\left(\frac{2.50\text{ g}}{\text{L}}\right)\left(\frac{22.4\text{ L}}{\text{mol}}\right) = 56.0\text{ g/mol (molar mass)}$$

The empirical formula mass is 14.0 $\frac{56.0}{14.0} = 4$

Therefore, the molecular formula is $(CH_2)_4 = C_4H_8$

115. $2\,CO(g) + O_2(g) \rightarrow 2\,CO_2(g)$ Determine the limiting reactant

$$(10.0\text{ mol CO})\left(\frac{2\text{ mol }CO_2}{2\text{ mol CO}}\right) = 10.0\text{ mol }CO_2\text{ (from CO)}$$

$$(8.0\text{ mol }O_2)\left(\frac{2\text{ mol }CO_2}{1\text{ mol }O_2}\right) = 16\text{ mol }CO_2\text{ (from }O_2)$$

CO: the limiting reactant, O_2: in excess, 3.0 mol O_2 unreacted.

(a) 10.0 mol CO react with 5.0 mol O_2

10.0 mol CO_2 and 3.0 mol O_2 are present, no CO will be present.

(b) $$P = \frac{nRT}{V} = \frac{(13\text{ mol})(0.0821\text{ L atm/mol K})(273\text{ K})}{10.\text{ L}} = 29\text{ atm}$$

116. $2\,KClO_3(s) \xrightarrow{\Delta} 2\,KCl(s) + 3O_2(g)$

First calculate the moles of O_2 produced. Then calculate the grams of $KClO_3$ required to produce the O_2. Then calculate the % $KClO_3$.

$$(0.25\text{ L }O_2)\left(\frac{1\text{ mol}}{22.4\text{ L}}\right) = 0.011\text{ mol }O_2$$

$$(0.011\text{ mol }O_2)\left(\frac{2\text{ mol }KClO_3}{3\text{ mol }O_2}\right)\left(\frac{122.6\text{ g}}{\text{mol}}\right) = 0.90\text{ g }KClO_3\text{ in the sample}$$

$$\left(\frac{0.90\text{ g}}{1.20\text{ g}}\right)(100) = 75\%\ KClO_3\text{ in the mixture}$$

117. Some ammonia gas dissolves in the water squirted into the flask, lowering the pressure inside the flask. The atomospheric pressure outside is greater than the pressure inside the flask and pushes water from the beaker up the tube and into the flask, filling the flask.

118. (a) The pressure of the helium is simply the difference between levels of Hg; 250 mm Hg (250 torr).

(b) The pressure of the oxygen is the difference between the pressure of the atmosphere and the difference in the levels of Hg.

$$P_{O_2} = P_{atm} + 300 \text{ mm Hg}$$
$$= 760 \text{ mm Hg} + 300 \text{ mm Hg}$$
$$= 1060 \text{ mm Hg } (1060 \text{ torr})$$

119. Assume 1.00 L of air. The mass of 1.00 L of air is 1.29 g.

$$\frac{P_1V_1}{T_1} = \frac{P_2V_2}{T_2}$$

$$V_2 = \frac{P_1V_1T_2}{P_2T_1} = \frac{(760 \text{ torr})(1.00 \text{ L})(290. \text{ K})}{(450 \text{ torr})(273 \text{ K})} = 1.8 \text{ L}$$

$$d = \frac{m}{V} = \frac{1.29 \text{ g}}{1.8 \text{ L}} = 0.72 \text{ g/L}$$

120. Each gas behaves as though it were alone in a 4.0 L system.

(a) After expansion: $P_1V_1 = P_2V_2$

For CO_2 $$P_2 = \frac{P_1V_1}{V_2} = \frac{(150. \text{ torr})(3.0 \text{ L})}{4.0 \text{ L}} = 1.1 \times 10^2 \text{ torr}$$

For H_2 $$P_2 = \frac{P_1V_1}{V_2} = \frac{(50. \text{ torr})(1.0 \text{ L})}{4.0 \text{ L}} = 13 \text{ torr}$$

(b) $P_{total} = P_{H_2} + P_{CO_2} = 110 \text{ torr} + 13 \text{ torr} = 120 \text{ torr}$ (2 sig. figures)

121. $P_1 = 40.0$ atm $\quad P_2 = P_2$

$V_1 = 50.0$ L $\quad V_2 = 50.0$ L

$T_1 = 25°C = 298$ K $\quad T_2 = 25°C + 152°C = 177°C = 450.$ K

Gas cylinders have constant volume, so pressure varies directly with temperature.

$$P_2 = \frac{P_1T_2}{T_1} = \frac{(40.0 \text{ atm})(450. \text{ K})}{298 \text{ K}} = 60.4 \text{ atm}$$

122. You can identify the gas by determining its density.

mass of gas $= 1.700\text{ g} - 0.500\text{ g} = 1.200\text{ g}$

volume of gas: Charles law problem. Correct volume to 273 K

$$\frac{V_1}{T_1} = \frac{V_2}{T_2} \qquad V_2 = \frac{V_1T_2}{T_1} = \frac{(0.4478\text{ L})(273\text{ K})}{323\text{ K}} = 0.3785\text{ L}$$

$$d = \frac{m}{V} = \frac{1.200\text{ g}}{0.3785\text{ L}} = 3.170\text{ g/L}$$

gas is chlorine (see Table 12.3)

CHAPTER 13

WATER AND THE PROPERTIES OF LIQUIDS

1. The potential energy is greater in the liquid water than in the ice. The heat necessary to melt the ice increases the potential energy of the liquid, thus allowing the molecules greater freedom of motion. The potential energy of steam (gas) is greater than that of liquid water.

2. At 0°C, all three substances, H_2S, H_2Se, and H_2Te, are gases, because they all have boiling points below 0°C.

3. Liquids are made of particles that are close together, they are not compressible and they have definite volume. Solids also exhibit similar properties.

4. Liquids take the shape of the container they are in. Gases also exhibit this property.

5. The pressure of the atmosphere must be 1.00 atmosphere, otherwise the water would be boiling at some other temperature.

6. 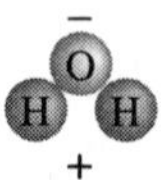

7. Melting point, boiling point, heat of fusion, heat of vaporization, density, and crystal structure in the solid state are some of the physical properties of water that would be very different, if the molecules were linear and nonpolar instead of bent and highly polar. For example, the boiling point, melting point, heat of fusion and heat of vaporization would be lower because linear molecules have no dipole moment and the attraction among molecules would be much less.

8. Prefixes preceding the word hydrate are used in naming hydrates, indicating the number of molecules of water present in the formulas. The prefixes used are:

mono = 1	di = 2	tri = 3	tetra = 4	penta = 5
hexa = 6	hepta = 7	octa = 8	nona = 9	deca = 10

9. The distillation setup in Figure 13.10 would be satisfactory for separating salt and water, but not for separating ethyl alcohol and water. In the first case, the water is easily vaporized, the salt is not, so the water boils off and condenses to a pure liquid. In the second case, both substances are easily vaporized, so both would vaporize (though not to and equal degree) and the condensed liquid would contain both substances.

10. The thermometer would be at about 70°C. The liquid is boiling, which means its vapor pressure equals the confining pressure. From Table 13.1, we find that ethyl alcohol has a vapor pressure of 543 torr at 70°C.

11. The water in both containers would have the same vapor pressure, for it is a function of the temperature of the liquid.

12. Vapor pressure is the pressure exerted by a vapor when it is in equilibrium with its liquid. The liquid molecules on the surface and the gas molecules above the liquid are the important players in vapor pressure. One liter of water will evaporate at the same rate as 100 mL of water if they are at the same temperature and have the same surface area. The liquid and vapor in both cases will come to equilibrium in the same amount of time and they will have the same vapor pressure. Vapor pressure is a characteristic of the type of liquid, not the number of molecules of a liquid.

13. In Figure 13.1, it would be case (b) in which the atmosphere would reach saturation. The vapor pressure of water is the same in both (a) and (b), but since (a) is an open container the vapor escapes into the atmosphere and doesn't reach saturation.

14. If ethyl ether and ethyl alcohol were both placed in a closed container, (a) both substances would be present in the vapor, for both are volatile liquids; (b) ethyl ether would have more molecules in the vapor because it has a higher vapor pressure at a given temperature.

15. The vapor pressure observed in (c) would remain unchanged. The presence of more water in (b) does not change the magnitude of the vapor pressure of the water. The temperature of the water determines the magnitude of the vapor pressure.

16. At 30 torr, H_2O would boil at approximately 29°C, ethyl alcohol at 14°C, and ethyl ether at some temperature below 0°C.

17. (a) At a pressure of 500 torr, water boils at 88°C.
    (b) The normal boiling point of ethyl alcohol is 78°C.
    (c) At a pressure of 0.50 atm (380 torr), ethyl ether boils at 16°C.

18. Based on Figure 13.5:
    (a) Line BC is horizontal because the temperature remains constant during the entire process of melting. The energy input is absorbed in changing from the solid to the liquid state.
    (b) During BC, both solid and liquid phases are present.
    (c) The line DE represents the change from liquid water to steam (vapor) at the boiling temperature of water.

19. Physical properties of water:
    (a) melting point, 0°C
    (b) boiling point, 100°C (at 1 atm pressure)
    (c) colorless
    (d) odorless
    (e) tasteless
    (f) heat of fusion, 335 J/g (80 cal/g)
    (g) heat of vaporization, 2.26 kJ/g (540 cal/g)
    (h) density = 1.0 g/mL (at 4°C)
    (i) specific heat = 4.184 J/g°C

20. For water, to have its maximum density, the temperature must be 4°C, and the pressure sufficient to keep it liquid. $d = 1.0$ g/mL

21. Apply heat to an ice-water mixture, the heat energy is absorbed to melt the ice (heat of fusion), rather than warm the water, so the temperature remains constant until all the ice has melted.

22. Ice at 0°C contains less heat energy than water at 0°C. Heat must be added to convert ice to water, so the water will contain that much additional heat energy.

23. Ice floats in water because it is less dense than water. The density of ice at 0°C is 0.915 g/mL. Liquid water, however, has a density of 1.00 g/mL. Ice will sink in ethyl alcohol, which has a density of 0.789 g/mL.

24. The heat of vaporization of water would be lower if water molecules were linear instead of bent. If linear, the molecules of water would be nonpolar. The relatively high heat of vaporization of water is a result of the molecule being highly polar and having strong dipole-dipole hydrogen bonding attraction for other water molecules.

25. Ethyl alcohol exhibits hydrogen bonding; ethyl ether does not. This is indicated by the high heat of vaporization of ethyl alcohol, even though its molar mass is much less than the molar mass of ethyl ether.

26. A linear water molecule, being nonpolar, would exhibit less hydrogen bonding than the highly polar, bent, water molecule. The polar molecule has a greater intermolecular attractive force than a nonpolar molecule.

27. Hydrogen bonding occurs between molecules that have hydrogen covalently bonded to a strongly electronegative element such as oxygen, fluorine or nitrogen.

28. Ammonia exhibits hydrogen bonding; methane does not. The ammonia molecule is polar; the methane molecule is not.

29. Water, at 80°C, will have fewer hydrogen bonds than water at 40°C. At the higher temperature, the molecules of water are moving faster than at the lower temperature. This results in less hydrogen bonding at the higher temperature.

30. $H_2NCH_2CH_2NH_2$ has two polar NH_2 groups. It should, therefore, show more hydrogen bonding and a higher boiling point (117°C) versus 49°C for $CH_3CH_2CH_2NH_2$.

31. Rubbing alcohol feels cold when applied to the skin, because the evaporation of the alcohol absorbs heat from the skin. The alcohol has a fairly high vapor pressure (low boiling point) and evaporates quite rapidly. This produces the cooling effect.

32. (a) Order of increasing rate of evaporation: mercury, acetic acid, water, toluene, benzene, carbon tetrachloride, methyl alcohol, bromine.
    (b) Highest boiling point is mercury. Lowest boiling point is bromine.

33. Water boils when its vapor pressure equals the prevailing atmospheric pressure over the water. In order for water to boil at 50°C, the pressure over the water would need to be reduced to a point equal to the vapor pressure of the water (92.5 torr).

34. In a pressure cooker, the temperature at which the water boils increases above its normal boiling point, because the water vapor (steam) formed by boiling cannot escape. This results in an increased pressure over water and, consequently, an increased boiling temperature.

35. Vapor pressure varies with temperature. The temperature at which the vapor pressure of a liquid equals the prevailing pressure is the boiling point of the liquid.

36. Adhesive forces are the forces found between a liquid and the walls of its container. Cohesive forces are the forces found between the individual molecules of a liquid.

37. As temperature increases, molecular velocities increase. At higher molecular velocities, it becomes easier for molecules to break away from the attractive forces in the liquid.

38. Water has a relatively high boiling point because there is a high attraction between molecules due to hydrogen bonding.

39. Ammonia would have a higher vapor pressure than SO_2 at -40°C because it has a lower boiling point (NH_3 is more volatile than SO_2).

40. As the temperature of a liquid increases, the kinetic energy of the molecules as well as the vapor pressure of the liquid increases. When the vapor pressure of the liquid equals the external pressure, boiling begins with many of the molecules having enough energy to

escape from the liquid. Bubbles of vapor are formed throughout the liquid and these bubbles rise to the surface, escaping as boiling continues.

41. HF has a higher boiling point than HCl because of the strong hydrogen bonding in HF (F is the most electronegative element). Neither F_2 nor Cl_2 will have hydrogen bonding, so the compound, F_2, with the lower molar mass, has the lower boiling point.

42. The boiling liquid remains at constant temperature because the added heat energy is being used to convert the liquid to a gas, i.e., to supply the heat of vaporization for the liquid at its boiling point.

43. 34.6°C, the boiling point of ethyl ether. (See Table 13.2)

44. The lake freezes from the top down because, as the temperature drops to freezing or below, the water on the surface tends to cool faster than the water that lies deeper. As the surface water freezes, the ice formed floats because the ice is less dense than the liquid water below it.

45. If the lake is in an area where the temperature is below freezing for part of the year, the expected temperature would be 4°C at the bottom of the lake. This is because the surface water would cool to 4°C (maximum density) and sink.

46. The formation of hydrogen and oxygen from water is an endothermic reaction, due to the following evidence:

    (a) Energy must continually be provided to the system for the reaction to proceed. The reaction will cease when the energy source is removed.

    (b) The reverse reaction, burning hydrogen in oxygen, releases energy as heat.

47. Acid anhydride: $[HClO_4, Cl_2O_7]$ $[H_2CO_3, CO_2]$ $[H_3PO_4, P_2O_5]$

48. Acid anhydride: $[H_2SO_3, SO_2]$ $[H_2SO_4, SO_3]$ $[HNO_3, N_2O_5]$

49. Basic anhydrides: $[LiOH, Li_2O]$ $[NaOH, Na_2O]$ $[Mg(OH)_2, MgO]$

50. Basic anhydrides: $[KOH, K_2O]$ $[Ba(OH)_2, BaO]$ $[Ca(OH)_2, CaO]$

51. (a) $Ba(OH)_2 \xrightarrow{\Delta} BaO + H_2O$

    (b) $2\,CH_3OH + 3O_2 \longrightarrow 2\,CO_2 + 4\,H_2O$

    (c) $2\,Rb + 2\,H_2O \longrightarrow 2\,RbOH + H_2$

    (d) $SnCl_2 \cdot 2\,H_2O \xrightarrow{\Delta} SnCl_2 + 2\,H_2O$

(e) $HNO_3 + NaOH \longrightarrow NaNO_3 + H_2O$

(f) $CO_2 + H_2O \longrightarrow H_2CO_3$

52. (a) $Li_2O + H_2O \longrightarrow 2\,LiOH$

(b) $2\,KOH \xrightarrow{\Delta} K_2O + H_2O$

(c) $Ba + 2\,H_2O \longrightarrow Ba(OH)_2 + H_2$

(d) $Cl_2 + H_2O \longrightarrow HCl + HClO$

(e) $SO_3 + H_2O \longrightarrow H_2SO_4$

(f) $H_2SO_3 + 2\,KOH \longrightarrow K_2SO_3 + 2\,H_2O$

53. (a) barium bromide dihydrate
(b) aluminum chloride hexahydrate
(c) iron(III) phosphate tetrahydrate

54. (a) magnesium ammonium phosphate hexahydrate
(b) iron(II) sulfate heptahydrate
(c) tin(IV) chloride pentahydrate

55. Deionized water is water from which the ions have been removed.
(a) Hard water contains dissolved calcium and magnesium salts.
(b) Soft water is free of ions that cause hardness (Ca^{2+} and Mg^{2+}) but it may contain other ions such as Na^+ and K^+.

56. Deionized water is water from which the ions have been removed.
(a) Distilled water has been vaporized by boiling and recondensed. It is free of nonvolatile impurities, but may still contain any volatile impurities that were initially present in the water.
(b) Natural waters are generally not pure, but contain dissolved minerals and suspended matter, and can even contain harmful bacteria.

57. $$(100.\text{ g CoCl}_2 \cdot 6\,\text{H}_2\text{O})\left(\frac{1\text{ mol}}{238.0\text{ g}}\right) = 0.420\text{ mol CoCl}_2 \cdot 6\,\text{H}_2\text{O}$$

58. $$(100.\text{ g FeI}_2 \cdot 4\,\text{H}_2\text{O})\left(\frac{1\text{ mol}}{381.7\text{ g}}\right) = 0.262\text{ mol FeI}_2 \cdot 4\,\text{H}_2\text{O}$$

59. $$(100.\text{ g CoCl}_2 \cdot 6\,\text{H}_2\text{O})\left(\frac{1\text{ mol}}{238.0\text{ g}}\right)\left(\frac{6\text{ mol H}_2\text{O}}{1\text{ mol CoCl}_2 \cdot 6\,\text{H}_2\text{O}}\right) = 2.52\text{ mol H}_2\text{O}$$

60. $(100.\text{ g FeI}_2 \cdot 4\text{ H}_2\text{O})\left(\dfrac{1\text{ mol}}{381.7\text{ g}}\right)\left(\dfrac{4\text{ mol H}_2\text{O}}{1\text{ mol FeI}_2 \cdot 4\text{ H}_2\text{O}}\right) = 1.05\text{ mol H}_2\text{O}$

61. Assume 1 mol of the compound which contains seven moles of water.

$$\%\text{ H}_2\text{O} = \left(\frac{\text{g H}_2\text{O}}{\text{g MgSO}_4 \cdot 7\text{ H}_2\text{O}}\right)(100) = \left(\frac{(7)(18.02\text{ g})}{246.5\text{ g}}\right)(100) = 51.17\%\text{ H}_2\text{O}$$

62. Assume 1 mol of the hydrate which contains 18 mol of water.

$$\%\text{ H}_2\text{O} = \frac{\text{g H}_2\text{O}}{\text{g Al}_2(\text{SO}_4)_3 \cdot 18\text{ H}_2\text{O}} = \left(\frac{(18)(18.02\text{ g})}{666.5\text{ g}}\right)(100) = 48.67\%\text{ H}_2\text{O}$$

63. Assume 100. g of the compound.

$(0.142)(100.\text{ g}) = 14.2\text{ g H}_2\text{O}$

$(0.858)(100.\text{ g}) = 85.8\text{ g Pb(C}_2\text{H}_3\text{O}_2)_2$

$(14.2\text{ g H}_2\text{O})\left(\dfrac{1\text{ mol}}{18.02\text{ g}}\right) = 0.788\text{ mol H}_2\text{O}$

$(85.8\text{ g Pb(C}_2\text{H}_3\text{O}_2)_2)\left(\dfrac{1\text{ mol}}{325.3\text{ g}}\right) = 0.264\text{ mol Pb(C}_2\text{H}_3\text{O}_2)_2$

In the formula for the hydrate, there is one mole of $\text{Pb(C}_2\text{H}_3\text{O}_2)_2$, so divide each of the moles calculated by 0.264.

$$\frac{0.264\text{ mol Pb(C}_2\text{H}_3\text{O}_2)_2}{0.264\text{ mol}} = 1\text{ Pb(C}_2\text{H}_3\text{O}_2)_2$$

$$\frac{0.788\text{ mol H}_2\text{O}}{0.264\text{ mol}} = 2.98\text{ H}_2\text{O}$$

Therefore, the formula is $\text{Pb(C}_2\text{H}_3\text{O}_2)_2 \cdot 3\text{ H}_2\text{O}$.

64. $25.0\text{ g hydrate} - 16.9\text{ g FePO}_4 = 8.1\text{ g H}_2\text{O}$ driven off

$(8.1\text{ g H}_2\text{O})\left(\dfrac{1\text{ mol}}{18.02\text{ g}}\right) = 0.45\text{ mol H}_2\text{O}$ $\qquad \dfrac{0.45}{0.112} = 4.0$

$(16.9\text{ g FePO}_4)\left(\dfrac{1\text{ mol}}{150.8\text{ g}}\right) = 0.112\text{ mol FePO}_4$ $\qquad \dfrac{0.112}{0.112} = 1.00$

The formula is $\text{FePO}_4 \cdot 4\text{ H}_2\text{O}$.

65. (a) Energy (E_a) to heat the water from 20.°C $\longrightarrow$ 100.°C

$$E_a = (\text{m})(\text{specific heat})(\Delta\text{T}) = (120.\text{ g})\left(\frac{4.184\text{ J}}{\text{g°C}}\right)(80.\text{°C}) = 4.0 \times 10^4\text{ J}$$

(b) Energy (E_b) to convert water to steam: heat of vaporization $= 2.26 \times 10^3$ J/g

$$E_b = \text{(m)(heat of vaporization)} = (120.\text{ g})(2.26 \times 10^3\text{ J/g}) = 2.71 \times 10^5\text{ J}$$

$$E_{total} = E_a + E_b = (4.0 \times 10^4\text{ J}) + (2.71 \times 10^5\text{ J}) = 3.11 \times 10^5\text{ J}$$

66. (a) Energy (E_a) to cool the water from 24°C $\longrightarrow$ 0°C

$$E_a = \text{(m)(specific heat)}(\Delta T) = (126\text{ g})\left(\frac{4.184\text{ J}}{\text{g°C}}\right)(24°\text{C}) = 1.3 \times 10^4\text{ J}$$

(b) Energy (E_b) to convert water to ice

$$E_b = \text{(m)(heat of fusion)} = (126\text{ g})(335\text{ J/g}) = 4.22 \times 10^4\text{ J}$$

$$E_{total} = E_a + E_b = (1.3 \times 10^4\text{ J}) + (4.22 \times 10^4\text{ J}) = 5.5 \times 10^4\text{ J}$$

67. Energy released in cooling the water: 25°C to 0°C

$$E = \text{(m)(specific heat)}(\Delta T) = (300.\text{ g})\left(\frac{1\text{ cal}}{\text{g°C}}\right)(25°\text{C}) = 7.5 \times 10^3\text{ cal}$$

Energy required to melt the ice

$$E = \text{(m)(heat of fusion)} = (100.\text{ g})(80.\text{ cal/g}) = 8.0 \times 10^3\text{ cal}$$

Less energy is released in cooling the water than is required to melt the ice. Ice will remain and the water will be at 0°C.

68. Energy to heat the water = energy to condense the steam

$$(300.\text{ g})\left(\frac{1\text{ cal}}{\text{g°C}}\right)(100.°\text{C} - 25°\text{C}) = \text{(m)}(540\text{ cal/g})$$

m = 42 g (grams of steam required to heat the water to 100.°C) 42 g of steam are required to heat 300. g of water to 100.°C. Since only 35 g of steam are added to the system, the final temperature will be less than 100.°C. Not sufficient steam.

69. Energy lost by warm water = energy gained by the ice

x = final temperature

$$\text{mass}(H_2O) = (1.5\text{ L }H_2O)\left(\frac{1000\text{ mL}}{\text{L}}\right)\left(\frac{1.0\text{ g}}{\text{mL}}\right) = 1500\text{ g}$$

$$(1500\text{ g})\left(\frac{1\text{ cal}}{\text{g°C}}\right)(75°\text{C} - x) = (75\text{ g})\left(80.\frac{\text{cal}}{\text{g}}\right) + (75\text{ g})\left(\frac{1\text{ cal}}{\text{g°C}}\right)(x - 0°\text{C})$$

$$(112{,}500\text{ cal}) - (1500x\text{ cal/°C}) = 6.0 \times 10^3\text{ cal} + 75x\text{ cal/°C}$$

$$106{,}500\text{ cal} = 1575x\text{ cal/°C}$$

$$68°\text{C} = x$$

70. E = (m)(heat of fusion)
(500. g)(335 J/g) = 167,000 J needed to melt the ice
9560 J < 167,500 J
Since 167,500 J are required to melt all the ice, and only 9560 J are available, the system will be at 0°C. It will be a mixture of ice and water.

71. (a) $2\,Na + 2\,H_2O \longrightarrow 2\,NaOH + H_2$

$$(1.00\text{ g Na})\left(\frac{1\text{ mol}}{22.99\text{ g}}\right)\left(\frac{2\text{ mol }H_2O}{2\text{ mol Na}}\right)\left(\frac{18.02\text{ g}}{\text{mol}}\right) = 0.784\text{ g }H_2O$$

(b) $MgO + H_2O \longrightarrow Mg(OH)_2$

$$(1.00\text{ g MgO})\left(\frac{1\text{ mol}}{40.31\text{ g}}\right)\left(\frac{1\text{ mol }H_2O}{1\text{ mol MgO}}\right)\left(\frac{18.02\text{ g}}{\text{mol}}\right) = 0.447\text{ g }H_2O$$

(c) $N_2O_5 + H_2O \longrightarrow 2\,HNO_3$

$$(1.00\text{ g }N_2O_5)\left(\frac{1\text{ mol}}{108.0\text{ g}}\right)\left(\frac{1\text{ mol }H_2O}{1\text{ mol }N_2O_5}\right)\left(\frac{18.02\text{ g}}{\text{mol}}\right) = 0.167\text{ g }H_2O$$

72. (a) $2\,K + 2\,H_2O \longrightarrow 2\,KOH + H_2$

$$(1.00\text{ mol K})\left(\frac{2\text{ mol }H_2O}{2\text{ mol K}}\right)\left(\frac{18.02\text{ g}}{\text{mol}}\right) = 18.0\text{ g }H_2O$$

(b) $Ca + 2\,H_2O \longrightarrow Ca(OH)_2 + H_2$

$$(1.00\text{ mol Ca})\left(\frac{2\text{ mol }H_2O}{1\text{ mol Ca}}\right)\left(\frac{18.02\text{ g}}{\text{mol}}\right) = 36.0\text{ g }H_2O$$

(c) $SO_3 + H_2O \longrightarrow H_2SO_4$

$$(1.00\text{ mol }SO_3)\left(\frac{1\text{ mol }H_2O}{1\text{ mol }SO_3}\right)\left(\frac{18.02\text{ g}}{\text{mol}}\right) = 18.0\text{ g }H_2O$$

73. Water forms droplets because of surface tension, or the desire for a droplet of water to minimize its ratio of surface area to volume. The molecules of water inside the drop are attracted to other water molecules all around them, but on the surface of the droplet the water molecules feel an inward attraction only. This inward attraction of surface water molecules for internal water molecules is what holds the droplets together and minimizes their surface area.

74. Steam molecules will cause a more severe burn. Steam molecules contain more energy at 100°C than water molecules at 100°C due to the energy absorbed during the vaporization stage (heat of vaporization).

75. The alcohol has a higher vapor pressure than water and thus evaporates faster than water. When the alcohol evaporates it absorbs energy from the water, cooling the water. Eventually the water will lose enough energy to change from a liquid to a solid (freeze).

76. When one leaves the swimming pool, water starts to evaporate from the skin of the body. Part of the energy needed for evaporation is absorbed from the skin, resulting in the cool feeling.

77.

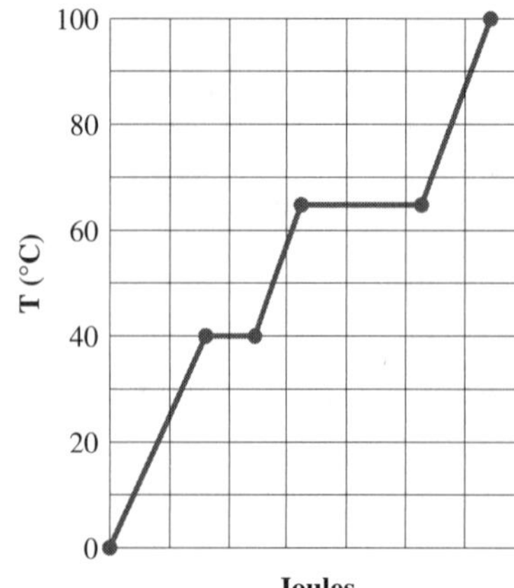

(a) From 0°C to 40.°C solid X warms until at 40.°C it begins to melt. The temperature remains at 40.0°C until all of X is melted. After that, liquid X will warm steadily to 65°C where it will boil and remain at 65°C until all of the liquid becomes vapor. Beyond 65°C, the vapor will warm steadily until 100°C.

(b)

Joules needed (0°C to 40°C)	=	(60. g)(3.5 J/g°C)(40.°C)	=	8400 J
Joules needed at 40°C	=	(60. g)(80. J/g)	=	4800 J
Joules needed (40°C to 65°C)	=	(60. g)(3.5 J/g°C)(25°C)	=	5300 J
Joules needed at 65°C	=	(60. g)(190 J/g)	=	11,000 J
Joules needed (65°C to 100°C)	=	(60. g)(3.5 J/g°C)(35°C)	=	7400 L
Total Joules needed				37,000 J

(each step rounded to two significant figures)

78. During phase changes (ice melting to liquid water or liquid water evaporating to steam), all the heat energy is used to cause the phase change. Once the phase change is complete the heat energy is once again used to increase the temperature of the substance.

79. As the temperature of a liquid increases, the molecules gain kinetic energy thereby increasing their escaping tendency (vapor pressure).

80. Since boiling occurs when vapor pressure equals atmospheric pressure, the graph in Figure 13.4 indicates that water will boil at about 75°C at 270 torr pressure.

81. $CuSO_4$ (anhydrous) is gray white. When exposed to moisture, it turns bright blue forming $CuSO_4 \cdot 5\,H_2O$. The color change is an indicator of moisture in the environment.

82. $MgSO_4 \cdot 7\,H_2O$ $\quad$ $Na_2HPO_4 \cdot 12\,H_2O$

83. Soap can soften hard water by forming a precipitate with, and thus removing, the calcium and magnesium ions. This precipitate is a greasy scum and is very objectionable, so it is a poor way to soften water.

84. Chlorine is commonly used to destroy bacteria in water. Ozone and ultraviolet radiation are also used in some places.

85. Ozone, O_3

86. When organic pollutants in water are oxidized by dissolved oxygen, there may not be sufficient dissolved oxygen to sustain marine life, such as fish. Most marine life forms depend on dissolved oxygen for cellular respiration.

87. Liquids that are stored in ceramic containers should never be drunk, for they are likely to have dissolved some of the lead from the ceramic. If the ceramic is glazed, the liquid is less apt to dissolve lead from the ceramic.

88. $Na_2\,\text{zeolite}(s) + Mg^{2+}(aq) \longrightarrow Mg\,\text{zeolite}(s) + 2\,Na^{+}(aq)$

89. Softening of hard water using sodium carbonate:
$CaCl_2(aq) + Na_2CO_3(aq) \longrightarrow CaCO_3(s) + 2\,NaCl(aq)$

90.

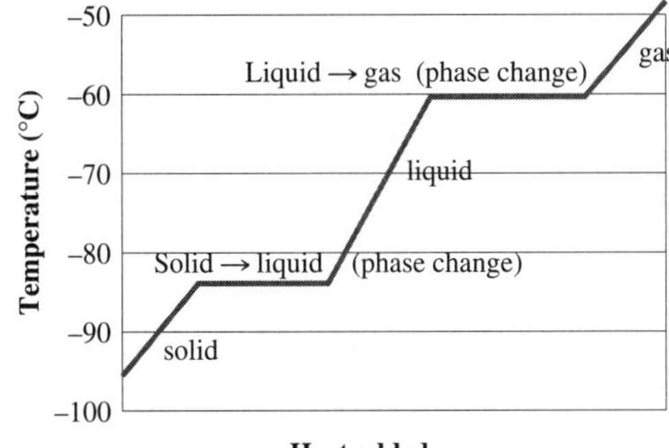

91. (a) Melt ice: $E_a = (m)(\text{heat of fusion}) = (225\ g)(80.\ cal/g) = 18{,}000\ cal$

(b) Warm the water: $E_b = (m)(\text{specific heat})(\Delta T)$

$$= (225\ g)\left(\frac{1\ cal}{g°C}\right)(100.°C) = 22{,}500\ cal$$

(c) Vaporize the water:

$E_c = (m)(\text{heat of vaporization}) = (225\ g)(540\ cal/g) = 121{,}500\ cal$

$E_{total} = E_a + E_b + E_c = 1.6 \times 10^5\ cal$

92. The heat of vaporization of water is 2.26 kJ/g.

$$(2.26\ kJ/g)\left(\frac{18.02\ g}{mol}\right) = 40.7\ kJ/mol$$

93. $$E = (m)(\text{specific heat})(\Delta T) = (250.\ g)\left(\frac{0.096\ cal}{g°C}\right)(150. - 20.0°C)$$

$$= 3.1 \times 10^3\ cal\ (3.1\ kcal)$$

94. Heat lost by warm water = heat gained by ice
m = grams of ice to lower temperature of water to 0.0°C.

$$(120.\ g)\left(\frac{1\ cal}{g°C}\right)(45°C - 0.0°C) = (m)(80.\ cal/g)$$

68 g = m (grams of ice melted)
68 g of ice melted. Therefore, 150. g − 68 g = 82 g ice remain.

95. Energy liberated when steam at 100.0°C condenses to water at 100.0°C

$$(50.0\ mol\ steam)\left(\frac{18.02\ g}{mol}\right)\left(\frac{2.26\ kJ}{g}\right)\left(\frac{1000\ J}{kJ}\right) = 2.04 \times 10^6\ J$$

Energy liberated in cooling water from 100.0°C to 30.0°C

$$(50.0\ mol\ H_2O)\left(\frac{18.02\ g}{mol}\right)\left(\frac{4.184\ J}{g°C}\right)(100.0°C - 30.0°C) = 2.64 \times 10^5\ J$$

Total energy liberated $= 2.04 \times 10^6\ J + 2.64 \times 10^5\ J = 2.30 \times 10^6\ J$

96. Energy to warm the ice from −10.0°C to 0°C

$$(100.\ g)\left(\frac{2.01\ J}{g°C}\right)(10.0°C) = 2010\ J$$

Energy to melt the ice at 0°C

$(100.\ g)(335\ J/g) = 33{,}500\ J$

Energy to heat the water from 0°C to 20.0°C

$$(100.\ g)\left(\frac{4.184\ J}{g°C}\right)(20.0°C) = 8370\ J$$

$E_{total} = 2010\ J + 33{,}500\ J + 8370\ J = 4.39 \times 10^4\ J = 43.9\ kJ$

97. $2\ H_2O(l) \longrightarrow 2\ H_2(g) + O_2(g)$

The conversion is $L\ O_2 \longrightarrow mol\ O_2 \longrightarrow mol\ H_2O \longrightarrow g\ H_2O$

$$(25.0\ L\ O_2)\left(\frac{1\ mol}{22.4\ L}\right)\left(\frac{2\ mol\ H_2O}{1\ mol\ O_2}\right)\left(\frac{18.02\ g}{mol}\right) = 40.2\ g\ H_2O$$

98. The conversion is

$$\frac{mol}{day} \longrightarrow \frac{molecules}{day} \longrightarrow \frac{molecules}{hr} \longrightarrow \frac{molecules}{min} \longrightarrow \frac{molecules}{s}$$

$$\left(\frac{1.00\ mol\ H_2O}{day}\right)\left(\frac{6.022 \times 10^{23}\ molecules}{mol}\right)\left(\frac{1.00\ day}{24\ hr}\right)\left(\frac{1\ hr}{60\ min}\right)\left(\frac{1\ min}{60\ s}\right) =$$

6.97×10^{18} molecules/s

99. Liquid water has a density of 1.00 g/mL.

$$d = \frac{m}{V} \qquad V = \frac{m}{d} = \frac{18.02\ g}{1.00\ g/mL} = 18.0\ mL \qquad \text{(volume of 1 mole)}$$

1.00 mole of water vapor at STP has a volume of 22.4 L (gas)

100. Mass solution − mass H_2O = mass H_2SO_4

$(122\ mL)(1.26\ g/mL) - (100.\ mL)(1.00\ g/mL) = 54\ g\ H_2SO_4$ added

101. $2\ H_2(g) + O_2(g) \longrightarrow 2\ H_2O(g)$

(a) $(80.0\ mL\ H_2)\left(\frac{1\ mL\ O_2}{2\ mL\ H_2}\right) = 40.0\ mL\ O_2$ react with 80.0 mL of H_2

Since 60.0 mL of O_2 are available, some oxygen remains unreacted.

(b) $60.0\ mL - 40.0\ mL = 20.0\ mL\ O_2$ unreacted.

102. Energy absorbed by the student when steam at 100.°C changes to water at 100.°C

$$(1.5 \text{ g steam})\left(\frac{2.26 \text{ kJ}}{\text{g}}\right) = 3.4 \text{ kJ} \quad (3.4 \times 10^3 \text{ J})$$

Energy absorbed when water cools from 100.°C to 20.0°C

$$E = (\text{m})(\text{specific heat})(\Delta \text{t})$$

$$E = (1.5 \text{ g})\left(\frac{4.184 \text{ J}}{\text{g°C}}\right)(100.\text{°C} - 20.0\text{°C}) = 5.0 \times 10^2 \text{ J}$$

$$E_{total} = 3.4 \times 10^3 \text{ J} + 5.0 \times 10^2 \text{ J} = 3.9 \times 10^3 \text{ J}$$

103. (a) won't hydrogen bond – no hydrogen in the molecule
(b) will hydrogen bond

$CH_3—O—H\cdots O—CH_3$ (with H bonded below the second O)

(c) won't hydrogen bond–no hydrogen covalently bonded to a strongly electronegative element
(d) will hydrogen bond

$H—O\cdots H$ (with H bonded below O and H bonded below the second H)

104. During the fusion (melting) of a substance the temperature remains constant so a temperature factor is not needed.

105. Energy needed to heat Cu to its melting point:

$$E = (50.0 \text{ g})(0.385 \text{ J/g°C})(1083\text{°C} - 25.0\text{°C}) = 2.04 \times 10^4 \text{ J}$$

Energy needed to melt the Cu.

$$E = (50.0 \text{ g})(134 \text{ J/g}) = 6.70 \times 10^3 \text{ J}$$

$$E_{total} = 2.04 \times 10^4 \text{ J} + 6.70 \times 10^3 \text{ J} = 2.71 \times 10^4 \text{ J}$$

CHAPTER 14

SOLUTIONS

1.

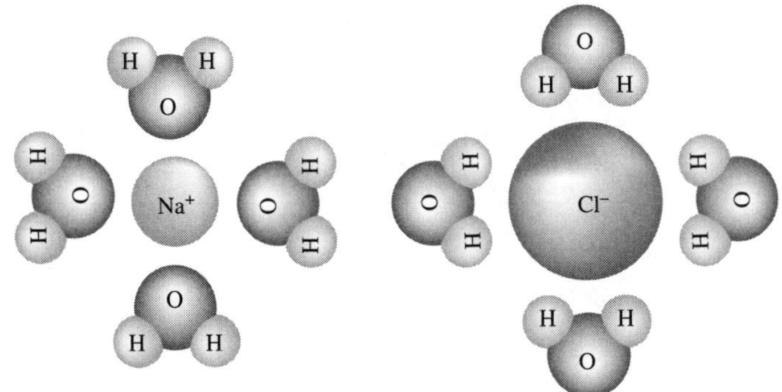

These diagrams are intended to illustrate the orientation of the water molecules about the ions, not the number of water molecules.

2. From Table 14.3, approximately 4.5 g of NaF would be soluble in 100 g of water at 50°C.

3. From Figure 14.3, solubilities in water at 25°C are:
   (a) KCl 35g/100 g H_2O
   (b) $KClO_3$ 9g/100 g H_2O
   (c) KNO_3 39g/100 g H_2O

4. Potassium fluoride has a relatively high solubility when compared to lithium or sodium fluoride. For lithium and sodium halides, the order of solubility (in order of increasing solubilities) is:

   F^- Cl^- Br^- I^-

   For potassium halides, the order of increasing solubilities is:

   Cl^- Br^- F^- I^-

5. (a) $KClO_3$ at 60°C, 25 g
   (b) HCl at 20°C, 72 g
   (c) Li_2SO_4 at 80°C, 31 g
   (d) KNO_3 at 0°C, 14 g

6. KNO_3

7. A one molal solution in camphor will show a greater freezing point depression than a 2 molal solution in benzene.

$$\Delta t_f = \left(\frac{1 \text{ mol solute}}{\text{kg camphor}}\right)\left(\frac{40°\text{C kg camphor}}{\text{mol solute}}\right) = 40°\text{C (freezing point depression)}$$

$$\Delta t_f = \left(\frac{2 \text{ mol solute}}{\text{kg benzene}}\right)\left(\frac{5.1°\text{C kg benzene}}{\text{mol solute}}\right) = 10.2°\text{C (freezing point depression)}$$

8.

Cube	1 cm	0.01 cm
Volume	1 cm^3	$1 \times 10^{-6}\,cm^3$
Number/1 cm cube	1	$10^6\,[(1\,cm^3)/(1 \times 10^{-6}\,cm^3) = 10^6\text{ cubes}]$
Area of face	1 cm^2	$1 \times 10^{-4}\,cm^2$
Total surface area	6 cm^2	$6 \times 10^2\,cm^2$

$$(1 \times 10^6\text{ cubes})(6\text{ faces/cube})(1 \times 10^{-4}\,cm^2/\text{face}) = 6 \times 10^2\,cm^2$$

9. $\frac{63\text{ g }NH_4Cl}{150\text{ g }H_2O} = \frac{42\text{ g }NH_4Cl}{100\text{ g }H_2O}$ From Figure 14.3, the solubility of NH_4Cl in water is approximately 42 g/100 g H_2O at 30°C, 46 g/100 g H_2O at 40°C. Therefore, the solution of 63 g/150 g of water would be saturated at 10°C, 20°C, and 30°C. The solution would be unsaturated at 40°C and 50°C.

10. The dissolving process involves solvent molecules attaching to the solute ions or molecules. This rate decreases as more of the solvent molecules are already attached to solute molecules. As the solution becomes more saturated, the number of unused solvent molecules decreases. Also, the rate of recrystallization increases as the concentration of dissolved solute increases.

11. A supersaturated solution of $NaC_2H_3O_2$ may be prepared in the following sequence:
   (a) Determine the mass of $NaC_2H_3O_2$ necessary to saturate a specific amount of water at room temperature.
   (b) Place a bit more $NaC_2H_3O_2$ in the water than the amount needed to saturate the solution.
   (c) Heat the solution until all the solid dissolves.
   (d) Cover the container and allow it to cool undisturbed. The cooled solution, which should contain no solid $NaC_2H_3O_2$, is supersaturated.

   To test for supersaturation, add one small crystal of $NaC_2H_3O_2$ to the solution. Immediate crystallization is an indication that the solution was supersaturated.

12. Because the concentration of water is greater in the thistle tube, the water will flow through the membrane from the thistle tube to the urea solution in the beaker. The solution level in the thistle tube will fall.

13. A true solution is one in which the size of the particles of solute are between $0.1 - 1\text{nm}$. True solutions are homogeneous and the ratio of solute to solvent can be varied. They can be colored or colorless but are transparent. The solute remains distributed evenly in the solution, it will not settle out.

14. The two components of a solution are the solute and the solvent. The solute is dissolved into the solvent or is the least abundant component. The solvent is the dissolving agent or the most abundant component.

15. It is not always apparent which component in a solution is the solute. For example, in a solution composed of equal volumes of two liquids, the designation of solute and solvent would be simply a matter of preference on the part of the person making the designation.

16. The ions or molecules of a dissolved solute do not settle out because the individual particles are so small that the force of molecular collisions is large compared to the force of gravity.

17. Yes. It is possible to have one solid dissolved in another solid. Metal alloys are of this type. Atoms of one metal are dissolved among atoms of another metal.

18. Orange. The three reference solutions are KCl, $KMnO_4$, and $K_2Cr_2O_7$. They all contain K^+ ions in solution. The different colors must result from the different anions dissolved in the solutions: MnO_4^- (purple) and $Cr_2O_7^{2-}$ (orange). Therefore, it is predictable that the $Cr_2O_7^{2-}$ ion present in an aqueous solution of $Na_2Cr_2O_7$ will impart an orange color to the solution.

19. Hexane and benzene are both nonpolar molecules. There are no strong intermolecular forces between molecules of either substance or with each other, so they are miscible. Sodium chloride consists of ions strongly attracted to each other by electrical attractions. The hexane molecules, being nonpolar, have no strong forces to pull the ions apart, so sodium chloride is insoluble in hexane.

20. Coca Cola has two main characteristics, taste and fizz (carbonation). The carbonation is due to a dissolved gas, carbon dioxide. Since dissolved gases become less soluble as temperature increases, warm Coca Cola would be flat, with little to no carbonation. It is, therefore, unappealing to most people.

21. Air is considered to be a solution because it is a homogeneous mixture of several gaseous substances and does not have a fixed composition.

22. A teaspoon of sugar would definitely dissolve more rapidly in 200 mL of hot coffee than in 200 mL of iced tea. The much greater thermal agitation of the hot coffee will help break the sugar molecules away from the undissolved solid and disperse them throughout the solution. Other solutes in coffee and tea would have no significant effect. The temperature difference is the critical factor.

23. The solubility of gases in liquids is greatly affected by the pressure of a gas above the liquid. The greater the pressure, the more soluble the gas. There is very little effect of pressure regarding the dissolution of solids in liquids.

24. For a given mass of solute, the smaller the particles, the faster the dissolution of the solute. This is due to the smaller particles having a greater surface area exposed to the dissolving action of the solvent.

25. In a saturated solution, the net rate of dissolution is zero. There is no further increase in the amount of dissolved solute, even though undissolved solute is continuously dissolving, because dissolved solute is continuously coming out of solution, crystallizing at a rate equal to the rate of dissolving.

26. When crystals of $AgNO_3$ and NaCl are mixed, the contact between the individual ions is not intimate enough for the double displacement reaction to occur. When solutions of the two chemicals are mixed, the ions are free to move and come into intimate contact with each other, allowing the reaction to occur easily. The AgCl formed is insoluble.

27. A nonvolatile solute (such as salt) lowers the freezing point of water. Adding salt to icy roads in winter melts the ice because the salt lowers the freezing point of water.

28. A 16 molar solution of nitric acid is a solution that contains 16 moles HNO_3 per liter of solution.

29. The two solutions contain the same number of chloride ions. One liter of 1 M NaCl contains 1 mole of NaCl, therefore 1 mole of chloride ions. 0.5 liter of 1 M $MgCl_2$ contains 0.5 mol of $MgCl_2$ and 1 mole of chloride ions.

$$(0.5\text{ L})\left(\frac{1\text{ mol MgCl}_2}{\text{L}}\right)\left(\frac{2\text{ mol Cl}^-}{1\text{ mol MgCl}_2}\right) = 1\text{ mol Cl}^-$$

30. The champagne would spray out of the bottle all over the place. The rise in temperature and the increase in kinetic energy of the molecules by shaking both act to decrease the solubility of gas within the liquid. The pressure inside the bottle would be great. As the cork is popped, much of the gas would escape from the liquid very rapidly, causing the champagne to spray.

31. The number of grams of NaCl in 750 mL of 5.0 molar solution is

$$(0.75\ \text{L})\left(\frac{5.0\ \text{mol NaCl}}{\text{L}}\right)\left(\frac{58.44\ \text{g}}{1\ \text{mol}}\right) = 2.2 \times 10^2\ \text{g NaCl}$$

Dissolve the 220 g of NaCl in a minimum amount of water, then dilute the resulting solution to a final volume of 750 mL (0.75 L).

32. A semipermeable membrane will allow water molecules to pass through in both directions. If it has pure water on one side and 10% sugar solutions on the other side of the membrane, there is a higher concentration of water molecules on the pure water side. Therefore, there are more water molecule impacts per second on the pure water side of the membrane. The net result is more water molecules pass from the pure water to the sugar solution. Osmotic pressure effect.

33. The urea solution will have the greater osmotic pressure because it has 1.67 mol solute/kg H_2O, while the glucose solution has only 0.83 mol solute/kg H_2O.

34. A lettuce leaf immersed in salad dressing containing salt and vinegar will become limp and wilted as a result of osmosis. As the water inside the leaf flows into the dressing where the solute concentration is higher the leaf becomes limp from fluid loss. In water, osmosis proceeds in the opposite direction flowing into the lettuce leaf maintaining a high fluid content and crisp leaf.

35. The concentration of solutes (such as salts) is higher in seawater than in body fluids. The survivors who drank seawater suffered further dehydration from the transfer of water by osmosis from body tissues to the intestinal tract.

36. Ranking of the specified bases in descending order of the volume of each required to react with 1 liter of 1 M HCl. The volume of each required to yield 1 mole of OH^- ion is shown.

(a) 1 M NaOH — 1 liter
(b) 0.6 M $Ba(OH)_2$ — 0.83 liter
(c) 2 M KOH — 0.50 liter
(d) 1.5 M $Ca(OH)_2$ — 0.33 liter

37. The boiling point of a liquid or solution is the temperature at which the vapor pressure of the liquid equals the pressure of the atmosphere. Since a solution containing a nonvolatile solute has a lower vapor pressure than the pure solvent, the boiling point of the solution must be at a higher temperature than for the pure solvent. At the higher boiling temperature the vapor pressure of the solution equals the atmospheric pressure.

38. The freezing point is the temperature at which a liquid changes to a solid. The vapor pressure of a solution is lower than that of a pure solvent. Therefore, the vapor pressure curve of the solution intersects the vapor pressure curve of the pure solvent, at a temperature lower than the freezing point of the pure solvent. (See Figure 14.8a) At this point of intersection, the vapor pressure of the solution equals the vapor pressure of the pure solvent.

39. Water and ice are different phases of the same substance in equilibrium at the freezing point of water, 0°C. The presence of the methanol lowers the vapor pressure and hence the freezing point of water. If the ratio of alcohol to water is high, the freezing point can be lowered as much as 10°C or more.

40. Effectiveness in lowering the freezing point of 500. g water:

    (a) 100. g (2.17 mol) of ethyl alcohol is more effective than 100. g (0.292 mol) of sucrose.

    (b) 20.0 g (0.435 mol) of ethyl alcohol is more effective than 100. g (0.292 mol) of sucrose.

    (c) 20.0 g (0.625 mol) of methyl alcohol is more effective than 20.0 g (0.435 mol) of ethyl alcohol.

41. Both molarity and molality describe the concentration of a solution. However, molarity is the ratio of moles of solute per liter of solution, and molality is the ratio of moles of solute per kilogram of solvent.

42. 5 molal NaCl = 5 mol NaCl/kg H_2O; 5 molar NaCl = 5 mol NaCl/L of solution. The volume of the 5 molal solution will be larger than 1 liter (1 L H_2O + 5 mol NaCl). The volume of the 5 molar solution is exactly 1 L (5 mol NaCl + sufficient H_2O to produce 1 L of solution). The molarity of a 5 molal solution is therefore, less than 5 molar.

43. Reasonably soluble: (a) KOH (b) $NiCl_2$ (d) $AgC_2H_3O_2$ (e) Na_2CrO_4
    Insoluble: (c) ZnS

44. Reasonably soluble: (c) $CaCl_2$ (d) $Fe(NO_3)_3$
    Insoluble: (a) PbI_2 (b) $MgCO_3$ (e) $BaSO_4$

45. Mass percent calculations.

    (a) 25.0 g NaBr + 100. g H_2O = 125 g solution

$$\left(\frac{25.0 \text{ g NaBr}}{125 \text{ g solution}}\right)(100) = 20.0\% \text{ NaBr}$$

(b) $1.20 \text{ g } K_2SO_4 + 10.0 \text{ g } H_2O = 11.2 \text{ g solution}$

$$\left(\frac{1.20 \text{ g } K_2SO_4}{11.2 \text{ g solution}}\right)(100) = 10.7\% \; K_2SO_4$$

46. (a) $40.0 \text{ g } Mg(NO_3)_2 + 500. \text{ g } H_2O = 540. \text{ g solution}$

$$\left(\frac{40.0 \text{ g } Mg(NO_3)_2}{540. \text{ g solution}}\right)(100) = 7.41\% \; Mg(NO_3)_2$$

(b) $17.5 \text{ g } NaNO_3 + 250. \text{ g } H_2O = 268 \text{ g solution}$

$$\left(\frac{17.5 \text{ g } NaNO_3}{268 \text{ g solution}}\right)(100) = 6.53\% \; NaNO_3$$

47. A 12.5% $AgNO_3$ solution contains 12.5 g $AgNO_3$ per 100. g solution

$$(30.0 \text{ g } AgNO_3)\left(\frac{100. \text{ g solution}}{12.5 \text{ g } AgNO_3}\right) = 240. \text{ g solution}$$

48. A 12.5% $AgNO_3$ solution contains 12.5 g $AgNO_3$ per 100. g solution

$$(0.400 \text{ mol } AgNO_3)\left(\frac{169.9 \text{ g}}{\text{mol}}\right)\left(\frac{100. \text{ g solution}}{12.5 \text{ g } AgNO_3}\right) = 544 \text{ g solution}$$

49. Mass percent calculations.

(a) $60.0 \text{ g NaCl} + 200.0 \text{ g } H_2O = 260.0 \text{ g solution}$

$$\left(\frac{60.0 \text{ g NaCl}}{260.0 \text{ g solution}}\right)(100) = 23.1\% \text{ NaCl}$$

(b) $$(0.25 \text{ mol } HC_2H_3O_2)\left(\frac{60.03 \text{ g}}{\text{mol}}\right) = 15 \text{ g } HC_2H_3O_2$$

$$(3.0 \text{ mol } H_2O)\left(\frac{18.02 \text{ g}}{\text{mol}}\right) = 54 \text{ g } H_2O$$

$$\left(\frac{15 \text{ g } HC_2H_3O_2}{69 \text{ g solution}}\right)(100) = 22\% \; HC_2H_3O_2$$

50. Mass percent calculation.

(a) $145.0 \text{ g NaOH} + 1500 \text{ g } H_2O = 1645 \text{ g solution}$

$$\left(\frac{145.0 \text{ g NaOH}}{1645 \text{ g solution}}\right)(100) = 8.815\% \text{ NaOH}$$

(b) 1.0 molal solution of $C_6H_{12}O_6 = \left(\dfrac{1 \text{ mol } C_6H_{12}O_6}{1000. \text{ g } H_2O}\right)$

$$(1.0 \text{ mol } C_6H_{12}O_6)\left(\frac{180.2 \text{ g}}{\text{mol}}\right) = 180 \text{ g } C_6H_{12}O_6$$

$$1000. \text{ g } H_2O + 180 \text{ g } C_6H_{12}O_6 = 1180 \text{ g solution}$$

$$\left(\frac{180 \text{ g } C_6H_{12}O_6}{1180 \text{ g solution}}\right)(100) = 15\% \; C_6H_{12}O_6$$

51. $(65 \text{ g solution})\left(\dfrac{5.0 \text{ g KCl}}{100. \text{ g solution}}\right) = 3.3 \text{ g KCl}$

52. $(250. \text{ g solution})\left(\dfrac{15.0 \text{ g } K_2CrO_4}{100. \text{ g solution}}\right) = 37.5 \text{ g } K_2CrO_4$

53. Mass/volume percent.

$$\left(\frac{22.0 \text{ g } CH_3OH}{100. \text{ mL solution}}\right)(100) = 22.0\% \; CH_3OH$$

54. Mass/volume percent.

$$\left(\frac{4.20 \text{ g NaCl}}{12.5 \text{ mL solution}}\right)(100) = 33.6\% \text{ NaCl}$$

55. Volume percent.

$$\left(\frac{10.0 \text{ mL } CH_3OH}{40.0 \text{ mL solution}}\right)(100) = 25.0\% \; CH_3OH$$

56. Volume percent.

$$\left(\frac{2.0 \text{ mL } C_6H_{14}}{9.0 \text{ mL solution}}\right)(100) = 22\% \; C_6H_{14}$$

57. Molarity problems $\left(\text{M} = \dfrac{\text{mol}}{\text{L}}\right)$

(a) $\left(\dfrac{0.10 \text{ mol}}{250 \text{ mL}}\right)\left(\dfrac{1000 \text{ mL}}{\text{L}}\right) = 0.40 \text{ M}$

(b) $\left(\dfrac{2.5 \text{ mol NaCl}}{0.650 \text{ L}}\right) = 3.8 \text{ M NaCl}$

(c) $\left(\dfrac{53.0 \text{ g } Na_2CrO_4}{1.00 \text{ L}}\right)\left(\dfrac{1 \text{ mol}}{162.0 \text{ g}}\right) = 0.327 \text{ M } Na_2CrO_4$

(d) $\left(\dfrac{260 \text{ g } C_6H_{12}O_6}{800. \text{ mL}}\right)\left(\dfrac{1000 \text{ mL}}{\text{L}}\right)\left(\dfrac{1 \text{ mol}}{180.2 \text{ g}}\right) = 1.8 \text{ M } C_6H_{12}O_6$

58. (a) $\left(\dfrac{0.025 \text{ mol HCl}}{10. \text{ mL}}\right)\left(\dfrac{1000 \text{ mL}}{\text{L}}\right) = 2.5 \text{ M HCl}$

(b) $\left(\dfrac{0.35 \text{ mol } BaCl_2 \cdot 2H_2O}{593 \text{ mL}}\right)\left(\dfrac{1000 \text{ mL}}{\text{L}}\right) = 0.59 \text{ M } BaCl_2 \cdot 2H_2O$

(c) $\left(\dfrac{1.5 \text{ g } Al_2(SO_4)_3}{2.00 \text{ L}}\right)\left(\dfrac{1 \text{ mol}}{342.2 \text{ g}}\right) = 2.19 \times 10^{-3} \text{ M } Al_2(SO_4)_3$

(d) $\left(\dfrac{0.0282 \text{ g } Ca(NO_3)_2}{1.00 \text{ mL}}\right)\left(\dfrac{1000 \text{ mL}}{\text{L}}\right)\left(\dfrac{1 \text{ mol}}{164.1 \text{ g}}\right) = 0.172 \text{ M } Ca(NO_3)_2$

59. $\text{Molarity} = \dfrac{\text{mol solute}}{\text{L solution}}$ or $\text{mol solute} = (\text{L solution})(\text{Molarity})$

(a) $(40.0 \text{ L})\left(\dfrac{1.0 \text{ mol LiCl}}{\text{L}}\right) = 40. \text{ mol LiCl}$

(b) $(25.0 \text{ mL})\left(\dfrac{1 \text{ L}}{1000 \text{ mL}}\right)\left(\dfrac{3.0 \text{ mol } H_2SO_4}{\text{L}}\right) = 0.0750 \text{ mol } H_2SO_4$

60. $\text{Molarity} = \dfrac{\text{mol solute}}{\text{L solution}}$ or $\text{mol solute} = (\text{L solution})(\text{Molarity})$

(a) $(349 \text{ mL})\left(\dfrac{1 \text{ L}}{1000 \text{ mL}}\right)\left(\dfrac{0.0010 \text{ mol NaOH}}{\text{L}}\right) = 3.5 \times 10^{-4} \text{ mol NaOH}$

(b) $(5000. \text{ mL})\left(\dfrac{1 \text{ L}}{1000 \text{ mL}}\right)\left(\dfrac{3.1 \text{ mol } CoCl_2}{\text{L}}\right) = 16 \text{ mol } CoCl_2$

61. (a) $(150 \text{ L})\left(\dfrac{1.0 \text{ mol NaCl}}{\text{L}}\right)\left(\dfrac{58.44 \text{ g}}{\text{mol}}\right) = 8.8 \times 10^3 \text{ g NaCl}$

(b) $(260 \text{ mL})\left(\dfrac{18 \text{ mol } H_2SO_4}{1000 \text{ mL}}\right)\left(\dfrac{98.09 \text{ g}}{\text{mol}}\right) = 4.6 \times 10^2 \text{ g } H_2SO_4$

62. (a) $(0.035\ \text{L})\left(\frac{10.0\ \text{mol HCl}}{\text{L}}\right)\left(\frac{36.46\ \text{g}}{\text{mol}}\right) = 13\ \text{g HCl}$

(b) $(8.00\ \text{mL})\left(\frac{1\ \text{L}}{1000\ \text{mL}}\right)\left(\frac{8.00\ \text{mol Na}_2\text{C}_2\text{O}_4}{\text{L}}\right)\left(\frac{134.0\ \text{g}}{\text{mol}}\right) = 8.58\ \text{g Na}_2\text{C}_2\text{O}_4$

63. (a) $(0.430\ \text{mol})\left(\frac{1\ \text{L}}{0.256\ \text{mol}}\right)\left(\frac{1000\ \text{mL}}{\text{L}}\right) = 1.68 \times 10^3\ \text{mL}$

(b) $(20.0\ \text{g KCl})\left(\frac{1\ \text{mol}}{74.55\ \text{g}}\right)\left(\frac{1\ \text{L}}{0.256\ \text{mol}}\right)\left(\frac{1000\ \text{mL}}{\text{L}}\right) = 1.05 \times 10^3\ \text{mL}$

64. (a) $(10.0\ \text{mol})\left(\frac{1\ \text{L}}{0.256\ \text{mol}}\right)\left(\frac{1000\ \text{mL}}{\text{L}}\right) = 3.91 \times 10^4\ \text{mL}$

(b) The conversion is: $\text{g Cl}^- \longrightarrow \text{mol Cl}^- \longrightarrow \text{mol KCl} \longrightarrow \text{L} \longrightarrow \text{mL}$

$$(71.0\ \text{g Cl}^-)\left(\frac{1\ \text{mol}}{35.45\ \text{g}}\right)\left(\frac{1\ \text{mol KCl}}{1\ \text{mol Cl}^-}\right)\left(\frac{1\ \text{L}}{0.256\ \text{mol KCl}}\right)\left(\frac{1000\ \text{mL}}{\text{L}}\right) = 7.82 \times 10^3\ \text{mL}$$

65. (a) First calculate the moles of HCl in each solution. Then calculate the molarity.

$(100.\ \text{mL})\left(\frac{1\ \text{L}}{1000\ \text{mL}}\right)\left(\frac{1.0\ \text{mol}}{\text{L}}\right) = 0.10\ \text{mol HCl}$

$(150.\ \text{mL})\left(\frac{1\ \text{L}}{1000\ \text{mL}}\right)\left(\frac{2.0\ \text{mol}}{\text{L}}\right) = 0.30\ \text{mol HCl}$

Total mol = 0.40 mol HCl

Total volume = 100. mL + 150. mL = 250. mL (0.250 L)

$\frac{0.40\ \text{mol HCl}}{0.250\ \text{L}} = 1.6\ \text{M HCl}$

(b) First calculate the moles of NaCl in each solution. Then calculate the molarity.

$(25.0\ \text{mL})\left(\frac{1\ \text{L}}{1000\ \text{mL}}\right)\left(\frac{1.25\ \text{mol}}{\text{L}}\right) = 0.0313\ \text{mol NaCl}$

$(75.0\ \text{mL})\left(\frac{1\ \text{L}}{1000\ \text{mL}}\right)\left(\frac{2.0\ \text{mol}}{\text{L}}\right) = 0.150\ \text{mol NaCl}$

Total mol = 0.181 mol NaCl

Total volume = 25.0 mL + 75.0 mL = 100. mL = 0.100 L

$$\frac{0.181 \text{ mol NaCl}}{0.100 \text{ L}} = 1.81 \text{ M NaCl}$$

66. Dilution problem

$V_1M_1 = V_2M_2$

(a) $V_1 = 200.\text{ mL} \qquad V_2 = 400.\text{ mL}$

$M_1 = 12\text{ M} \qquad M_2 = M_2$

$(200.\text{ mL})(12\text{ M}) = (400.\text{ mL})(M_2)$

$$M_2 = \frac{(200.\text{ mL})(12\text{ M})}{400.\text{ mL}} = 6.0\text{ M HCl}$$

(b) $V_1 = 60.0\text{ mL} \qquad V_2 = 560.\text{ mL}$

$M_1 = 0.60\text{ M} \qquad M_2 = M_2$

$(60.0\text{ mL})(0.60\text{ M}) = (560.\text{ mL})(M_2)$

$$M_2 = \frac{(60.0\text{ mL})(0.60\text{ M})}{560.\text{ mL}} = 0.064\text{ M } ZnSO_4$$

67. $V_1M_1 = V_2M_2$

(a) $(V_1)(15\text{ M}) = (50.\text{ mL})(6.0\text{ M})$

$$V_1 = \frac{(50.0\text{ mL})(6.0\text{ M})}{15\text{ M}} = 20.\text{ mL } 15\text{ M } NH_3$$

(b) $(V_1)(18\text{ M}) = (250\text{ mL})(10.00\text{ M})$

$$V_1 = \frac{(250\text{ mL})(10.00\text{ M})}{18\text{ M}} = 140\text{ mL } 18\text{ M } H_2SO_4$$

68. $V_1M_1 = V_2M_2$

(a) $(V_1)(12\text{ M}) = (400.\text{ mL})(6.0\text{ M})$

$$V_1 = \frac{(400.0\text{ mL})(6.0\text{ M})}{12\text{ M}} = 2.0 \times 10^2\text{ mL } 12\text{ M HCl}$$

(b) $(V_1)(16\text{ M}) = (100.\text{ mL})(2.5\text{ M})$

$$V_1 = \frac{(100.\text{ mL})(2.5\text{ M})}{16\text{ M}} = 16\text{ mL } 16\text{ M } HNO_3$$

69. $(0.25\ \text{L})\left(\dfrac{0.75\ \text{mol}}{\text{L}}\right) = 0.19\ \text{mol}\ H_2SO_4$

(a) Final volume after mixing

$250\ \text{mL} + 150\ \text{mL} = 400\ \text{mL} = 0.40\ \text{L}$

$$\frac{0.19\ \text{mol}\ H_2SO_4}{0.40\ \text{L}} = 0.48\ \text{M}\ H_2SO_4$$

(b) $(250\ \text{mL})\left(\dfrac{1\ \text{L}}{1000\ \text{mL}}\right)\left(\dfrac{0.70\ \text{mol}\ H_2SO_4}{\text{L}}\right) = 0.18\ \text{mol}\ H_2SO_4$

Total moles $= 0.19\ \text{mol} + 0.18\ \text{mol} = 0.37\ \text{mol}\ H_2SO_4$

Final volume $= 250.\ \text{mL} + 250.\ \text{mL} = 500.\ \text{mL} = 0.500\ \text{L}$

$$\frac{0.37\ \text{mol}\ H_2SO_4}{0.500\ \text{L}} = 0.74\ \text{M}\ H_2SO_4$$

70. $(0.250\ \text{L})\left(\dfrac{0.75\ \text{mol}}{\text{L}}\right) = 0.19\ \text{mol}\ H_2SO_4$

(a) $(400.\ \text{mL})\left(\dfrac{1\ \text{L}}{1000\ \text{mL}}\right)\left(\dfrac{2.50\ \text{mol}\ H_2SO_4}{\text{L}}\right) = 1.00\ \text{mol}\ H_2SO_4$

Total moles $= 0.19\ \text{mol} + 1.00\ \text{mol} = 1.19\ \text{mol}\ H_2SO_4$

Final volume $= 250.\ \text{mL} + 400.\ \text{mL} = 650.\ \text{mL} = 0.650\ \text{L}$

$$\frac{1.19\ \text{mol}\ H_2SO_4}{0.650\ \text{L}} = 1.83\ \text{M}\ H_2SO_4$$

(b) Final volume after mixing

$250.\ \text{mL} + 375\ \text{mL} = 625\ \text{mL} = 0.625\ \text{L}$

$$\frac{0.19\ \text{mol}\ H_2SO_4}{0.625\ \text{L}} = 0.30\ \text{M}\ H_2SO_4$$

71. $BaCl_2(aq) + K_2CrO_4(aq) \longrightarrow BaCrO_4(s) + 2KCl(aq)$

(a) mL $BaCl_2 \longrightarrow$ mol $BaCl_2 \longrightarrow$ mol $BaCrO_4 \longrightarrow$ g $BaCrO_4$

$$(100.0\ \text{mL}\ BaCl_2)\left(\frac{0.300\ \text{mol}}{1000\ \text{mL}}\right)\left(\frac{1\ \text{mol}\ BaCrO_4}{1\ \text{mol}\ BaCl_2}\right)\left(\frac{253.3\ \text{g}}{\text{mol}}\right) = 7.60\ \text{g}\ BaCrO_4$$

(b) mL $K_2CrO_4 \longrightarrow$ mol $K_2CrO_4 \longrightarrow$ mol $BaCl_2 \longrightarrow$ mL $BaCl_2$

$$(50.0 \text{ mL } K_2CrO_4)\left(\frac{0.300 \text{ mol}}{1000 \text{ mL}}\right)\left(\frac{1 \text{ mol } BaCl_2}{1 \text{ mol } K_2CrO_4}\right)\left(\frac{1000 \text{ mL}}{1.0 \text{ mol}}\right) = 15 \text{ mL of } 1.0 \text{ M } BaCl_2$$

72. $3\,MgCl_2(aq) + 2\,Na_3PO_4(aq) \longrightarrow Mg_3(PO_4)_2(s) + 6\,NaCl(aq)$

(a) mL $MgCl_2 \longrightarrow$ mol $MgCl_2 \longrightarrow$ mol $Na_3PO_4 \longrightarrow$ mL Na_3PO_4

$$(50.0 \text{ mL } MgCl_2)\left(\frac{0.250 \text{ mol}}{1000 \text{ mL}}\right)\left(\frac{2 \text{ mol } Na_3PO_4}{3 \text{ mol } MgCl_2}\right)\left(\frac{1000 \text{ mL}}{0.250 \text{ mol}}\right)$$
$$= 33.3 \text{ mL of } 0.250 \text{ M } Na_3PO_4$$

(b) mL $MgCl_2 \longrightarrow$ mol $MgCl_2 \longrightarrow$ mol $Mg_3(PO_4)_2 \longrightarrow$ g $Mg_3(PO_4)_2$

$$(50.0 \text{ mL } MgCl_2)\left(\frac{0.250 \text{ mol}}{1000 \text{ mL}}\right)\left(\frac{1 \text{ mol } Mg_3(PO_4)_2}{3 \text{ mol } MgCl_2}\right)\left(\frac{262.9 \text{ g}}{\text{mol}}\right) = 1.10 \text{ g } Mg_3(PO_4)_2$$

73. The balanced equation is
$6\,FeCl_2 + K_2Cr_2O_7 + 14\,HCl \longrightarrow 6\,FeCl_3 + 2\,CrCl_3 + 2\,KCl + 7\,H_2O$

(a) $$(2.0 \text{ mol } FeCl_2)\left(\frac{2 \text{ mol KCl}}{6 \text{ mol } FeCl_2}\right) = 0.67 \text{ mol KCl}$$

(b) $$(1.0 \text{ mol } FeCl_2)\left(\frac{2 \text{ mol } CrCl_3}{6 \text{ mol } FeCl_2}\right) = 0.33 \text{ mol } CrCl_3$$

(c) $$(0.050 \text{ mol } K_2Cr_2O_7)\left(\frac{6 \text{ mol } FeCl_2}{1 \text{ mol } K_2Cr_2O_7}\right) = 0.30 \text{ mol } FeCl_2$$

(d) $$(0.025 \text{ mol } FeCl_2)\left(\frac{1 \text{ mol } K_2Cr_2O_7}{6 \text{ mol } FeCl_2}\right)\left(\frac{1000 \text{ mL}}{0.060 \text{ mol}}\right) = 69 \text{ mL of } 0.060 \text{ M } K_2Cr_2O_7$$

(e) $$(15.0 \text{ mL } FeCl_2)\left(\frac{6.0 \text{ mol}}{1000 \text{ mL}}\right)\left(\frac{14 \text{ mol HCl}}{6 \text{ mol } FeCl_2}\right)\left(\frac{1000 \text{ mL}}{6.0 \text{ mol}}\right) = 35 \text{ mL of } 6.0 \text{ M HCl}$$

74. $2\,KMnO_4 + 16\,HCl \longrightarrow 2\,MnCl_2 + 5\,Cl_2 + 8\,H_2O + 2\,KCl$

(a) $$(0.050 \text{ mol } KMnO_4)\left(\frac{5 \text{ mol } Cl_2}{2 \text{ mol } KMnO_4}\right) = 0.13 \text{ mol } Cl_2$$

(b) $$(1.0 \text{ L } KMnO_4)\left(\frac{2.0 \text{ mol}}{\text{L}}\right)\left(\frac{16 \text{ mol HCl}}{2 \text{ mol } KMnO_4}\right) = 16 \text{ mol HCl}$$

(c) $(200.\ \text{mL KMnO}_4)\left(\dfrac{0.50\ \text{mol}}{1000\ \text{mL}}\right)\left(\dfrac{16\ \text{mol HCl}}{2\ \text{mol KMnO}_4}\right)\left(\dfrac{1000\ \text{mL}}{6.0\ \text{mol}}\right)$

$= 1.3 \times 10^2$ mL of 6 MHCl

(d) $(75.0\ \text{mL HCl})\left(\dfrac{6.0\ \text{mol}}{1000\ \text{mL}}\right)\left(\dfrac{5\ \text{mol Cl}_2}{16\ \text{mol HCl}}\right)\left(\dfrac{22.4\ \text{L}}{\text{mol}}\right) = 3.2\ \text{L Cl}_2$

75. Molality $= m = \dfrac{\text{mol solute}}{\text{kg solvent}}$

(a) $\left(\dfrac{14.0\ \text{g CH}_3\text{OH}}{100.\ \text{g H}_2\text{O}}\right)\left(\dfrac{1000\ \text{g}}{\text{kg}}\right)\left(\dfrac{1\ \text{mol}}{32.04\ \text{g}}\right) = \left(\dfrac{4.37\ \text{mol CH}_3\text{OH}}{\text{kg H}_2\text{O}}\right) = 4.37\ m\ \text{CH}_3\text{OH}$

(b) $\left(\dfrac{2.50\ \text{mol C}_6\text{H}_6}{250\ \text{g C}_6\text{H}_{14}}\right)\left(\dfrac{1000\ \text{g}}{\text{kg}}\right) = \left(\dfrac{10.\ \text{mol C}_6\text{H}_6}{\text{kg C}_6\text{H}_{14}}\right) = 10.\ m\ \text{C}_6\text{H}_6$

76. Molality $= m = \dfrac{\text{mol solute}}{\text{kg solvent}}$

(a) $\left(\dfrac{1.0\ \text{g C}_6\text{H}_{12}\text{O}_6}{1.0\ \text{g H}_2\text{O}}\right)\left(\dfrac{1000\ \text{g}}{\text{kg}}\right)\left(\dfrac{1\ \text{mol}}{180.2\ \text{g}}\right) = \left(\dfrac{5.5\ \text{mol C}_6\text{H}_{12}\text{O}_6}{\text{kg H}_2\text{O}}\right) = 5.5\ m\ \text{C}_6\text{H}_{12}\text{O}_6$

(b) $\left(\dfrac{0.250\ \text{mol I}_2}{1.0\ \text{kg H}_2\text{O}}\right) = 0.25\ m\ \text{I}_2$

77. (a) $\left(\dfrac{2.68\ \text{g C}_{10}\text{H}_8}{38.4\ \text{g C}_6\text{H}_6}\right)\left(\dfrac{1\ \text{mol}}{128.2\ \text{g C}_{10}\text{H}_8}\right)\left(\dfrac{1000\ \text{g C}_6\text{H}_6}{\text{kg}}\right) = 0.544\ m$

(b) K_f (for benzene) $= \dfrac{5.1°\text{C}}{m}$ Freezing point of benzene = 5.5°C

$\Delta t_f = (0.544\ m)\left(\dfrac{5.1°\text{C}}{m}\right) = 2.8°\text{C}$

Freezing point of solution = 5.5°C − 2.8°C = 2.7°C

(c) K_b (for benzene) $= \dfrac{2.53°\text{C}}{m}$ Boiling point of benzene = 80.1°C

$\Delta t_b = (0.544\ m)\left(\dfrac{2.53°\text{C}}{m}\right) = 1.38°\text{C}$

Boiling point of solution = 80.1°C + 1.38°C = 81.5°C

78. (a) $\left(\frac{100.0 \text{ g } C_2H_6O_2}{150.0 \text{ g } H_2O}\right)\left(\frac{1 \text{ mol}}{62.07 \text{ g}}\right)\left(\frac{1000 \text{ g}}{\text{kg}}\right) = 10.74\ m$

(b) $\Delta t_b = mK_b = (10.74\ m)\left(\frac{0.512°C}{m}\right) = 5.50°C$ (Increase in boiling point)

Boiling point $= 100.00°C + 5.50°C = 105.50°C$

(c) $\Delta t_f = mK_f = (10.74\ m)\left(\frac{1.86°C}{m}\right) = 20.0°C$ (Decrease in freezing point)

Freezing point $= 0.00°C - 20.0°C = -20.0°C$

79. Freezing point of acetic acid is 16.6°C K_f acetic acid $= \frac{3.90°C}{m}$

$\Delta t_f = 16.6°C - 13.2°C = 3.4°C$

$\Delta t_f = mK_f$

$m = \frac{3.4°C}{3.90°C/m} = 0.87\ m$

Convert 8.00 g unknown/60.0 g $HC_2H_3O_2$ to g/mol (molar mass)

Conversion: $\frac{\text{g unknown}}{\text{g } HC_2H_3O_2} \longrightarrow \frac{\text{g unknown}}{\text{kg } HC_2H_3O_2} \longrightarrow \frac{\text{g}}{\text{mol}}$

$\left(\frac{8.00 \text{ g unknown}}{60.0 \text{ g } HC_2H_3O_2}\right)\left(\frac{1000 \text{ g}}{\text{kg}}\right)\left(\frac{1 \text{ kg } HC_2H_3O_2}{0.87 \text{ mol unknown}}\right) = 153 \text{ g/mol}$

80. $\Delta t_f = 2.50°C \qquad K_f(\text{for } H_2O) = \frac{1.86°C}{m}$

$\Delta t_f = mK_f$

$m = \frac{2.50°C}{1.86°C/m} = 1.34\ m$

Convert 4.80 g unknown/22.0 g H_2O to g/mol (molar mass)

$\left(\frac{4.80 \text{ g unknown}}{22.0 \text{ g } H_2O}\right)\left(\frac{1000 \text{ g}}{\text{kg}}\right)\left(\frac{1 \text{ kg } H_2O}{1.34 \text{ mol unknown}}\right) = 163 \text{ g/mol}$

81. Salt (NaCl) is an ionic compound. When it is dissolved in water the sodium and chloride ions "break apart" or dissociate. The polar water molecules are attracted to the polar sodium and chloride ions and are hydrated (surrounded by water molecules). The sodium and chloride ions are separated from one another and

distributed throughout the water in this way. Na^+ and Cl^- are hydrated in aqueous solution: See Question 1.

82. Sugar molecules are not ionic; therefore they do not dissociate when dissolved in water. However, sugar molecules are polar, so water molecules are attracted to them and the sugar becomes hydrated. The water molecules help separate the sugar molecules from each other and distribute them throughout the water.

83. Sugar and salt behave differently when dissolved in water because salt is an ionic compound and sugar is a molecular compound.

84. mass of solute = mass of container & solute − mass of container − mass of water
mass of water = (5.549 moles)(18.02 g/mol) = 100.0 g
mass of solute = 563 g − 375 g − 100.0 g = 88 g
solubility in water = g solute/100 g H_2O
Using this data solubility is 88 g solute/100.0 g water = $NaNO_3$ (see Table 14.3)

85. An isotonic sodium chloride solution has the same osmotic pressure as human blood plasma. When blood cells are placed in an isotonic solution the osmotic pressure inside the cells is equal to the osmotic pressure outside the cells so there is no change in the appearance of the blood cells.

86. The $KMnO_4$ crystals give the solution its purple color. The purple streaks are formed because the solute has not been evenly distributed throughout that solvent yet. The MnO_4^- has a purple color in solution.

87. The line for KNO_3 slopes upward, because the solubility increases as the temperature increases. KNO_3 has the steepest slope of all the compounds given in the diagram. It exhibits the greatest increase in the number of grams of solute that is able to dissolve in 100 g of water than any other compound in the diagram as the temperature increases.

88. First calculate the g NaOH to neutralize the HCl.

$$\text{NaOH} + \text{HCl} \longrightarrow \text{NaCl} + \text{H}_2\text{O}$$

$$(0.15\ \text{L HCl})\left(\frac{1.0\ \text{mol}}{\text{L}}\right)\left(\frac{1\ \text{mol NaOH}}{1\ \text{mol HCl}}\right)\left(\frac{40.00\ \text{g}}{\text{mol}}\right) = 6.0\ \text{g NaOH required to neutralize the acid}$$

Now calculate the grams of 10% NaOH solution that contains 6.0 g NaOH

$$\frac{6.0\ \text{g NaOH}}{x} = \frac{10.0\ \text{g NaOH}}{100.0\ \text{g}\ 10.0\%\ \text{NaOH solution}}$$

$x = 60.$ g 10% NaOH solution

89. $1.0\ m\ \text{HCl} = \dfrac{1\ \text{mol HCl}}{1\ \text{kg}\ H_2O} = \dfrac{36.46\ \text{g HCl}}{1000\ \text{g}\ H_2O}$

Total mass of solution $= 1000\ \text{g} + 36.46\ \text{g} = 1036.46\ \text{g}$

Therefore, $1.0\ m\ \text{HCl} = \dfrac{1\ \text{mol HCl}}{1036.46\ \text{g HCl solution}}$

$NaOH + HCl \longrightarrow NaCl + H_2O$

Calculate the grams NaOH to neutralize HCl

$$(250.0\ \text{g solution})\left(\frac{1\ \text{mol HCl}}{1036.46\ \text{solution}}\right)\left(\frac{1\ \text{mol NaOH}}{1\ \text{mol HCl}}\right)\left(\frac{40.00\ \text{g}}{\text{mol}}\right) = 9.648\ \text{g NaOH}$$

Calculate the grams of 10.0% NaOH solution that contains 9.648 g NaOH.

$$\frac{9.648\ \text{g NaOH}}{x} = \frac{10.0\ \text{g NaOH}}{100.0\ \text{g}\ 10.0\%\ \text{NaOH solution}}$$

$x = 96.5\ \text{g}\ 10\%\ \text{NaOH solution}$

90. (a) $$(1.0\ \text{L syrup})\left(\frac{1000\ \text{mL}}{\text{L}}\right)\left(\frac{1.06\ \text{g}}{\text{mL}}\right)\left(\frac{15.0\ \text{g sugar}}{100.\ \text{g syrup}}\right) = 1.6 \times 10^2\ \text{g sugar}$$

(b) $$\left(\frac{1.6 \times 10^2\ \text{g}\ C_{12}H_{22}O_{11}}{\text{L}}\right)\left(\frac{1\ \text{mol}}{342.3\ \text{g}}\right) = 0.47\ \text{M}$$

(c) $m = \dfrac{\text{mol sugar}}{\text{kg}\ H_2O}$ 15% sugar by mass $= 15.0\ \text{g}\ C_{12}H_{22}O_{11} + 85.0\ \text{g}\ H_2O$

$$\left(\frac{15.0\ \text{g}\ C_{12}H_{22}O_{11}}{85.0\ \text{g}\ H_2O}\right)\left(\frac{1000\ \text{g}\ H_2O}{1\ \text{kg}\ H_2O}\right)\left(\frac{1\ \text{mol}}{342.3\ \text{g}\ C_{12}H_{22}O_{11}}\right) = 0.516\ m$$

91. $K_f = \dfrac{5.1^\circ\text{C}}{m}$ $\quad \Delta t_f = 0.614^\circ\text{C}$

$$\left(\frac{3.84\ \text{g}\ C_4H_2N}{250.0\ \text{g}\ C_6H_6}\right)\left(\frac{1000\ \text{g}}{\text{kg}}\right) = \frac{15.4\ \text{g}\ C_4H_2N}{\text{kg}\ C_6H_6}$$

$\Delta t_f = mK_f$

$$m = \frac{0.614^\circ\text{C}}{5.1^\circ\text{C}/m} = 0.12\ m = \frac{0.12\ \text{mol}\ C_4H_2N}{\text{kg}\ C_6H_6}$$

$$\left(\frac{15.4\ \text{g}\ C_4H_2N}{\text{kg}\ C_6H_6}\right)\left(\frac{1\ \text{kg}\ C_6H_6}{0.12\ \text{mol}\ C_4H_2N}\right) = 128\ \text{g/mol} = 1.3 \times 10^2\ \text{g/mol}$$

Empirical mass (C_4H_2N) $= 64.07$ g

$$\frac{130 \text{ g}}{64.07 \text{ g}} = 2.0 \text{ (number of empirical formulas per molecular formula)}$$

Therefore, the molecular formula is twice the empirical formula, or $C_8H_4N_2$.

92. $$(12.0 \text{ mol HCl})\left(\frac{36.46 \text{ g}}{\text{mol}}\right) = 438 \text{ g HCl in } 1.00 \text{ L solution}$$

$$(1.00 \text{ L})\left(\frac{1.18 \text{ g solution}}{\text{mL}}\right)\left(\frac{1000 \text{ mL}}{\text{L}}\right) = 1180 \text{ g solution}$$

$$1180 \text{ g solution} - 438 \text{ g HCl} = 742 \text{ g } H_2O \text{ } (0.742 \text{ kg } H_2O)$$

$$\text{Since molality} = \frac{\text{mol HCl}}{\text{kg } H_2O} = \frac{12.0 \text{ mol HCl}}{0.742 \text{ kg } H_2O} = 16.2\ m \text{ HCl}$$

93. First calculate the g KNO_3 in the solution.

$$\text{The conversion is: } \frac{\text{mg } K^+}{\text{mL}} \longrightarrow \frac{\text{g } K^+}{\text{mL}} \longrightarrow \frac{\text{g } KNO_3}{\text{mL}} \longrightarrow \text{g } KNO_3$$

$$\left(\frac{5.5 \text{ mg } K^+}{\text{mL}}\right)\left(\frac{1 \text{ g}}{1000 \text{ mg}}\right)\left(\frac{101.1 \text{ g } KNO_3}{39.10 \text{ g } K^+}\right)(450 \text{ mL}) = 6.4 \text{ g } KNO_3$$

Now calculate the mol KNO_3 and the molarity.

$$(6.4 \text{ g } KNO_3)\left(\frac{1 \text{ mol}}{101.1 \text{ g}}\right) = 0.063 \text{ mol } KNO_3$$

$$\frac{0.063 \text{ mol } KNO_3}{0.450 \text{ L}} = 0.14 \text{ M}$$

94. $$(25.0 \text{ g KCl})\left(\frac{100. \text{ g solution}}{5.50 \text{ g KCl}}\right) = 455 \text{ g solution}$$

Alternate solution:

$$\left(\frac{25.0 \text{ g KCl}}{x}\right) = \left(\frac{5.50 \text{ g KCl}}{100. \text{ g solution}}\right)$$

$x = 455$ g solution

95. Verification of K_b for water

$$\Delta t_b = mK_b \quad \Delta t_b = 101.62°\text{C} - 100°\text{C} = 1.62°\text{C} \qquad K_b = \frac{\Delta t_b}{m}$$

First calculate the molality of the solution.

$$m = \frac{16.10 \text{ g } C_2H_6O_2}{(62.07 \text{ g/mol})(0.0820 \text{ kg } H_2O)} = \frac{3.16 \text{ mol } C_2H_6O_2}{\text{kg } H_2O}$$

$$K_b = \frac{\Delta t_b}{m} = \frac{1.62°C}{3.16 \text{ mol/kg } H_2O} = \frac{0.513°C \text{ kg } H_2O}{\text{mol}}$$

96. (a) $(500.0 \text{ mL solution})\left(\frac{0.90 \text{ g NaCl}}{100. \text{ mL solution}}\right) = 4.5 \text{ g NaCl}$

(b) $\left(\frac{4.5 \text{ g NaCl}}{x \text{ mL}}\right)(100) = 9.0\%$ $\quad x$ = volume of 9.0% solution

$$x = \frac{4.5 \text{ g NaCl}}{9.0\%} = 50. \text{ mL (4.5 g NaCl in solution)}$$

$500. \text{ mL} - 50. \text{ mL} = 450. \text{ mL } H_2O$ must evaporate

97. From Figure 14.4, the solubility of KNO_3 in H_2O at 20°C is 32 g per 100. g H_2O.

$(50.0 \text{ g } KNO_3)\left(\frac{100. \text{ g } H_2O}{32.0 \text{ g } KNO_3}\right) = 156 \text{ g } H_2O$ to produce a saturated solution.

$175 \text{ g } H_2O - 156 \text{ g } H_2O = 19 \text{ g } H_2O$ must be evaporated.

98. $(150 \text{ mL alcohol})\left(\frac{100. \text{ mL solution}}{70.0 \text{ mL alcohol}}\right) = 210 \text{ mL solution}$

99. (a) $(1.00 \text{ L solution})\left(\frac{1000 \text{ mL solution}}{\text{L solution}}\right)\left(\frac{1.21 \text{ g}}{\text{mL}}\right)\left(\frac{35.0 \text{ g } HNO_3}{100. \text{ g solution}}\right) = 424 \text{ g } HNO_3$

(b) $(500. \text{ g } HNO_3)\left(\frac{1000 \text{ mL solution}}{424 \text{ g } HNO_3}\right)\left(\frac{1.00 \text{ L}}{1000 \text{ mL}}\right) = 1.18 \text{ L solution}$

100. Assume 1.000 L (1000. mL) of solution

$$\left(\frac{1000. \text{ mL}}{\text{L}}\right)\left(\frac{1.21 \text{ g solution}}{\text{mL}}\right)\left(\frac{35.0 \text{ g } HNO_3}{100. \text{ g solution}}\right)\left(\frac{1 \text{ mol}}{63.02 \text{ g}}\right) = 6.72 \text{ M } HNO_3$$

101. First calculate the molarity of the solution

$$\left(\frac{80.0 \text{ g } H_2SO_4}{500. \text{ mL}}\right)\left(\frac{1000 \text{ mL}}{\text{L}}\right)\left(\frac{1 \text{ mol}}{98.09 \text{ g}}\right) = 1.63 \text{ M } H_2SO_4$$

$M_1V_1 = M_2V_2$

$$(1.63\text{ M})(500.\text{ mL}) = (0.10\text{ M})(V_2)$$

$$V_2 = \frac{(1.63\text{ M})(500.\text{ mL})}{0.10\text{ M}} = 8.2 \times 10^3\text{ mL} = 8.2\text{ L}$$

102. Note that the problem asks for the volume of water to be added, not the final volume of the solution.

$$M_1V_1 = M_2V_2$$

$$(1.40\text{ M})(300.\text{ mL}) = (0.500\text{ M})(V_2)$$

$$V_2 = \frac{(1.40\text{ M})(300.\text{ mL})}{0.500\text{ M}} = 840.\text{ mL (final volume)}$$

840. mL − 300. mL = 540. mL water to be added

103. $M_1V_1 = M_2V_2$

$$(16\text{ M})(10.0\text{ mL}) = (M_2)(500.\text{ mL})$$

$$M_2 = \frac{(16\text{ M})(10.0\text{ mL})}{500.0\text{ M}} = 0.32\text{ M } HNO_3$$

104. $(V_1)(5.00\text{ M}) = (250.\text{ mL})(0.625\text{ M})$

$$V_1 = \frac{(250.\text{ mL})(0.625\text{ M})}{5.00\text{ M}} = 31.3\text{ mL } 5.00\text{ M KOH}$$

To make 250. mL of 0.625 M KOH, take 31.3 mL of 5.00 M KOH and dilute with water to a volume of 250. mL.

105. $Mg + 2\,HCl \longrightarrow MgCl_2 + H_2(g)$

(a) mL HCl $\longrightarrow$ mol HCl $\longrightarrow$ mol H_2

$$(200.0\text{ mL HCl})\left(\frac{3.00\text{ mol}}{1000\text{ mL}}\right)\left(\frac{1\text{ mol } H_2}{2\text{ mol HCl}}\right) = 0.300\text{ mol } H_2$$

(b) $PV = nRT$

$$P = (720\text{ torr})\left(\frac{1\text{ atm}}{760\text{ torr}}\right) = 0.95\text{ atm}$$

$$T = 27^\circ\text{C} = 300.\text{ K}$$

$$n = 0.300\text{ mol}$$

$$V = \frac{nRT}{P} = \frac{(0.300\text{ mol})(0.0821\text{ L atm/mol K})(300.\text{ K})}{0.95\text{ atm}} = 7.8\text{ L } H_2$$

106. $Mg + 2\,HCl \longrightarrow MgCl_2 + H_2(g)$

$L\ H_2 \longrightarrow mol\ H_2 \longrightarrow mol\ HCl \longrightarrow M\ HCl$

$$(3.50\ L\ H_2)\left(\frac{1\ mol}{22.4\ L}\right)\left(\frac{2\ mol\ HCl}{1\ mol\ H_2}\right)\left(\frac{1}{0.150\ L}\right) = 2.08\ M\ HCl$$

107. Use $M_1V_1 = M_2V_2$

$$1\ drop = \frac{1}{20.}\ mL = 0.050\ mL$$

$100.\ mL + 0.050\ mL = 100.050\ mL$

$(17.8\ M)(0.050\ mL) = (M)(100.050\ mL)$

$$M = \frac{(17.8\ M)(0.050\ mL)}{100.050\ mL} = 8.9 \times 10^{-3}\ M$$

108. $Mg(OH)_2 + 2\,HCl \longrightarrow MgCl_2 + 2\,H_2O$

$Al(OH)_3 + 3\,HCl \longrightarrow AlCl_3 + 3\,H_2O$

Calculate the moles of HCl neutralized by each base.

$$(12.0\ g\ Mg(OH)_2)\left(\frac{1\ mol}{58.33\ g}\right)\left(\frac{2\ mol\ HCl}{1\ mol\ Mg(OH)_2}\right) = 0.400\ mol\ HCl$$

$$(10.0\ g\ Al(OH)_3)\left(\frac{1\ mol}{78.00\ g}\right)\left(\frac{3\ mol\ HCl}{1\ mol\ Al(OH)_3}\right) = 0.385\ mol\ HCl$$

12.0 g $Mg(OH)_2$ reacts with more HCl than 10.0 g $Al(OH)_3$. Therefore, $Mg(OH)_2$ is more effective in neutralizing stomach acid.

109. (a) With equal masses of CH_3OH and C_2H_5OH, the substance with the lower molar mass will represent more moles of solute in solution. Therefore, the CH_3OH will be more effective than C_2H_5OH as an antifreeze.

(b) Equal molal solutions will lower the freezing point of the solution by the same amount.

110. Calculate molarity and molality. Assume 1000 mL of solution to calculate the amounts of H_2SO_4 and H_2O in the solution.

$$(1000\ mL\ solution)\left(\frac{1.29\ g}{mL}\right) = 1.29 \times 10^3\ g\ solution$$

$$(1.29 \times 10^3\ g\ solution)\left(\frac{38\ g\ H_2SO_4}{100\ g\ solution}\right) = 4.9 \times 10^2\ g\ H_2SO_4$$

1.29×10^3 g solution $-$ 4.9×10^2 g H_2SO_4 $=$ 8.0×10^2 g H_2O in the solution

$$m = \left(\frac{490 \text{ g } H_2SO_4}{8.0 \times 10^2 \text{ g } H_2O}\right)\left(\frac{1000 \text{ g}}{\text{kg}}\right)\left(\frac{1 \text{ mol}}{98.09 \text{ g}}\right) = 6.2\ m\ H_2SO_4$$

$$\text{M} = \left(\frac{4.9 \times 10^2 \text{ g } H_2SO_4}{\text{L}}\right)\left(\frac{1 \text{ mol}}{98.09 \text{ g}}\right) = 5.0 \text{ M } H_2SO_4$$

111. 1.00 lb $=$ 453.6 g sugar ($C_{12}H_{22}O_{11}$)

$$(4.00 \text{ lb } H_2O)\left(\frac{453.6 \text{ g}}{\text{lb}}\right) = 1.81 \times 10^3 \text{ g } H_2O \text{ (1.81 kg } H_2O)$$

$$(453.6 \text{ g } C_{12}H_{22}O_{11})\left(\frac{1 \text{ mol}}{342.3 \text{ g}}\right) = 1.33 \text{ mol } C_{12}H_{22}O_{11}$$

K_f (for H_2O) $=$ 1.86°C kg solvent/mol solute

$$\Delta t_f = mK_f = \left(\frac{1.33 \text{ mol } C_{12}H_{22}O_{11}}{1.81 \text{ kg } H_2O}\right)\left(\frac{1.86°\text{C kg } H_2O}{\text{mol } C_{12}H_{22}O_{11}}\right) = 1.37°\text{C}$$

Freezing point of solution $=$ 0°C $-$ 1.37°C $=$ -1.37°C $=$ 29.5°F

If the sugar solution is placed outside, where the temperature is 20°F, the solution will freeze.

112. Freezing point depression is 5.4°C

(a) $\Delta t_f = mK_f$

$$m = \frac{\Delta t_f}{K_f} = \frac{5.4°\text{C}}{1.86°\text{C kg solvent/mol solute}} = 2.9\ m$$

(b) $$K_b \text{ (for } H_2O) = \frac{0.512°\text{C kg solvent}}{\text{mol solute}} = \frac{0.512°\text{C}}{m}$$

$$\Delta t_b = mK_b = (2.9\ m)\left(\frac{0.512°\text{C}}{m}\right) = 1.5°\text{C}$$

Boiling point $=$ 100°C $+$ 1.5°C $=$ 101.5°C

113. Freezing point depression $=$ 0.372°C $\quad K_f = \dfrac{1.86°\text{C}}{m}$

$$\Delta t_f = mK_f$$

$$m = \frac{0.372°\text{C}}{1.86°\text{C}/m} = 0.200\ m$$

$$(6.20 \text{ g } C_2H_6O_2)\left(\frac{1 \text{ mol}}{62.07 \text{ g}}\right) = 0.100 \text{ mol } C_2H_6O_2$$

$$(0.100 \text{ mol } C_2H_6O_2)\left(\frac{1 \text{ kg } H_2O}{0.200 \text{ mol } C_2H_6O_2}\right)\left(\frac{1000 \text{ g } H_2O}{\text{kg } H_2O}\right) = 500. \text{ g } H_2O$$

114. (a) Freezing point depression = 20.0°C

$$12.0 \text{ L } H_2O\left(\frac{1000 \text{ mL}}{\text{L}}\right)\left(\frac{1.00 \text{ g}}{\text{mL}}\right) = 1.20 \times 10^4 \text{ g } H_2O$$

$$\Delta t_f = mK_f$$

$$m = \frac{20.0°\text{C}}{1.86°\text{C}/m} = 10.8\ m$$

$$(1.20 \times 10^4 \text{ g } H_2O)\left(\frac{10.8 \text{ mol } C_2H_6O_2}{1000 \text{ g } H_2O}\right)\left(\frac{62.07 \text{ g}}{\text{mol}}\right) = 8.04 \times 10^3 \text{ g } C_2H_6O_2$$

(b) $$(8.04 \times 10^3 \text{ g } C_2H_6O_2)\left(\frac{1.00 \text{ mL}}{1.11 \text{ g}}\right) = 7.24 \times 10^3 \text{ mL } C_2H_6O_2$$

(c) °F = 1.8 (°C) + 32 = 1.8 (−20.0) + 32 = −4.0°F

115. Yes, a saturated solution can also be a dilute solution. For example, the solubility of AgCl in water at 25°C is 1.3×10^{-5} mol/L. Thus, the solution formed by dissolving AgCl in water is both saturated and very dilute.

116. $HCl + NaOH \longrightarrow NaCl + H_2O$

1 mol 1 mol

g NaOH ⟶ mol NaOH ⟶ mol HCl ⟶ L HCl

$$(12 \text{ g NaOH})\left(\frac{1 \text{ mol}}{40.00 \text{ g}}\right)\left(\frac{1 \text{ mol HCl}}{1 \text{ mol NaOH}}\right)\left(\frac{1 \text{ L HCl}}{0.65 \text{ mol HCl}}\right) = 0.46 \text{ L HCl (460 mL)}$$

117. $HNO_3 + NaHCO_3 \longrightarrow NaNO_3 + H_2O + CO_2$

First calculate the grams of $NaHCO_3$ in the sample.

mL HNO_3 ⟶ L HNO_3 ⟶ mol HNO_3 ⟶ mol $NaHCO_3$ ⟶ g $NaHCO_3$

$$(150 \text{ mL } HNO_3)\left(\frac{1 \text{ L}}{1000 \text{ mL}}\right)\left(\frac{0.055 \text{ mol}}{\text{L}}\right)\left(\frac{1 \text{ mol } NaHCO_3}{1 \text{ mol } HNO_3}\right)\left(\frac{84.01 \text{ g}}{\text{mol}}\right)$$

$$= 0.69 \text{ g } NaHCO_3 \text{ in the sample}$$

$$\left(\frac{0.69 \text{ g NaHCO}_3}{1.48 \text{ g sample}}\right)(100) = 47\% \text{ NaHCO}_3$$

118. (a) Dilution problem: $M_1V_1 = M_2V_2$

$$(1.5 \text{ M})(8.4 \text{ L}) = (17.8 \text{ M})(V_2)$$

$$V_2 = \frac{(1.5 \text{ M})(8.4 \text{ L})}{17.8 \text{ M}} = 0.71 \text{ L}$$

0.71 L of 17.8 M H_2SO_4 is to be diluted to 8.4 L.

8.4 L − 0.71 L = 7.7 L H_2O must be added (assume volumes are additive)

(b) $\left(\frac{17.8 \text{ mol}}{1000. \text{ mL}}\right)(1.00 \text{ mL}) = 0.0178 \text{ mol}$

(c) $\left(\frac{1.5 \text{ mol}}{1000. \text{ mL}}\right)(1.00 \text{ mL}) = 0.0015 \text{ mol } H_2SO_4$ in each mL

119. Freezing point depression is 3.6°C

$$\Delta t_f = mK_f$$

$$m = \frac{\Delta t_f}{K_f} = \frac{3.6°\text{C}}{1.86°\text{C}/m} = 1.9\ m \text{ solution}$$

$$\Delta t_b = mK_b = (1.9\ m)\left(\frac{0.512°\text{C}}{m}\right) = 0.97°\text{C}$$

Boiling point = 100.00°C + 0.97°C = 100.97°C

120. moles HNO_3 total = moles HNO_3 from 3.00 M + moles HNO_3 from 12.0 M

$$M_TV_T = M_{3.00\text{ M}}V_{3.00\text{ M}} + M_{12.0\text{ M}}V_{12.0\text{ M}}$$

Assume preparation of 1000. mL of 6 M solution

Let y = volume of 3.00 M solution; volume of 12.0 M = 1000. mL − y

$$(6.00 \text{ M})(1000. \text{ mL}) = (3.00 \text{ M})(y) + (12.0 \text{ M})(1000. \text{ mL} - y)$$

$$6000. \text{ mL} = 3.00\ y \text{ mL} + 12{,}000 \text{ mL} - 12.0\ y$$

$$6000. \text{ mL} = 9.00\ y \qquad y = \frac{6000. \text{ mL}}{9.00} = 667 \text{ mL } 3 \text{ M}$$

1000. mL − 667 mL = 333 mL 12 M

Mix together 667 mL 3.00 M HNO_3 and 333 mL of 12.0 M HNO_3 to get 1000. mL of 6.00 M HNO_3.

121. $HBr + NaOH \longrightarrow NaBr + H_2O$
First calculate the molarity of the diluted HBr solution.
The reaction is 1 mol HBr to 1 mol NaOH, so

$$M_AV_A = M_BV_B$$

$$(M_A)(100.0\text{ mL}) = (0.37\text{ M})(88.4\text{ mL})$$

$$M_A = \frac{(0.37\text{ M})(88.4\text{ mL})}{100.00\text{ mL}} = 0.33\text{ M HBr (diluted solution)}$$

Now calculate the molarity of the HBr before dilution.

$$M_1V_1 = M_2V_2$$

$$(M_1)(20.0\text{ mL}) = (0.33\text{ M})(240.\text{ mL})$$

$$M_1 = \frac{(0.33\text{ M})(240.\text{ mL})}{20.0\text{ mL}} = 4.0\text{ M HBr (original solution)}$$

122. $Ba(NO_3)_2 + 2\ KOH \longrightarrow Ba(OH)_2 + 2\ KNO_3$

This is a limiting reactant problem. First calculate the moles of each reactant and determine the limiting reactant.

$$M \times L = \left(\frac{\text{moles}}{L}\right)(L) = \text{moles}$$

$$\left(\frac{0.642\text{ mol}}{L}\right)(0.0805\text{ L}) = 0.0517\text{ mol Ba(NO}_3)_2$$

$$\left(\frac{0.743\text{ mol}}{L}\right)(0.0445\text{ L}) = 0.0331\text{ mol KOH}$$

According to the equation, twice as many moles of KOH as $Ba(NO_3)_2$ are needed, so KOH is the limiting reactant.

$$(0.0331\text{ mol KOH})\left(\frac{1\text{ mol Ba(OH)}_2}{2\text{ mol KOH}}\right)\left(\frac{171.3\text{ g}}{\text{mol}}\right) = 2.84\text{ g Ba(OH)}_2\text{ is formed}$$

123. $(300.\text{ g solution})\left(\frac{5.0\text{ g sucrose}}{100.\text{ g solution}}\right) = 15\text{ g sucrose}$

$$(y\text{ g 2.0\% solution})\left(\frac{2.0\text{ g sucrose}}{100.\text{ g solution}}\right) = 15\text{ g sucrose}$$

$$y = \left(\frac{100.\text{ g solution}}{2.0\text{ g sucrose}}\right)(15\text{ g sucrose}) = 750\text{ g 2\% solution}$$

124. (a) $\left(\frac{0.25 \text{ mol}}{\text{L}}\right)(0.0458 \text{ L}) = 0.011 \text{ mol } Li_2CO_3$

(b) $\left(\frac{0.25 \text{ mol}}{\text{L}}\right)(0.75 \text{ L})\left(\frac{73.89 \text{ g}}{\text{mol}}\right) = 14 \text{ g } Li_2CO_3$

(c) $(6.0 \text{ g } Li_2CO_3)\left(\frac{1 \text{ mol}}{73.89 \text{ g}}\right)\left(\frac{1000. \text{ mL}}{0.25 \text{ mol}}\right) = 3.2 \times 10^2 \text{ mL solution}$

(d) Assume 1000. mL solution

$$\left(\frac{1.22 \text{ g}}{\text{mL}}\right)(1000. \text{ mL}) = 1220 \text{ g solution}$$

$$\left(\frac{0.25 \text{ mol}}{\text{L}}\right)\left(\frac{73.89 \text{ g } Li_2CO_3}{\text{mol}}\right) = 18 \text{ g } Li_2CO_3 \text{ per L solution}$$

$$\% = \left(\frac{\text{g solute}}{\text{g solution}}\right)(100) = \left(\frac{18 \text{ g}}{1220 \text{ g}}\right)(100) = 1.5\%$$

125. Calculate the total moles of HCl and divide by the total volume.

$$\left(\frac{0.35 \text{ mol}}{\text{L}}\right)(0.4000 \text{ L}) = 0.14 \text{ mol HCl}$$

$$\left(\frac{0.65 \text{ mol}}{\text{L}}\right)(1.1 \text{ L}) = \frac{0.72 \text{ mol HCl}}{0.86 \text{ mol HCl (total mol HCl)}}$$

Total volume $= 1.1 \text{ L} + 0.40 \text{ L} = 1.5 \text{ L}$

$$\text{Molarity} = \frac{0.86 \text{ mol HCl}}{1.5 \text{ L}} = 0.57 \text{ M HCl}$$

CHAPTER 15

ACIDS, BASES, AND SALTS

1. The Arrhenius definition is restricted to aqueous solutions, while the Brønsted-Lowry definition is not.

2. An electrolyte must be present in the solution for the bulb to glow.

3. Electrolytes include acids, bases, and salts.

4. First, the orientation of the polar water molecules about the Na^+ and Cl^- is different. The positive end (hydrogen) of the water molecule is directed towards Cl^-, while the negative end (oxygen) of the water molecule is directed towards the Na^+. Second, more water molecules will fit around Cl^-, since it is larger than the Na^+ ion.

5. The pH for a solution with a hydrogen ion concentration of 0.003 M will be between 2 and 3.

6. Tomato juice is more acidic than blood, since its pH is lower.

7. By the Arrhenius theory, an acid is a substance that produces hydrogen ions in aqueous solution. A base is a substance that produces hydroxide ions in aqueous solution.

   By the Brønsted-Lowry theory, an acid is a proton donor, while a base accepts protons. Since a proton is a hydrogen ion, then the two theories are very similar for acids, but not bases. A chloride ion can accept a proton (producing HCl), so it is a Brønsted-Lowry base, but would not be a base by the Arrhenius theory, since it does not produce hydroxide ions.

   By the Lewis theory, an acid is an electron pair acceptor, and a base is an electron pair donor. Many individual substances would be similarly classified as bases by Brønsted-Lowry or Lewis theories, since a substance with an electron pair to donate, can accept a proton. But, the Lewis definition is almost exclusively applied to reactions where the acid and base combine into a single molecule. The Brønsted-Lowry definition is usually applied to reactions that involve a transfer of a proton from the acid to the base. The Arrhenius definition is most often applied to individual substances, not to reactions. According to the Arrhenius theory, neutralization involves the reaction between a hydrogen ion and a hydroxide ion to form water.

   Neutralization, according to the Brønsted-Lowry theory, involves the transfer of a proton to a negative ion. The formation of a covalent bond constitutes a Lewis neutralization.

8. Neutralization reactions:

   Arrhenius: $HCl + NaOH \longrightarrow NaCl + H_2O$ $\quad (H^+ + OH^- \longrightarrow H_2O)$
   Brønsted-Lowry: $HCl + KCN \longrightarrow HCN + KCl$ $\quad (H^+ + CN^- \longrightarrow HCN)$
   Lewis: $AlCl_3 + NaCl \longrightarrow AlCl_4^- + Na^+$

```
  ••                             ••
 :Cl:                           :Cl:      -
  ••  ••        ••      -    ••  ••  ••
  Al:Cl:  +  [:Cl:]   →   [:Cl:Al:Cl:]
  ••  ••        ••           ••  ••  ••
 :Cl:                           :Cl:
  ••                             ••
```

9. (a) $[:\ddot{\underset{\cdot\cdot}{Br}}:]^-$ (b) $[:\ddot{\underset{\cdot\cdot}{O}}:H]^-$ (c) $[:C:::N:]^-$

   These ions are considered to be bases according to the Brønsted-Lowry theory, because they can accept a proton at any of their unshared pairs of electrons. They are considered to be bases according to the Lewis acid-base theory, because they can donate an electron pair.

10. The electrolytic compounds are acids, bases, and salts.

11. Names of the compounds in Table 15.3

H_2SO_4	sulfuric acid	$HC_2H_3O_2$	acetic acid
HNO_3	nitric acid	H_2CO_3	carbonic acid
HCl	hydrochloric acid	HNO_2	nitrous acid
HBr	hydrobromic acid	H_2SO_3	sulfurous acid
$HClO_4$	perchloric acid	H_2S	hydrosulfuric acid
$NaOH$	sodium hydroxide	$H_2C_2O_4$	oxalic acid
KOH	potassium hydroxide	H_3BO_3	boric acid
$Ca(OH)_2$	calcium hydroxide	$HClO$	hypochlorous acid
$Ba(OH)_2$	barium hydroxide	NH_3	ammonia
		HF	hydrofluoric acid

12. Hydrogen chloride dissolved in water conducts an electric current. HCl reacts with polar water molecules to produce H_3O^+ and Cl^- ions, which conduct electric current. Hexane is a nonpolar solvent, so it cannot pull the HCl molecules apart. Since there are no ions in the hexane solution, it does not conduct an electric current. HCl does not ionize in hexane.

13. In their crystalline structure, salts exist as positive and negative ions in definite geometric arrangement to each other, held together by the attraction of the opposite charges. When dissolved in water, the salt dissociates as the ions are pulled away from each other by the polar water molecules.

14. Testing the electrical conductivity of the solutions shows that CH_3OH is a nonelectrolyte, while NaOH is an electrolyte. This indicates that the OH group in CH_3OH must be covalently bonded to the CH_3 group.

15. Molten NaCl conducts electricity because the ions are free to move. In the solid state, however, the ions are immobile and do not conduct electricity.

16. Dissociation is the separation of already existing ions in an ionic compound. Ionization is the formation of ions from molecules. The dissolving of NaCl is a dissociation, since the ions already exist in the crystalline compound. The dissolving of HCl in water is an ionization process, because ions are formed from HCl molecules and H_2O.

17. Strong electrolytes are those which are essentially 100% ionized or dissociated in water. Weak electrolytes are those which are only slightly ionized in water.

18. Ions are hydrated in solution because there is an electrical attraction between the charged ions and the polar water molecules.

19. The main distinction between water solutions of strong and weak electrolytes is the degree of ionization of the electrolyte. A solution of an electrolyte contains many more ions than a solution of a nonelectrolyte. Strong electrolytes are essentially 100% ionized. Weak electrolytes are only slightly ionized in water.

20. (a) In a neutral solution, the concentration of H^+ and OH^- are equal.
    (b) In an acid solution, the concentration of H^+ is greater than the concentration of OH^-.
    (c) In a basic solution, the concentration of OH^- is greater than the concentration of H^+.

21. The net ionic equation for an acid-base reaction in aqueous solutions is:
    $H^+ + OH^- \longrightarrow H_2O$.

22. The HCl molecule is polar and, consequently, is much more soluble in the polar solvent, water, than in the nonpolar solvent, hexane. There is also a chemical reaction between HCl and H_2O molecules. $HCl + H_2O \longrightarrow H_3O^+ + Cl^-$

23. Pure water is neutral because when it ionizes it produces equal molar concentrations of acid $[H^+]$ and base $[OH^-]$ ions.

24. The fundamental difference between a colloidal dispersion and a true solution lies in the size of the particles. In a true solution particles are usually ions or hydrated molecules and are less than 1 nm in size. In colloidals the particles are aggregates of ions or molecules, ranging in size from 1-1000 nm.

25. Dialysis is the process of removing dissolved solutes from a colloidal dispersion by use of a dialyzing membrane. The dissolved solutes pass through the membrane leaving the colloidal dispersion behind. Dialysis is used in artificial kidneys to remove soluble waste products from the blood.

26. A neutral solution is one in which the concentration of acid is equal to the concentration of base ($[H^+] = [OH^-]$). An acidic solution is one in which the concentration of acid is greater than the concentration of base ($[H^+] > [OH^-]$). A basic solution is one in which the concentration of base is greater than the concentration of acid ($[H^+] < [OH^-]$).

27. Acid rain is caused by the release of nitrogen and sulfur oxides into the air. When these oxides are carried through the atmosphere they react with water and form sulfuric acid (H_2SO_4) and nitric acid (HNO_3). Precipitation (rain or snow) carries the acids to the ground.

28. A titration is used to determine the concentration of a specific substance (often an acid or a base) in a sample. A titration determines the volume of a reagent of known concentration that is required to completely react with a volume of a sample of unknown concentration. An indicator is used to help visualize the endpoint of a titration. The endpoint is the point at which enough of the reagent of known concentration has been added to the sample of unknown concentration to completely react with the unknown solution. An indicator color change is visible when the endpoint has been reached.

29. (a) $H_2SO_4 - HSO_4^-$; $H_2C_2H_3O_2^+ - HC_2H_3O_2$
    (b) step 1: $H_2SO_4 - HSO_4^-$; $H_3O^+ - H_2O$
    step 2: $HSO_4^- - SO_4^{2-}$; $H_3O^+ - H_2O$
    (c) $HClO_4 - ClO_4^-$; $H_3O^+ - H_2O$
    (d) $H_3O^+ - H_2O$; $CH_3OH - CH_3O^-$

30. Conjugate acid-base pairs:
    (a) $HCl - Cl^-$; $NH_4^+ - NH_3$
    (b) $HCO_3^- - CO_3^{2-}$; $H_2O - OH^-$
    (c) $H_3O^+ - H_2O$; $H_2CO_3 - HCO_3^-$
    (d) $HC_2H_3O_2 - C_2H_3O_2^-$; $H_3O^+ - H_2O$

31. Balancing equations
    (a) $Mg(s) + 2\,HCl(aq) \longrightarrow MgCl_2(aq) + H_2(g)$
    (b) $BaO(s) + 2\,HBr(aq) \longrightarrow BaBr_2(aq) + H_2O(l)$

(c) $2\,Al(s) + 3\,H_2SO_4(aq) \longrightarrow Al_2(SO_4)_3(aq) + 3\,H_2(g)$
(d) $Na_2CO_3(aq) + 2\,HCl(aq) \longrightarrow 2\,NaCl(aq) + H_2O(l) + CO_2(g)$
(e) $Fe_2O_3(s) + 6\,HBr(aq) \longrightarrow 2\,FeBr_3(aq) + 3\,H_2O(l)$
(f) $Ca(OH)_2(aq) + H_2CO_3(aq) \longrightarrow CaCO_3(s) + 2\,H_2O(l)$

32. (a) $NaOH(aq) + HBr(aq) \longrightarrow NaBr(aq) + H_2O(l)$
(b) $KOH(aq) + HCl(aq) \longrightarrow KCl(aq) + H_2O(l)$
(c) $Ca(OH)_2(aq) + 2\,HI(aq) \longrightarrow CaI_2(aq) + 2\,H_2O(l)$
(d) $Al(OH)_3(s) + 3\,HBr(aq) \longrightarrow AlBr_3(aq) + 3\,H_2O(l)$
(e) $Na_2O(s) + 2\,HClO_4(aq) \longrightarrow 2\,NaClO_4(aq) + H_2O(l)$
(f) $3\,LiOH(aq) + FeCl_3(aq) \longrightarrow Fe(OH)_3(s) + 3\,LiCl(aq)$

33. The following compounds are electrolytes:
(a) HCl, acid in water
(b) CO_2, acid in water
(c) $CaCl_2$, salt

34. The following compounds are electrolytes:
(a) $NaHCO_3$, salt
(c) $AgNO_3$, salt
(d) HCOOH, acid
(e) RbOH, base
(f) K_2CrO_4, salt

35. Calculation of molarity of ions.

(a) $$(0.015\text{ M NaCl})\left(\frac{1\text{ mol Na}^+}{1\text{ mol NaCl}}\right) = 0.015\text{ M Na}^+$$

$$(0.015\text{ M NaCl})\left(\frac{1\text{ mol Cl}^-}{1\text{ mol NaCl}}\right) = 0.015\text{ M Cl}^-$$

(b) $$(4.25\text{ M NaKSO}_4)\left(\frac{1\text{ mol Na}^+}{1\text{ mol NaKSO}_4}\right) = 4.25\text{ M Na}^+$$

$$(4.25\text{ M NaKSO}_4)\left(\frac{1\text{ mol K}^+}{1\text{ mol NaKSO}_4}\right) = 4.25\text{ M K}^+$$

$$(4.25\text{ M NaKSO}_4)\left(\frac{1\text{ mol SO}_4{}^{2-}}{1\text{ mol NaKSO}_4}\right) = 4.25\text{ M SO}_4{}^{2-}$$

(c) $$(0.20\ \text{M CaCl}_2)\left(\frac{1\ \text{mol Ca}^{2+}}{1\ \text{mol CaCl}_2}\right) = 0.20\ \text{M Ca}^{2+}$$

$$(0.20\ \text{M CaCl}_2)\left(\frac{2\ \text{mol Cl}^{-}}{1\ \text{mol CaCl}_2}\right) = 0.40\ \text{M Cl}^{-}$$

(d) $$\left(\frac{22.0\ \text{g KI}}{500.\ \text{mL}}\right)\left(\frac{1\ \text{mol}}{166.0\ \text{g}}\right)\left(\frac{1000\ \text{mL}}{\text{L}}\right) = 0.265\ \text{M KI}$$

$$(0.265\ \text{M KI})\left(\frac{1\ \text{mol K}^{+}}{1\ \text{mol KI}}\right) = 0.265\ \text{M K}^{+}$$

$$(0.265\ \text{M KI})\left(\frac{1\ \text{mol I}^{-}}{1\ \text{mol KI}}\right) = 0.265\ \text{M I}^{-}$$

36. (a) $$(0.75\ \text{M ZnBr}_2)\left(\frac{1\ \text{mol Zn}^{2+}}{1\ \text{mol ZnBr}_2}\right) = 0.75\ \text{M Zn}^{2+}$$

$$(0.75\ \text{M ZnBr}_2)\left(\frac{2\ \text{mol Br}^{-}}{1\ \text{mol ZnBr}_2}\right) = 1.5\ \text{M Br}^{-}$$

(b) $$(1.65\ \text{M Al}_2(\text{SO}_4)_3)\left(\frac{3\ \text{mol SO}_4{}^{2-}}{1\ \text{mol Al}_2(\text{SO}_4)_3}\right) = 4.95\ \text{M SO}_4{}^{2-}$$

$$(1.65\ \text{M Al}_2(\text{SO}_4)_3)\left(\frac{2\ \text{mol Al}^{3+}}{1\ \text{mol Al}_2(\text{SO}_4)_3}\right) = 3.30\ \text{M Al}^{3+}$$

(c) $$\left(\frac{900.\ \text{g(NH}_4)_2\text{SO}_4}{20.0\ \text{L}}\right)\left(\frac{1\ \text{mol}}{132.2\ \text{g}}\right) = 0.340\ \text{M (NH}_4)_2\text{SO}_4$$

$$(0.340\ \text{M (NH}_4)_2\text{SO}_4)\left(\frac{2\ \text{mol NH}_4{}^{+}}{1\ \text{mol (NH}_4)_2\text{SO}_4}\right) = 0.680\ \text{M NH}_4{}^{+}$$

$$(0.340\ \text{M (NH}_4)_2\text{SO}_4)\left(\frac{1\ \text{mol SO}_4{}^{2-}}{1\ \text{mol (NH}_4)_2\text{SO}_4}\right) = 0.340\ \text{M SO}_4{}^{2-}$$

(d) $$\left(\frac{0.0120\text{ g Mg(ClO}_3)_2}{0.00100\text{ L}}\right)\left(\frac{1\text{ mol}}{191.2\text{ g}}\right) = 0.0628\text{ M Mg(ClO}_3)_2$$

$$(0.0628\text{ M Mg(ClO}_3)_2)\left(\frac{1\text{ mol Mg}^{2+}}{1\text{ mol Mg(ClO}_3)_2}\right) = 0.0628\text{ M Mg}^{2+}$$

$$(0.0628\text{ M Mg(ClO}_3)_2)\left(\frac{2\text{ mol ClO}_3^{\,-}}{1\text{ mol Mg(ClO}_3)_2}\right) = 0.126\text{ M ClO}_3^{\,-}$$

37. The molarity of each ion, as calculated in Exercise 35 will be used to calculate the mass of each ion present in 100. mL of solution. 100. mL = 0.100 L

(a) $$(0.100\text{ L})\left(\frac{0.015\text{ mol Na}^+}{\text{L}}\right)\left(\frac{22.99\text{ g}}{\text{mol}}\right) = 0.034\text{ g Na}^+$$

$$(0.100\text{ L})\left(\frac{0.015\text{ mol Cl}^-}{\text{L}}\right)\left(\frac{35.45\text{ g}}{\text{mol}}\right) = 0.053\text{ g Cl}^-$$

(b) $$(0.100\text{ L})\left(\frac{4.25\text{ mol Na}^+}{\text{L}}\right)\left(\frac{22.99\text{ g}}{\text{mol}}\right) = 9.77\text{ g Na}^+$$

$$(0.100\text{ L})\left(\frac{4.25\text{ mol K}^+}{\text{L}}\right)\left(\frac{39.10\text{ g}}{\text{mol}}\right) = 16.6\text{ g K}^+$$

$$(0.100\text{ L})\left(\frac{4.25\text{ mol SO}_4^{\,2-}}{\text{L}}\right)\left(\frac{96.07\text{ g}}{\text{mol}}\right) = 40.8\text{ g SO}_4^{\,2-}$$

(c) $$(0.100\text{ L})\left(\frac{0.20\text{ mol Ca}^{2+}}{\text{L}}\right)\left(\frac{40.08\text{ g}}{\text{mol}}\right) = 0.80\text{ g Ca}^{2+}$$

$$(0.100\text{ L})\left(\frac{0.40\text{ mol Cl}^-}{\text{L}}\right)\left(\frac{35.45\text{ g}}{\text{mol}}\right) = 1.4\text{ g Cl}^-$$

(d) $$(0.100\text{ L})\left(\frac{0.265\text{ mol K}^+}{\text{L}}\right)\left(\frac{39.10\text{ g}}{\text{mol}}\right) = 1.04\text{ g K}^+$$

$$(0.100\text{ L})\left(\frac{0.265\text{ mol I}^-}{\text{L}}\right)\left(\frac{126.9\text{ g}}{\text{mol}}\right) = 3.36\text{ g I}^-$$

38. The molarity of each ion, as calculated in Exercise 36, will be used to calculate the mass of each ion present in 100 mL of solution. 100. mL = 0.100 L

(a) $$(0.100\ \text{L})\left(\frac{0.75\ \text{mol Zn}^{2+}}{\text{L}}\right)\left(\frac{65.39\ \text{g}}{\text{mol}}\right) = 4.9\ \text{g Zn}^{2+}$$

$$(0.100\ \text{L})\left(\frac{1.5\ \text{mol Br}^{-}}{\text{L}}\right)\left(\frac{79.90\ \text{g}}{\text{mol}}\right) = 12\ \text{g Br}^{-}$$

(b) $$(0.100\ \text{L})\left(\frac{3.30\ \text{mol Al}^{3+}}{\text{L}}\right)\left(\frac{26.98\ \text{g}}{\text{mol}}\right) = 8.90\ \text{g Al}^{3+}$$

$$(0.100\ \text{L})\left(\frac{4.95\ \text{mol SO}_4{}^{2-}}{\text{L}}\right)\left(\frac{96.07\ \text{g}}{\text{mol}}\right) = 47.6\ \text{g SO}_4{}^{2-}$$

(c) $$(0.100\ \text{L})\left(\frac{0.680\ \text{mol NH}_4{}^{+}}{\text{L}}\right)\left(\frac{18.04\ \text{g}}{\text{mol}}\right) = 1.23\ \text{g NH}_4{}^{+}$$

$$(0.100\ \text{L})\left(\frac{0.340\ \text{mol SO}_4{}^{2-}}{\text{L}}\right)\left(\frac{96.07\ \text{g}}{\text{mol}}\right) = 3.27\ \text{g SO}_4{}^{2-}$$

(d) $$(0.100\ \text{L})\left(\frac{0.0628\ \text{mol Mg}^{2+}}{\text{L}}\right)\left(\frac{24.31\ \text{g}}{\text{mol}}\right) = 0.153\ \text{g Mg}^{2+}$$

$$(0.100\ \text{L})\left(\frac{0.126\ \text{mol ClO}_3{}^{-}}{\text{L}}\right)\left(\frac{83.45\ \text{g}}{\text{mol}}\right) = 1.05\ \text{g ClO}_3{}^{-}$$

39. (a) $\text{pH} = -\log[\text{H}^+]$ $\quad [\text{H}^+] = 10^{-\text{pH}}$
$[\text{H}^+] = 1 \times 10^{-7.0}$
(b) $[\text{H}^+] = 3.98 \times 10^{-7}$
(c) $[\text{H}^+] = 3.16 \times 10^{-5}$

40. (a) $\text{pH} = -\log[\text{H}^+]$ $\quad [\text{H}^+] = 10^{-\text{pH}}$
$[\text{H}^+] = 6.31 \times 10^{-9}$
(b) $[\text{H}^+] = 3.98 \times 10^{-4}$
(c) $[\text{H}^+] = 1 \times 10^{-7.0}$

41. (a) $$(30.0\ \text{mL})\left(\frac{1.0\ \text{mol NaCl}}{1000\ \text{mL}}\right) = 0.030\ \text{mol NaCl}$$

$$(40.0\ \text{mL})\left(\frac{1.0\ \text{mol NaCl}}{1000\ \text{mL}}\right) = 0.040\ \text{mol NaCl}$$

Total mol NaCl = 0.030 mol + 0.040 mol = 0.070 mol NaCl

$$\frac{0.070 \text{ mol NaCl}}{0.070 \text{ L}} = 1.0 \text{ M NaCl}$$

$$(1.0 \text{ M NaCl})\left(\frac{1.0 \text{ mol Na}^+}{1.0 \text{ mol NaCl}}\right) = 1.0 \text{ M Na}^+$$

$$(1.0 \text{ M NaCl})\left(\frac{1.0 \text{ mol Cl}^-}{1.0 \text{ mol NaCl}}\right) = 1.0 \text{ M Cl}^-$$

(b) $HCl + NaOH \longrightarrow NaCl + H_2O$

$$(30.0 \text{ mL HCl})\left(\frac{1 \text{ L}}{1000 \text{ mL}}\right)\left(\frac{1.0 \text{ mol}}{\text{L}}\right) = 0.030 \text{ mol HCl}$$

$$(30.0 \text{ mL NaOH})\left(\frac{1 \text{ L}}{1000 \text{ mL}}\right)\left(\frac{1.0 \text{ mol}}{\text{L}}\right) = 0.030 \text{ mol NaOH}$$

0.030 mol HCl reacts with 0.030 mol NaOH and produces 0.030 mol NaCl. The final volume is 0.060 L. 0.030 mol NaCl/0.060 L = 0.50 M NaCl. Since there is one mole each of sodium and chloride ions per mole of NaCl, the molar concentration of Na^+ and Cl^- will be 0.50 M Na^+ and 0.50 M Cl^-.

(c) $KOH + HCl \longrightarrow KCl + H_2O$

$$(100.0 \text{ mL})\left(\frac{1 \text{ L}}{1000 \text{ mL}}\right)\left(\frac{0.40 \text{ mol KOH}}{\text{L}}\right) = 0.040 \text{ mol KOH}$$

$$(100.0 \text{ mL})\left(\frac{1 \text{ L}}{1000 \text{ mL}}\right)\left(\frac{0.80 \text{ mol HCl}}{\text{L}}\right) = 0.080 \text{ mol HCl}$$

0.040 mol KOH reacts with 0.040 mol HCl. 0.040 mol HCl remains unreacted and 0.040 mol KCl is produced. The final volume is 200.0 mL and contains 0.040 mol HCl and 0.040 mol KCl. Moles of ions are: 0.040 mol H^+, 0.040 mol K^+, and 0.080 mol Cl^-. Concentrations of ions are:

$$\frac{0.040 \text{ mol H}^+}{0.200 \text{ L}} = 0.20 \text{ M H}^+ \qquad \text{molarity K}^+ = \text{molarity H}^+$$

$$\frac{0.080 \text{ mol Cl}^-}{0.200 \text{ L}} = 0.40 \text{ M Cl}^-$$

42. (a) 100.0 mL of 2.0 M KCl and 100.0 mL of 1.0 M $CaCl_2$ are mixed, giving a final volume of 200.0 mL and concentrations of 1.0 M KCl and 0.50 M $CaCl_2$. The concentration of K^+ will be 1.0 M and the concentration of Ca^{2+} will be 0.50 M. The chloride ion concentration will be 2.0 M (1.0 M from the KCl and 2(0.50 M) from the $CaCl_2$).

(b) $(35.0\text{ mL})\left(\frac{1\text{ L}}{1000\text{ mL}}\right)\left(\frac{0.20\text{ mol Ba(OH)}_2}{\text{L}}\right) = 0.0070\text{ mol Ba(OH)}_2$

$(35.0\text{ mL})\left(\frac{1\text{ L}}{1000\text{ mL}}\right)\left(\frac{0.20\text{ mol H}_2\text{SO}_4}{\text{L}}\right) = 0.0070\text{ mol H}_2\text{SO}_4$

Final volume $= 35.0\text{ mL} + 35.0\text{ mL} = 70.0\text{ mL}$

$$\begin{array}{llll} H_2SO_4 & + \; Ba(OH)_2 & \longrightarrow BaSO_4(s) & + \; 2\,H_2O \\ 0.0070\text{ mol} & 0.0070\text{ mol} & 0.0070\text{ mol} & 0.014\text{ mol} \end{array}$$

The H_2SO_4 and the $Ba(OH)_2$ react completely producing insoluble $BaSO_4$ and H_2O. No ions are present in solution.

(c) $(0.500\text{ L NaCl})\left(\frac{2.0\text{ mol}}{\text{L}}\right) = 1.0\text{ mol NaCl}$

$(1.00\text{ L AgNO}_3)\left(\frac{1.00\text{ mol}}{\text{L}}\right) = 1.0\text{ mol AgNO}_3$

$$\begin{array}{llll} NaCl(aq) & + \; AgNO_3(aq) & \longrightarrow AgCl(s) & + \; NaNO_3(aq) \\ 1.0\text{ mol} & 1.0\text{ mol} & 1.0\text{ mol} & 1.0\text{ mol} \end{array}$$

The AgCl is insoluble and produces no ions. The 1.0 mol $NaNO_3$ will produce 1.0 mol Na^+ ions and 1.0 mol NO_3^- ions. The final volume of the solution is 1.5 L. The concentration of the ions are:

$$\frac{1.0\text{ mol Na}^+}{1.5\text{ L}} = 0.67\text{ M Na}^+ \qquad \frac{1.0\text{ mol NO}_3^-}{1.5\text{ L}} = 0.67\text{ M NO}_3^-$$

43. The reaction of HCl and NaOH occurs on a 1 : 1 mole ratio.

$$HCl + NaOH \longrightarrow NaCl + H_2O$$

At the endpoint in these titration reactions, equal moles of HCl and NaOH will have reacted. Moles = (molarity)(volume). At the endpoint, mol HCl = mol NaOH. Therefore, at the endpoint,

$$M_AV_A = M_BV_B \qquad M_A = \frac{M_BV_B}{V_A}$$

(a) $(37.70\text{ mL})(0.728\text{ M}) = (40.13\text{ mL})(\text{M HCl})$

$$\text{M HCl} = \frac{(37.70\text{ mL})(0.728\text{ M})}{40.13\text{ mL}} = 0.684\text{ M HCl}$$

(b) $\frac{(33.66\text{ mL})(0.306\text{ M})}{19.00\text{ mL}} = 0.542\text{ M HCl}$

(c) $\frac{(18.00\text{ mL})(0.555\text{ M})}{27.25\text{ mL}} = 0.367\text{ M HCl}$

44. The reaction of HCl and NaOH occurs on a 1 : 1 mole ratio.

$$HCl + NaOH \longrightarrow NaCl + H_2O$$

At the endpoint in these titration reactions, equal moles of HCl and NaOH will have reacted. Moles = (molarity)(volume). At the endpoint, mol HCl = mol NaOH. Therefore, at the endpoint,

$$M_AV_A = M_BV_B \qquad M_B = \frac{M_AV_A}{V_B}$$

(a) $\frac{(37.19\text{ mL})(0.126\text{ M})}{31.91\text{ mL}} = 0.147\text{ M NaOH}$

(b) $\frac{(48.04\text{ mL})(0.482\text{ M})}{24.02\text{ mL}} = 0.964\text{ M NaOH}$

(c) $\frac{(13.13\text{ mL})(1.425\text{ M})}{39.39\text{ mL}} = 0.4750\text{ M NaOH}$

45. (a) $SO_4^{2-}(aq) + Ba^{2+}(aq) \longrightarrow BaSO_4(s)$

(b) $CaCO_3(s) + 2\,H^+(aq) \longrightarrow Ca^{2+}(aq) + CO_2(g) + H_2O(l)$

(c) $Mg(s) + 2\,HC_2H_3O_2(aq) \longrightarrow Mg^{2+}(aq) + 2\,C_2H_3O_2^-(aq) + H_2(g)$

46. (a) $H_2S(g) + Cd^{2+}(aq) \longrightarrow CdS(s) + 2\,H^+(aq)$

(b) $Zn(s) + 2\,H^+(aq) \longrightarrow Zn^{2+}(aq) + H_2(g)$

(c) $Al^{3+}(aq) + PO_4^{3-}(aq) \longrightarrow AlPO_4(s)$

47. The more acidic solution is listed followed by an explanation.

(a) 1 molar H_2SO_4. The concentration of H^+ in 1 M H_2SO_4 is greater than 1 M since there are two ionizable hydrogens per mole of H_2SO_4. In HCl the concentration of H^+ will be 1 M, since there is only one ionizable hydrogen per mole HCl.

(b) 1 molar HCl. HCl is a strong electrolyte, producing more H^+ than $HC_2H_3O_2$ which is a weak electrolyte.

48. The more acidic solution is listed followed by an explanation.

(a) 2 molar HCl. 2 M HCl will yield 2 M H^+ concentration. 1 M HCl will yield 1 M H^+ concentration.

(b) 1 molar H_2SO_4. Both are strong acids. The concentration of H^+ in 1 M H_2SO_4 is greater than in 1 M HNO_3 because H_2SO_4 has two ionizable hydrogens per mole whereas HNO_3 has only one ionizable hydrogen per mole.

49. $3\ HCl + Al(OH)_3 \longrightarrow AlCl_3 + 3\ H_2O$

$g\ Al(OH)_3 \longrightarrow mol\ Al(OH)_3 \longrightarrow mol\ HCl \longrightarrow mL\ HCl$

0.245 M HCl contains 0.245 mol HCl/1000 mL

$$(10.0\ g\ Al(OH)_3)\left(\frac{1\ mol}{78.00\ g}\right)\left(\frac{3\ mol\ HCl}{1\ mol\ Al(OH)_3}\right)\left(\frac{1000\ mL}{0.245\ mol}\right)$$

$= 1.57 \times 10^3$ mL of 0.245 M HCl

50. $2\ HCl + Ca(OH)_2 \longrightarrow CaCl_2 + 2\ H_2O$

$M\ Ca(OH)_2 \longrightarrow mol\ Ca(OH)_2 \longrightarrow mol\ HCl \longrightarrow mL\ HCl$

0.245 M HCl contains 0.245 mol HCl/1000 mL

$$(0.0500\ L\ Ca(OH)_2)\left(\frac{0.100\ mol}{L}\right)\left(\frac{2\ mol\ HCl}{1\ mol\ Ca(OH)_2}\right)\left(\frac{1000\ mL}{0.245\ mol}\right)$$

$= 40.8$ mL of 0.245 M HCl

51. $NaOH + HCl \longrightarrow NaCl + H_2O$

First calculate the grams of NaOH in the sample.

$L\ HCl \longrightarrow mol\ HCl \longrightarrow mol\ NaOH \longrightarrow g\ NaOH$

$$(0.01825\ L\ HCl)\left(\frac{0.2406\ mol}{L}\right)\left(\frac{1\ mol\ NaOH}{1\ mol\ HCl}\right)\left(\frac{40.00\ g}{mol}\right) = 0.1756\ g\ NaOH$$

in the sample

$$\left(\frac{0.1756\ g\ NaOH}{0.200\ g\ sample}\right)(100) = 87.8\%\ NaOH$$

52. $NaOH + HCl \longrightarrow NaCl\ H_2O$

$L\ HCl \longrightarrow mol\ HCl \longrightarrow mol\ NaOH \longrightarrow g\ NaOH$

$$(0.04990\ L\ HCl)\left(\frac{0.466\ mol}{L}\right)\left(\frac{1\ mol\ NaOH}{1\ mol\ HCl}\right)\left(\frac{40.00\ g}{mol}\right) = 0.930\ g\ NaOH\ in\ the\ sample$$

1.00 g sample − 0.930 g NaOH = 0.070 g NaCl in the sample

$$\left(\frac{0.070\ g\ NaCl}{1.00\ g\ sample}\right)(100) = 7.0\%\ NaCl\ in\ the\ sample$$

53. $Zn + 2\ HCl \longrightarrow ZnCl_2 + H_2$

This is a limiting reactant problem. First find the moles of Zn and HCl from the given data and then identify the limiting reactant.

$$g\ Zn \longrightarrow mol\ Zn \qquad (5.00\ g\ Zn)\left(\frac{1\ mol}{65.39\ g}\right) = 0.0765\ mol\ Zn$$

$$(0.100\ L\ HCl)\left(\frac{0.350\ mol}{L}\right) = 0.0350\ mol\ HCl$$

Therefore Zn is in excess and HCl is the limiting reactant.

$$(0.0350\ mol\ HCl)\left(\frac{1\ mol\ H_2}{2\ mol\ HCl}\right) = 0.0175\ mol\ H_2 \text{ produced in the reaction}$$

$$T = 27°C = 300.\ K \qquad P = (700.\ torr)\left(\frac{1\ atm}{760\ torr}\right) = 0.921\ atm$$

$$PV = nRT$$

$$V = \frac{nRT}{P} = \frac{(0.0175\ mol\ H_2)(0.0821\ L\ atm/mol\ K)(300.\ K)}{0.921\ atm} = 0.468\ L\ H_2$$

54. $Zn + 2\,HCl \longrightarrow ZnCl_2 + H_2$

This is a limiting reactant problem. First find moles of Zn and HCl from the given data and then identify the limiting reactant.

$$g\ Zn \longrightarrow mol\ Zn \qquad (5.00\ g\ Zn)\left(\frac{1\ mol}{65.39\ g}\right) = 0.0765\ mol\ Zn$$

$$(0.200\ L\ HCl)\left(\frac{0.350\ mol}{L}\right) = 0.0700\ mol\ HCl$$

Zn is in excess and HCl is the limiting reactant.

$$(0.0700\ mol\ HCl)\left(\frac{1\ mol\ H_2}{2\ mol\ HCl}\right) = 0.0350\ mol\ H_2$$

$$T = 27°C = 300.\ K \qquad P = (700.\ torr)\left(\frac{1\ atm}{760\ torr}\right) = 0.921\ atm$$

$$PV = nRT$$

$$V = \frac{nRT}{P} = \frac{(0.0350\ mol\ H_2)(0.0821\ L\ atm/mol\ K)(300.\ K)}{0.921\ atm} = 0.936\ L\ H_2$$

55. Calculation of the pH solutions:

(a) $H^+ = 0.01\ M = 1 \times 10^{-2}\ M; \quad pH = -\log(1 \times 10^{-2}) = 2.0$

(b) $H^+ = 1.0\ M; \quad pH = -\log 1.0 = 0$

(c) $H^+ = 6.5 \times 10^{-9}\ M; \quad pH = -\log(6.5 \times 10^{-9}) = 8.19$

56. (a) $H^+ = 1 \times 10^{-7}\,M$; $pH = -\log(1 \times 10^{-7}) = 7.0$

(b) $H^+ = 0.50\,M$; $pH = -\log(5.0 \times 10^{-1}) = 0.30$

(c) $H^+ = 0.00010\,M = 1.0 \times 10^{-4}\,M$; $pH = -\log(1.0 \times 10^{-4}) = 4.00$

57. (a) Orange juice $= 3.7 \times 10^{-4}\,M\,H^+$
$pH = -\log(3.7 \times 10^{-4}) = 3.43$

(b) Vinegar $= 2.8 \times 10^{-3}\,M\,H^+$
$pH = -\log(2.8 \times 10^{-3}) = 2.55$

58. (a) Black coffee $= 5.0 \times 10^{-5}\,M\,H^+$
$pH = -\log(5.0 \times 10^{-5}) = 4.30$

(b) Limewater $= 3.4 \times 10^{-11}\,M\,H^+$
$pH = -\log(3.4 \times 10^{-11}) = 10.47$

59. (a) $CaCl_2(s) \longrightarrow Ca^{2+}(aq) + 2\,Cl^-(aq)$

For each $CaCl_2$ ionic compound, 1 calcium ion and 2 chloride ions result.

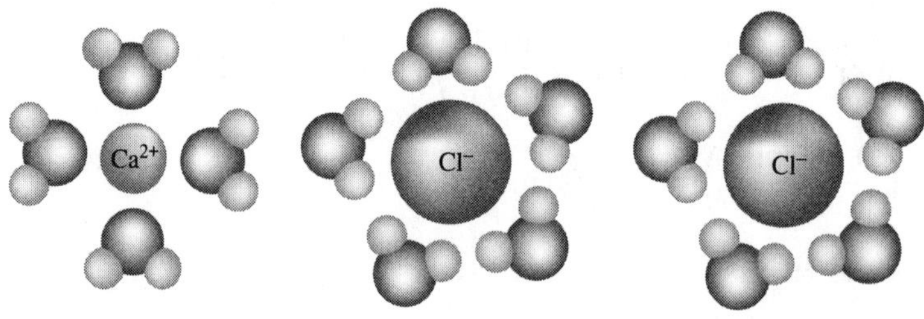

(b) $KF(s) \longrightarrow K^+(aq) + F^-(aq)$

For each KF ionic compound, 1 potassium ion and 1 fluoride ion result.

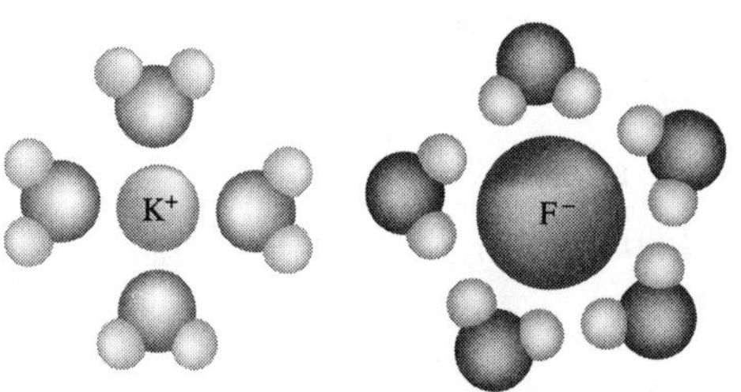

(c) $AlBr_3(s) \longrightarrow Al^{3+}(aq) + 3\,Br^-(aq)$

For each $AlBr_3$ ionic compound, 1 aluminum ion and 3 bromide ions result.

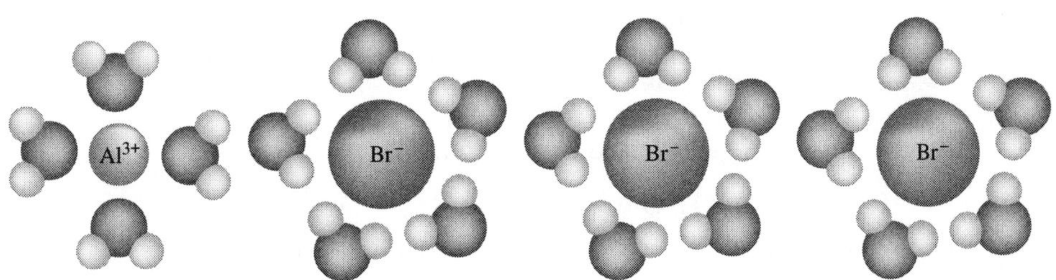

60. $CaI_2 \longrightarrow Ca^{2+} + 2\,I^-$

$$\left(\frac{0.520\text{ mol I}^-}{\text{L}}\right)\left(\frac{1\text{ mol Ca}^{2+}}{2\text{ mol I}}\right) = \left(\frac{0.260\text{ mol Ca}^{2+}}{\text{L}}\right) = 0.260\text{ M Ca}^{2+}$$

61. Dilution problem

$V_1M_1 = V_2M_2$

$(100.\text{ mL})(12\text{ M}) = (V_2)(0.40\text{ M})$

$$V_2 = \frac{(100.\text{ mL})(12\text{ M})}{0.40\text{ M}} = 3.0 \times 10^3\text{ mL}$$

62. $Ba(OH)_2 + 2\,HCl \longrightarrow BaCl_2 + 2\,H_2O$

$\text{M HCl} \longrightarrow \text{mol HCl} \longrightarrow \text{mol Ba(OH)}_2 \longrightarrow \text{M Ba(OH)}_2$

$$\left(\frac{0.430\text{ mol HCl}}{\text{L}}\right)\left(\frac{1\text{ L}}{1000\text{ mL}}\right)(29.26\text{ mL}) = 0.0126\text{ mol HCl}$$

$$(0.0126\text{ mol HCl})\left(\frac{1\text{ mol Ba(OH)}_2}{2\text{ mol HCl}}\right) = 0.00630\text{ mol Ba(OH)}_2$$

$$\frac{0.00630\text{ mol Ba(OH)}_2}{0.02040\text{ L}} = 0.309\text{ M Ba(OH)}_2$$

63. The acetic acid solution freezes at a lower temperature than the alcohol solution. The acetic acid ionizes slightly while the alcohol does not. The ionization of the acetic acid increases its particle concentration in solution above that of the alcohol solution, resulting in a lower freezing point for the acetic acid solution.

64. It is more economical to purchase CH_3OH at the same cost per pound as C_2H_5OH. Because CH_3OH has a lower molar mass than C_2H_5OH, the CH_3OH solution will contain more particles per pound in a given solution and therefore, have a greater effect on the freezing point of the radiator solution.

Assume 100. g of each compound.

$$CH_3OH: \quad \frac{100.\ g}{34.04\ g/mol} = 2.84\ mol$$

$$CH_3CH_2OH: \quad \frac{100.\ g}{46.07\ g/mol} = 2.17\ mol$$

65. A hydronium ion is a hydrated hydrogen ion.

$$\underset{\text{(hydrogen ion)}}{H^+} \quad + \quad H_2O \quad \longrightarrow \quad \underset{\text{(hydronium ion)}}{H_3O^+}$$

66. Freezing point depression is directly related to the concentration of particles in the solution.

$C_{12}H_{22}O_{11}$	>	$HC_2H_3O_2$	>	HCl	>	$CaCl_2$	
1 mol		1+ mol		2 mol		3 mol	(particles in solution)

67. (a) 100°C $pH = -\log(1 \times 10^{-6}) = 6.0$

25°C $pH = -\log(1 \times 10^{-7}) = 7.0$ pH of H_2O is greater at 25°C

(b) $1 \times 10^{-6} > 1 \times 10^{-7}$ so, H^+ concentration is higher at 100°C.

(c) The water is neutral at both temperatures, because the H_2O ionizes into equal concentrations of H^+ and OH^- at any temperature.

68. As the pH changes by 1 unit, the concentration of H^+ in solution changes by a factor of 10. For example, the pH of 0.10 M HCl is 1.00, while the pH of 0.0100 M HCl is 2.00.

69. First determine the molarity of the two HCl solutions. Take the antilog of the pH value to obtain the $[H^+]$.

pH 0.300; $H^+ = 2.00\ M = 2.00\ M\ HCl$

pH 0.150; $H^+ = 1.41\ M = 1.41\ M\ HCl$

Now treat the calculation as a dilution problem.

$$V_1M_1 = V_2M_2 \quad V_2 = \frac{V_1M_1}{M_2}$$

$$\frac{(200\ mL\ HCl)(2.00\ M)}{1.41\ M} = 284\ mL\ solution$$

$$284\ mL - 200\ mL = 84\ mL\ H_2O \text{ to be added}$$

70. First calculate the moles of HCl in 100.0 mL of HCl of pH -1.5. Take the antilog of the pH value -1.5 to obtain the $[H^+]$.

pH -1.5; $H^+ = 3.16 \times 10^{-2}\,M = 3.16 \times 10^{-2}\,M\,HCl$

Now calculate the moles of HCl in 100.0 mL HCl

$$\left(\frac{3.16 \times 10^{-2}\,\text{mol}}{\text{L}}\right)(0.1000\,\text{L}) = 3.16 \times 10^{-3}\,\text{mol HCl}$$

Now determine the volume of 12.0 M HCl that contains 3.16×10^{-3} mol HCl

$$(3.16 \times 10^{-3}\,\text{mol HCl})\left(\frac{1000\,\text{mL}}{12.0\,\text{mol HCl}}\right) = 0.263\,\text{mL } 12.0\,\text{M HCl}$$

$$100.0\,\text{mL } H_2O - 0.263\,\text{mL} = 99.7\,\text{mL } H_2O$$

Add 0.3 mL 12.0 M HCl to 99.7 mL H_2O to make 100.0 mL of HCl with pH -1.5.

71. First determine the molarity of the HCl solution. Take the antilog of the pH 0.250 to obtain the $[H^+]$.

pH 0.250; $H^+ = 1.78\,M = 1.78\,M\,HCl$

HCl reacts in a 1 : 1 molar ratio with NaOH

$$M_A V_A = M_B V_B$$

$$\frac{(1.78\,\text{M})(35.00\,\text{mL})}{2.00\,\text{M}} = 31.2\,\text{mL NaOH required for neutralization}$$

72. A 1.00 *m* solution contains 1 mol solute plus 1000 g H_2O. We need to find the total number of moles and then calculate the mole percent of each component.

$$\left(\frac{1000\,\text{g } H_2O}{18.02\,\text{g/mol}}\right) = 55.49\,\text{mol } H_2O$$

$$55.49\,\text{mol } H_2O + 1.00\,\text{mol solute} = 56.49\,\text{total moles}$$

$$\left(\frac{1.00\,\text{mol solute}}{56.49\,\text{mol}}\right)(100) = 1.77\%\,\text{solute}$$

$$\left(\frac{55.49\,\text{mol } H_2O}{56.49\,\text{mol}}\right)(100) = 98.22\%\,H_2O$$

73. Saturated NaCl solution contains dissolved NaCl in equilibrium with solid NaCl.

$$NaCl(s) \rightleftharpoons Na^+(aq) + Cl^-(aq)$$

Increasing the Na^+ or Cl^- concentration in a saturated NaCl solution will cause solid NaCl to precipitate. For example, adding HCl(*g*) to the solution will increase the Cl^- concentration and cause precipitation of NaCl.

74. $Na_2CO_3 + 2\,HCl \longrightarrow 2\,NaCl + CO_2 + H_2O$

$g\ Na_2CO_3 \longrightarrow mol\ Na_2CO_3 \longrightarrow mol\ HCl \longrightarrow M\ HCl$

$$(0.452\text{ g }Na_2CO_3)\left(\frac{1\text{ mol}}{106.0\text{ g}}\right)\left(\frac{2\text{ mol HCl}}{1\text{ mol }Na_2CO_3}\right)\left(\frac{1}{0.0424\text{ L}}\right) = 0.201\text{ M HCl}$$

75. $2\,HCl + Ca(OH)_2 \longrightarrow CaCl_2 + 2\,H_2O$

$g\ Ca(OH)_2 \longrightarrow mol\ Ca(OH)_2 \longrightarrow mol\ HCl \longrightarrow mL\ HCl$

$$(2.00\text{ g }Ca(OH)_2)\left(\frac{1\text{ mol}}{74.10\text{ g}}\right)\left(\frac{2\text{ mol HCl}}{1\text{ mol }Ca(OH)_2}\right)\left(\frac{1000\text{ mL}}{0.1234\text{ mol}}\right) = 437\text{ mL of }0.1234\text{ M HCl}$$

76. $KOH + HNO_3 \longrightarrow KNO_3 + H_2O$

$L\ HNO_3 \longrightarrow mol\ HNO_3 \longrightarrow mol\ KOH \longrightarrow g\ KOH$

$$(0.05000\text{ L }HNO_3)\left(\frac{0.240\text{ mol}}{\text{L}}\right)\left(\frac{1\text{ mol KOH}}{1\text{ mol }HNO_3}\right)\left(\frac{56.11\text{ g}}{\text{mol}}\right) = 0.673\text{ g KOH}$$

77. pH of 1.0 L solution containing 0.1 mL of 1.0 M HCl

$$(0.1\text{ mL})\left(\frac{1.0\text{ L}}{1000\text{ mL}}\right)\left(\frac{1\text{ mol HCl}}{\text{L}}\right) = 1 \times 10^{-4}\text{ mol HCl added}$$

$$\frac{1 \times 10^{-4}\text{ mol HCl}}{1.0\text{ L}} = 1 \times 10^{-4}\text{ M HCl}$$

1×10^{-4} M HCl produces 1×10^{-4} M H^+

$pH = -\log(1 \times 10^{-4}) = 4.0$

78. Dilution problem

$$V_1M_1 = V_2M_2 \qquad V_1 = \frac{V_2M_2}{M_1} = \frac{(50.0\text{ L})(5.00\text{ M})}{18.0\text{ M}} = 13.9\text{ L of }18.0\text{ M }H_2SO_4$$

79. $NaOH + HCl \longrightarrow NaCl + H_2O$

$$(3.0\text{ g NaOH})\left(\frac{1\text{ mol}}{40.00\text{ g}}\right) = 0.075\text{ mol NaOH}$$

$$(500.\text{ mL HCl})\left(\frac{1\text{ L}}{1000\text{ mL}}\right)\left(\frac{0.10\text{ mol}}{\text{L}}\right) = 0.050\text{ mol HCl}$$

This solution is basic. The NaOH will neutralize the HCl with an excess of 0.025 mol of NaOH remaining unreacted.

80. $Ba(OH)_2(aq) + 2\,HCl(aq) \longrightarrow BaCl_2(aq) + 2\,H_2O(l)$

$$(0.380\text{ L Ba(OH)}_2)\left(\frac{0.35\text{ mol}}{\text{L}}\right) = 0.13\text{ mol Ba(OH)}_2$$

$$0.13\text{ mol Ba(OH)}_2 \longrightarrow 0.26\text{ mol OH}^-$$

$$(0.5000\text{ L HCl})\left(\frac{0.65\text{ mol}}{\text{L}}\right) = 0.33\text{ mol HCl}$$

$$0.33\text{ mol HCl} \longrightarrow 0.33\text{ mol H}^+$$

0.33 mol H^+ will neutralize 0.26 mol OH^- and leave 0.07 mol $H^+(0.33 - 0.26)$ remaining in solution.

Total volume = 500.0 mL + 380 mL = 880 mL (0.88 L)

$$[\text{H}^+]\text{ in solution} = \frac{0.07\text{ mol H}^+}{0.88\text{ L}} = 0.08\text{ M H}^+$$

$$\text{pH} = -\log[\text{H}^+] = -\log(8 \times 10^{-2}) = 1.1$$

81. $$(0.05000\text{ L HCl})\left(\frac{0.2000\text{ mol}}{\text{L}}\right) = 0.01000\text{ mol HCl} = 0.01000\text{ mol H}^+\text{ in 50.00 mL HCl}$$

(a) no base added: pH = $-\log(0.2000) = 0.700$

(b) 10.00 mL base added: $(0.01000\text{ L})\left(\frac{0.2000\text{ mol}}{\text{L}}\right) = 0.002000\text{ mol NaOH}$

$= 0.002000\text{ mol OH}^-$

$(0.01000\text{ mol H}^+) - (0.002000\text{ mol OH}^-) = 0.00800\text{ mol H}^+$ in 60.00 mL solution

$$[\text{H}^+] = \frac{0.00800\text{ mol}}{0.06000\text{ L}} \qquad \text{pH} = -\log\left(\frac{0.00800}{0.06000}\right) = 0.880$$

(c) 25.00 mL base added:

$$(0.02500\text{ L})\left(\frac{0.2000\text{ mol}}{\text{L}}\right) = 0.005000\text{ mol NaOH} = \text{mol OH}^-$$

$(0.01000\text{ mol H}^+) - (0.005000\text{ mol OH}^-) = 0.00500\text{ mol H}^+$ in 75.00 mL solution

$$[\text{H}^+] = \frac{0.00500\text{ mol}}{0.07500\text{ L}} \qquad \text{pH} = -\log\left(\frac{0.00500}{0.07500}\right) = 1.2$$

(d) 49.00 mL base added:

$$(0.04900\text{ L})\left(\frac{0.2000\text{ mol}}{\text{L}}\right) = 0.009800\text{ mol NaOH} = \text{mol OH}^-$$

$$(0.01000\text{ mol H}^+) - (0.009800\text{ mol OH}^-) = 0.00020\text{ mol H}^+\text{ in 99.00 mL solution}$$

$$[\text{H}^+] = \frac{0.00020\text{ mol}}{0.09900\text{ L}} \qquad \text{pH} = -\log\left(\frac{0.00020}{0.09900}\right) = 2.69$$

(e) 49.90 mL base added:

$$(0.04990\text{ L})\left(\frac{0.2000\text{ mol}}{\text{L}}\right) = 0.009980\text{ mol NaOH} = \text{mol OH}^-$$

$(0.01000\text{ mol H}^+) - (0.009980\text{ mol OH}^-) = 2 \times 10^{-5}\text{ mol H}^+$ in 99.90 mL solution

$$[\text{H}^+] = \frac{2 \times 10^{-5}\text{ mol}}{0.09990\text{ L}} \qquad \text{pH} = -\log\left(\frac{2 \times 10^{-5}}{0.09990}\right) = 3.7$$

(f) 49.99 mL base added:

$$(0.04999\text{ L})\left(\frac{0.2000\text{ mol}}{\text{L}}\right) = 0.009998\text{ mol NaOH} = \text{mol OH}^-$$

$(0.01000\text{ mol H}^+) - (0.009998\text{ mol OH}^-) = 2 \times 10^{-6}\text{ mol H}^+$ in 99.99 mL solution

$$[\text{H}^+] = \frac{2 \times 10^{-6}\text{ mol}}{0.09999\text{ L}} \qquad \text{pH} = -\log\left(\frac{2 \times 10^{-6}}{9.999 \times 10^{-2}}\right) = 4.7$$

(g) 50.00 mL of 0.2000 M NaOH neutralizes 50.00 mL of 0.2000 M HCl. No excess acid or base is in the solution. Therefore, the solution is neutral with a pH = 7.0

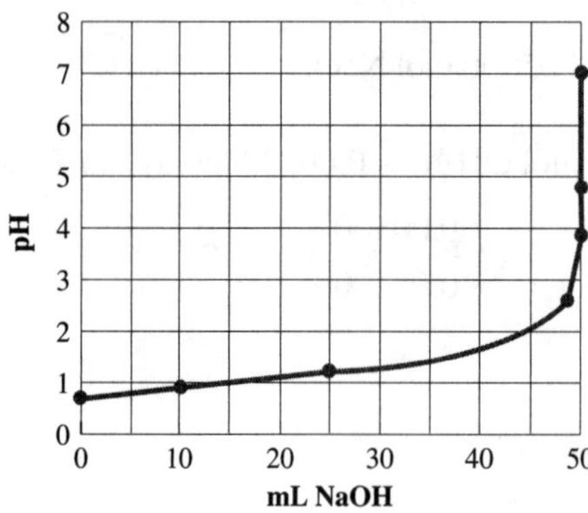

82. (a) $2\,NaOH(aq) + H_2SO_4(aq) \longrightarrow Na_2SO_4(aq) + 2\,H_2O(l)$

(b) mol $H_2SO_4 \longrightarrow$ mol NaOH $\longrightarrow$ mL NaOH

$$(0.0050\text{ mol }H_2SO_4)\left(\frac{2\text{ mol NaOH}}{1\text{ mol }H_2SO_4}\right)\left(\frac{1000\text{ mL}}{0.10\text{ mol}}\right) = 1.0 \times 10^2\text{ mL NaOH}$$

(c) $$(0.0050\text{ mol }H_2SO_4)\left(\frac{1\text{ mol }Na_2SO_4}{1\text{ mol }H_2SO_4}\right)\left(\frac{142.1\text{ g}}{\text{mol}}\right) = 0.71\text{ g }Na_2SO_4$$

83. $HNO_3 + KOH \longrightarrow KNO_3 + H_2O$

$M_AV_A = M_BV_B$

$(M_A)(25\text{ mL}) = (0.60\text{ M})(50.0\text{ mL})$

$M_A = 1.2$ M (diluted solution)

Dilution problem $M_1V_1 = M_2V_2$

$(M_1)(10.0\text{ mL}) = (1.2\text{ M})(100.00\text{ mL})$

$M_1 = 12$ M HNO_3 (original solution)

84. Yes, adding water changes the concentration of the acid, which changes the concentration of the $[H^+]$, and changes the pH.

No, the solution theoretically will never reach a pH of 7, but it will approach pH 7 as water is added.

85. $pH = -\log[H^+]$

$pH = 2, \quad [H^+] = 1 \times 10^{-2}$ (solution X)

$pH = 4, \quad [H^+] = 1 \times 10^{-4}$ (solution Y)

statement (c) is correct. The $[H^+]$ of X is 100 times that of Y.

86. (a) The solution is neutral, neither acidic nor basic ($pH = 7.0$).

(b) The solution is basic, $[OH^-] > 10^{-7}$; $pH = 12$.

(c) The solution is acidic.

(d) Cannot determine whether the solution is acidic or basic.

87. mol acid = mol base (lactic acid has one acidic H)

$$\frac{1.0\text{ g acid}}{\text{molar mass}} = (0.017\text{ L})\left(\frac{0.65\text{ mol}}{\text{L}}\right) = 90.\text{ g/mol (molar mass)}$$

mass of empirical formula ($HC_3H_5O_3$) = 90.17 g/mol

molar mass = mass of empirical formula

Therefore the molecular formula is $HC_3H_5O_3$

CHAPTER 16

CHEMICAL EQUILIBRIUM

1. At 25°C both tubes would appear the same and contain more molecules in the gaseous state than the tube at 0°C, and less molecules in the gaseous state than the tube at 80°C.

2. The reaction is endothermic because the increased temperature increases the concentration of product (NO_2) present at equilibrium.

3. In an endothermic process heat is absorbed or used by the system so it should be placed on the reactant side of a chemical equation. In an exothermic process heat is given off by the system so it belongs on the product side of a chemical equation.

4. At equilibrium, the rate of the forward reaction equals the rate of the reverse reaction.

5. Free protons (H^+) do not exist in water because they are hydrated forming H_3O^+.

6. The sum of the pH and the pOH is 14. A solution whose pH is -1 would have a pOH of 15.

7. Acids stronger than acetic acid are: benzoic, cyanic, formic, hydrofluoric, and nitrous acids (all equilibrium constants are greater than the equilibrium constant for acetic acid). Acids weaker than acetic acid are: carbolic, hydrocyanic, and hypochlorous acids (all have equilibrium constants smaller than the equilibrium constant for acetic acid). All have one ionizable hydrogen atom.

8. The order of solubility will correspond to the order of the values of the solubility product constants of the salts being compared. This occurs because each salt in the comparison produces the same number of ions (two in this case) for each formula unit of salt that dissolves. This type of comparison would not necessarily be valid if the salts being compared gave different numbers of ions per formula unit of salt dissolving. The order is: $AgC_2H_3O_2$, $PbSO_4$, $BaSO_4$, AgCl, $BaCrO_4$, AgBr, AgI, PbS.

9. (a) $K_{sp}\ Mn(OH)_2 = 2.0 \times 10^{-13}$; $K_{sp}\ Ag_2CrO_4 = 1.9 \times 10^{-12}$.
   Each salt gives 3 ions per formula unit of salt dissolving. Therefore, the salt with the largest K_{sp} (in this case Ag_2CrO_4) is more soluble.

   (b) $K_{sp}\ BaCrO_4 = 8.5 \times 10^{-11}$; $K_{sp}\ Ag_2CrO_4 = 1.9 \times 10^{-12}$. Ag_2CrO_4 has a greater molar solubility than $BaCrO_4$, even though its K_{sp} is smaller, because the Ag_2CrO_4 produces more ions per formula unit of salt dissolving than $BaCrO_4$.

$BaCrO_4(s) \rightleftharpoons Ba^{2+} + CrO_4^{2-}$ $\qquad K_{sp} = [Ba^{2+}][CrO_4^{2-}]$

Let y = molar solubility of $BaCrO_4$

$K_{sp} = [y][y] = 8.5 \times 10^{-11}$

$y = \sqrt{8.5 \times 10^{-11}} = 9.2 \times 10^{-6}$ mol $BaCrO_4$/L

$Ag_2CrO_4(s) \rightleftharpoons 2\,Ag^+ + CrO_4^{2-}$ $\qquad K_{sp} = [Ag^+]^2[CrO_4^{2-}]$

Let y = molar solubility of Ag_2CrO_4

$K_{sp} = [2y]^2[y] = 1.9 \times 10^{-12}$

$$y = \sqrt[3]{\frac{1.9 \times 10^{-12}}{4}} = 7.8 \times 10^{-5} \text{ mol } Ag_2CrO_4/L$$

Ag_2CrO_4 has the greater solubility.

10. $HC_2H_3O_2 \rightleftharpoons H^+ + C_2H_3O_2^-$

Initial Concentrations		Added	Concentration After Equilibrium Shifts
$HC_2H_3O_2$	1.00 M	-----	1.01 M
H^+	1.8×10^{-5}M	0.010 mol	1.9×10^{-5}
$C_2H_3O_2^-$	1.00 M	-----	0.99 M

The initial concentration of H^+ in the buffer solution is very low (1.8×10^{-5} M) because of the large excess of acetate ions. 0.010 mole of HCl is added to one liter of the buffer solution. This will supply 0.010 M H^+. The added H^+ creates a stress on the right side of the equation. The equilibrium shifts to the left, using up almost all the added H^+, reducing the acetate ion by approximately 0.010 M, and increasing the acetic acid by approximately 0.010 M. The concentration of H^+ will not increase significantly and the pH is maintained relatively constant.

11. In a saturated sodium chloride solution, the equilibrium is

$Na^+(aq) + Cl^-(aq) \rightleftharpoons NaCl(s)$

Bubbling in HCl gas increases the concentration of Cl^-, creating a stress, which will cause the equilibrium to shift to the right, precipitating solid NaCl.

12. The rate of a reaction increases when the concentration of one of the reactants increases. The increase in concentration causes the number of collisions between the reactants to increase. The rate of a reaction, being proportional to the frequency of such collisions, as a result, will increase.

13. If pure HI is placed in a vessel at 700 K, some of it will decompose. Since the reaction is reversible ($H_2 + I_2 \rightleftharpoons 2\,HI$) HI molecules will react to produce H_2 and I_2.

14. An increase in temperature causes the rate of reaction to increase, because it increases the velocity of the molecules. Faster moving molecules increase the number and effectiveness of the collisions between molecules resulting in an increase in the rate of the reaction.

15. A catalyst speeds up the rate of a reaction by lowering the activation energy. A catalyst is not used up in the reaction.

16. $A + B \rightleftharpoons C + D$

    When A and B are initially mixed, the rate of the forward reaction to produce C and D is at its maximum. As the reaction proceeds, the rate of production of C and D decreases because the concentrations of A and B decrease. As C and D are produced, some of the collisions between C and D will result in the reverse reaction, forming A and B. Finally, an equilibrium is achieved in which the forward rate exactly equals the reverse rate.

17. $HC_2H_3O_2 + H_2O \rightleftharpoons H_3O^+ + C_2H_3O_2^-$

    As water is added (diluting the solution from 1.0 M to 0.10 M), the equilibrium shifts to the right, yielding a higher percent ionization.

18. The statement does not contradict Le Chatelier's Principle. The previous question deals with the case of dilution. If pure acetic acid is added to a dilute solution, the reaction will shift to the right, producing more ions in accordance with Le Chatelier's Principle. But, the concentration of the un-ionized acetic acid will increase faster than the concentration of the ions, thus yielding a smaller percent ionization.

19. At different temperatures, the degree of ionization of water varies, being higher at higher temperatures. Consequently, the pH of water can be different at different temperatures.

20. In pure water, H^+ and OH^- are produced in equal quantities by the ionization of the water molecules, $H_2O \rightleftharpoons H^+ + OH^-$. Since $pH = -\log[H^+]$, and $pOH = -\log[OH^-]$, they will always be identical for pure water. At 25°C, they each have the value of 7, but at higher temperatures, the degree of ionization is greater, so the pH and pOH would both be less than 7, but still equal.

21. In water the silver acetate dissociates until the equilibrium concentration of ions is reached. In nitric acid solution, the acetate ions will react with hydrogen ions to form acetic acid molecules. The HNO_3 removes acetate ions from the silver acetate equilibrium allowing more silver acetate to dissolve. If HCl is used, a precipitate of silver chloride would be formed, since silver chloride is less soluble than silver acetate. Thus, more silver acetate would dissolve in HCl than in pure water.

    $AgC_2H_3O_2(s) \rightleftharpoons Ag^+(aq) + C_2H_3O_2^-(aq)$

22. When the salt, sodium acetate, is dissolved in water, the solution becomes basic. The dissolving reaction is

$$NaC_2H_3O_2(s) \xrightarrow{H_2O} Na^+(aq) + C_2H_3O_2^-(aq)$$

The acetate ion reacts with water. The reaction does not go to completion, but some OH^- ions are produced and at equilibrium the solution is basic.

$$C_2H_3O_2^-(aq) + H_2O(l) \rightleftharpoons OH^-(aq) + HC_2H_3O_2(aq)$$

23. A buffer solution contains a weak acid or base plus a salt of that weak acid or base, such as dilute acetic acid and sodium acetate.

$$HC_2H_3O_2(aq) \rightleftharpoons H^+(aq) + C_2H_3O_2^-(aq)$$
$$NaC_2H_3O_2(aq) \longrightarrow Na^+(aq) + C_2H_3O_2^-(aq)$$

When a small amount of a strong acid (H^+) is added to this buffer solution, the H^+ reacts with the acetate ions to form un-ionized acetic acid, thus neutralizing the added acid. When a strong base, OH^-, is added it reacts with un-ionized acetic acid to neutralized the added base. As a result, in both cases, the approximate pH of the solution is maintained.

24. All four K expressions are equilibrium constants. They describe the ratio between the concentrations of products and reactants for different types of reactions when at equilibrium. K_a is the equilibrium constant expression for the ionization of a weak acid, K_b is the equilibrium constant expression for the ionization of a weak base, K_w is the equilibrium constant expression for the ionization of water and K_{sp} is the equilibrium constant expression for a slightly soluble salt.

25. Reversible systems

(a) $H_2O(s) \xrightleftharpoons{0°C} H_2O(l)$

(b) $Na_2SO_4(s) \rightleftharpoons 2\,Na^+(aq) + SO_4^{2-}(aq)$

26. Reversible systems

(a) $H_2O(l) \xrightleftharpoons{100°C} H_2O(g)$

(b) $SO_2(l) \rightleftharpoons SO_2(g)$

27. Equilibrium system

$$4\,NH_3(g) + 3\,O_2(g) \rightleftharpoons 2\,N_2(g) + 6\,H_2O(g) + 1531\text{ kJ}$$

(a) The reaction is exothermic with heat being evolved.

(b) The addition of O_2 will shift the reaction to the right until equilibrium is reestablished. The concentration of N_2, H_2O, and O_2 will be increased. The concentration of the NH_3 will be decreased.

28. Equilibrium system

$$4\,NH_3(g) + 3\,O_2(g) \rightleftharpoons 2\,N_2(g) + 6\,H_2O(g) + 1531\text{ kJ}$$

(a) The addition of N_2 will shift the reaction to the left until equilibrium is reestablished. The concentration of NH_3, O_2 and N_2 will be increased. The concentration of H_2O will be decreased.

(b) The addition of heat will shift the reaction to the left.

29. $N_2(g) + 3\,H_2(g) \rightleftharpoons 2\,NH_3(g) + 92.5\text{ kJ}$

Change or stress imposed on the system at equilibrium	Direction of reaction, left or right, to reestablish equilibrium	Changes in number of moles N_2	H_2	NH_3
(a) Add N_2	right	I	D	I
(b) Remove H_2	left	I	D	D
(c) Decrease volume of reaction vessel	right	D	D	I
(d) Increase temperature	left	I	I	D

I = Increase; D = Decrease; N = No Change;
? = insufficient information to determine

30. $N_2(g) + 3\,H_2(g) \rightleftharpoons 2\,NH_3(g) + 92.5\text{ kJ}$

Change or stress imposed on the system at equilibrium	Direction of reaction, left or right, to reestablish equilibrium	Changes in number of moles N_2	H_2	NH_3
(a) Add NH_3	left	I	I	I
(b) Increase volume of reaction vessel	left	I	I	D
(c) Add a catalyst	no change	N	N	N
(d) Add H_2 and NH_3	?	?	I	I

I = Increase; D = Decrease; N = No Change;
? = insufficient information to determine

31. Direction of shift in equilibrium:

Reaction	Increased Temperature	Increased Pressure (Volume Decreases)	Add Catalyst
(a)	right	right	no change
(b)	left	no change	no change
(c)	left	right	no change

32. Direction of shift in equilibrium:

Reaction	Increased Temperature	Increased Pressure (Volume Decreases)	Add Catalyst
(a)	right	left	no change
(b)	left	left	no change
(c)	left	left	no change

33. Equilibrium shifts
   (a) right
   (b) left
   (c) none

34. Equilibrium shifts
   (a) left
   (b) right
   (c) right

35. (a) $K_{eq} = \dfrac{[Cl_2]^2[H_2O]^2}{[HCl]^4[O_2]}$

   (b) $K_{eq} = \dfrac{[NH_3]^2}{[N_2][H_2]^3}$

   (c) $K_{eq} = \dfrac{[PCl_3][Cl_2]}{[PCl_5]}$

36. (a) $K_{eq} = \dfrac{[H^+][ClO_2^-]}{[HClO_2]}$

   (b) $K_{eq} = \dfrac{[H^+][C_2H_3O_2^-]}{[HC_2H_3O_2]}$

   (c) $K_{eq} = \dfrac{[NO]^4[H_2O]^6}{[NH_3]^4[O_2]^5}$

37. (a) $K_{sp} = [Cu^{2+}][S^{2-}]$
(b) $K_{sp} = [Ba^{2+}][SO_4^{2-}]$
(c) $K_{sp} = [Pb^{2+}][Br^-]^2$
(d) $K_{sp} = [Ag^+]^3[AsO_4^{3-}]$

38. (a) $K_{sp} = [Fe^{3+}][OH^-]^3$
(b) $K_{sp} = [Sb^{5+}]^2[S^{2-}]^5$
(c) $K_{sp} = [Ca^{2+}][F^-]^2$
(d) $K_{sp} = [Ba^{2+}]^3[PO_4^{3-}]^2$

39. If the H^+ ion concentration is decreased:
(a) pH is increased
(b) pOH is decreased
(c) $[OH^-]$ is increased
(d) K_w remains the same. K_w is a constant at a given temperature.

40. If the H^+ ion concentration is increased:
(a) pH is decreased (pH of 1 is more acidic than that of 4)
(b) pOH is increased
(c) $[OH^-]$ is decreased
(d) K_w remains unchanged. K_w is a constant at a given temperature.

41. The basis for deciding if a salt dissolved in water produces an acidic, a basic, or a neutral solution, is whether or not the salt reacts with water. Salts that contain an ion derived from a weak acid or base will produce an acidic or a basic solution.
(a) KCl, neutral
(b) Na_2CO_3, basic
(c) K_2SO_4, neutral
(d) $(NH_4)_2SO_4$, acidic

42. The basis for deciding if a salt dissolved in water produces an acidic, a basic, or a neutral solution, is whether or not the salt reacts with water. Salts that contain an ion derived from a weak acid or base will produce an acidic or a basic solution.
(a) $Ca(CN)_2$, basic
(b) $BaBr_2$, neutral
(c) $NaNO_2$, basic
(d) NaF, basic

43. (a) $NO_2^-(aq) + H_2O(l) \rightleftharpoons OH^-(aq) + HNO_2(aq)$
(b) $C_2H_3O_2^-(aq) + H_2O(l) \rightleftharpoons OH^-(aq) + HC_2H_3O_2(aq)$

44. (a) $NH_4^+(aq) + H_2O(l) \rightleftharpoons H_3O^+(aq) + NH_3(aq)$
(b) $SO_3^{2-}(aq) + H_2O(l) \rightleftharpoons OH^-(aq) + HSO_3^-(aq)$

45. (a) $HCO_3^-(aq) + H_2O(l) \rightleftharpoons OH^-(aq) + H_2CO_3(aq)$
(b) $NH_4^+(aq) + H_2O(l) \rightleftharpoons H_3O^+(aq) + NH_3(aq)$

46. (a) $OCl^-(aq) + H_2O(l) \rightleftharpoons OH^-(aq) + HOCl(aq)$

(b) $ClO_2^-(aq) + H_2O(l) \rightleftharpoons OH^-(aq) + HClO_2(aq)$

47. When excess acid (H^+) gets into the blood stream it reacts with HCO_3^- to form un-ionized H_2CO_3, thus neutralizing the acid and maintaining the approximate pH of the blood.

48. When excess base gets into the blood stream it reacts with H^+ to form water. Then H_2CO_3 ionizes to replace H^+, thus maintaining the approximate pH of the blood.

49. (a) $HC_2H_3O_2(aq) \rightleftharpoons H^+(aq) + C_2H_3O_2^-(aq)$

Let x = molarity of H^+

$[H^+] = [C_2H_3O_2^-] = x$

$[HC_2H_3O_2] = 0.25 - x = 0.25$ (since x is small)

$$K_a = \frac{[H^+][C_2H_3O_2^-]}{[HC_2H_3O_2]} = \frac{x^2}{0.25} = 1.8 \times 10^{-5}$$

$x^2 = (0.25)(1.8 \times 10^{-5})$

$x = \sqrt{(0.25)(1.8 \times 10^{-5})} = 2.1 \times 10^{-3}\,M = [H^+]$

(b) $pH = -\log[H^+] = -\log(2.1 \times 10^{-3}) = 2.68$

(c) Percent ionization

$$\frac{[H^+]}{[HC_2H_3O_2]}(100) = \left(\frac{2.1 \times 10^{-3}}{0.25}\right)(100) = 0.84\%$$

50. (a) $HC_6H_5O(aq) \rightleftharpoons H^+(aq) + C_6H_5O^-(aq)$

Let x = molarity of H^+

$[H^+] = [C_6H_5O^-] = x$

$[HC_6H_5O] = 0.25 - x = 0.25$ (since x is small)

$$K_a = \frac{[H^+][C_6H_5O^-]}{[HC_6H_5O]} = \frac{x^2}{0.25} = 1.3 \times 10^{-10}$$

$x^2 = (0.25)(1.3 \times 10^{-10})$

$x = \sqrt{(0.25)(1.3 \times 10^{-10})} = 5.7 \times 10^{-6} = [H^+]$

(b) $pH = -\log[H^+] = -\log(5.7 \times 10^{-6}) = 5.24$

(c) Percent ionization

$$\frac{[H^+]}{[HC_6H_5O]}(100) = \left(\frac{5.7 \times 10^{-6}}{0.25}\right)(100) = 2.3 \times 10^{-3}\,\%$$

51. $HA \rightleftharpoons H^+ + A^-$ $\quad 0.52\% = 0.0052$

$[H^+] = [A^-] = (1.000\ M)(0.0052) = 5.2 \times 10^{-3}\ M$

$[HA] = 1.000\ M - 0.0052\ M = 0.9948\ M$

$$K_a = \frac{[H^+][A^-]}{[HA]} = \frac{(5.2 \times 10^{-3})^2}{0.9948} = 2.7 \times 10^{-5}$$

52. $HA \rightleftharpoons H^+ + A^-$ $\quad pH = 5 = -\log[H^+]$ $\quad [H^+] = 1 \times 10^{-5}\ M$

$[H^+] = [A^-] = 1 \times 10^{-5}\ M$

$[HA] = 0.15\ M - 1 \times 10^{-5}\ M = 0.15\ M$

$$K_a = \frac{[H^+][A^-]}{[HA]} = \frac{(1 \times 10^{-5})^2}{0.15} = 7 \times 10^{-10}$$

53. $HC_2H_3O_2 \rightleftharpoons H^+ + C_2H_3O_2^-$

$$K_a = \frac{[H^+][C_2H_3O_2^-]}{[HC_2H_3O_2]} = 1.8 \times 10^{-5}$$

Let x = molarity of H^+

$[H^+] = [C_2H_3O_2^-] = x$

$[HC_2H_3O_2]$ = initial concentration $- x$ = initial concentration

Since K_a is small, the degree of ionization is small. Therefore, the approximation, initial concentration $- x$ = initial concentration, is valid.

(a) $[H^+] = [C_2H_3O_2^-] = x$ $\quad [HC_2H_3O_2] = 1.0\ M$

$$\frac{(x)(x)}{1.0} = 1.8 \times 10^{-5}$$

$$x^2 = (1.0)(1.8 \times 10^{-5})$$

$$x = \sqrt{1.8 \times 10^{-5}} = 4.2 \times 10^{-3}\ M$$

$$\left(\frac{4.2 \times 10^{-3}\ M}{1.0\ M}\right)(100) = 0.42\% \text{ ionized}$$

$$pH = -\log(4.2 \times 10^{-3}) = 2.38$$

(b) $[HC_2H_3O_2] = 0.10\ M$

$$\frac{(x)(x)}{0.10} = 1.8 \times 10^{-5}$$

$$x^2 = (0.10)(1.8 \times 10^{-5}) = 1.8 \times 10^{-6}$$

$x = \sqrt{1.8 \times 10^{-6}} = 1.3 \times 10^{-3}\,M$

$\left(\dfrac{1.3 \times 10^{-3}\,M}{0.10\,M}\right)(100) = 1.3\%$ ionized

$pH = -\log(1.3 \times 10^{-3}) = 2.89$

(c) $[HC_2H_3O_2] = 0.010\,M$

$\dfrac{(x)(x)}{0.010} = 1.8 \times 10^{-5}$

$x^2 = (0.010)(1.8 \times 10^{-5}) = 1.8 \times 10^{-7}$

$x = \sqrt{1.8 \times 10^{-7}} = 4.2 \times 10^{-4}\,M$

$\left(\dfrac{4.2 \times 10^{-4}\,M}{0.010\,M}\right)(100) = 4.2\%$ ionized

$pH = -\log(4.2 \times 10^{-4}) = 3.38$

54. $HClO \rightleftharpoons H^+ + ClO^-$

$K_a = \dfrac{[H^+][ClO^-]}{[HClO]} = 3.5 \times 10^{-8}$

Let x = molarity of H^+

$[H^+] = [ClO^-] = x$

$[HClO]$ = initial concentration $- x$ = initial concentration

Since K_a is small, the degree of ionization is small. Therefore, the approximation, initial concentration $- x$ = initial concentration, is valid.

(a) $[H^+] = [ClO^-] = x$ $\quad [HClO] = 1.0\,M$

$\dfrac{(x)(x)}{1.0} = 3.5 \times 10^{-8}$ $\qquad x^2 = (1.0)(3.5 \times 10^{-8})$

$x = \sqrt{3.5 \times 10^{-8}} = 1.9 \times 10^{-4}\,M$

$\left(\dfrac{1.9 \times 10^{-4}\,M}{1.0\,M}\right)(100) = 1.9 \times 10^{-2}\%$ ionized

$pH = -\log(1.9 \times 10^{-4}) = 3.72$

(b) $[HClO] = 0.10\,M$

$\dfrac{(x)(x)}{0.10} = 3.5 \times 10^{-8}$ $\qquad x^2 = (0.10)(3.5 \times 10^{-8})$

$x = \sqrt{3.5 \times 10^{-9}} = 5.9 \times 10^{-5}\,M$

$$\left(\frac{5.9 \times 10^{-5}\,\text{M}}{0.10\,\text{M}}\right)(100) = 0.059\%\ \text{ionized}$$

$$\text{pH} = -\log(5.9 \times 10^{-5}) = 4.23$$

(c) $[\text{HClO}] = 0.010\,\text{M}$

$$\frac{(x)(x)}{0.010} = 3.5 \times 10^{-8} \qquad x^2 = (0.010)(3.5 \times 10^{-8})$$

$$x = \sqrt{3.5 \times 10^{-10}} = 1.9 \times 10^{-5}\,\text{M}$$

$$\left(\frac{1.9 \times 10^{-5}\,\text{M}}{0.010\,\text{M}}\right)(100) = 0.19\%\ \text{ionized}$$

$$\text{pH} = -\log(1.9 \times 10^{-5}) = 4.72$$

55. $\text{HA} \rightleftharpoons \text{H}^+ + \text{A}^-$ $\qquad K_a = \dfrac{[\text{H}^+][\text{A}^-]}{[\text{HA}]}$

First, find the $[\text{H}^+]$. This is calculated from the pH expression, $\text{pH} = -\log[\text{H}^+] = 3.7$.

$[\text{H}^+] = 2 \times 10^{-4}$

$[\text{H}^+] = [\text{A}^-] = 2 \times 10^{-4} \qquad [\text{HA}] = 0.37$

$$K_a = \frac{[\text{H}^+][\text{A}^-]}{[\text{HA}]} = \frac{(2 \times 10^{-4})(2 \times 10^{-4})}{0.37} = 1 \times 10^{-7}$$

56. See problem 51 for a discussion of calculating $[\text{H}^+]$ from pH.

$\text{HA} \rightleftharpoons \text{H}^+ + \text{A}^- \qquad K_a = \dfrac{[\text{H}^+][\text{A}^-]}{[\text{HA}]} \qquad \text{pH} = 2.89$

$-\log[\text{H}^+] = 2.89 \qquad [\text{H}^+] = 1.3 \times 10^{-3}$

$[\text{H}^+] = 1.3 \times 10^{-3} = [\text{A}^-] \qquad [\text{HA}] = 0.23$

$$K_a = \frac{[\text{H}^+][\text{A}^-]}{[\text{HA}]} = \frac{(1.3 \times 10^{-3})(1.3 \times 10^{-3})}{0.23} = 7.3 \times 10^{-6}$$

57. 6.0 M HCl yields $[\text{H}^+] = 6.0\,\text{M}$ (100% ionized)

$\text{pH} = -\log 6.0 = -0.78$

$\text{pOH} = 14 - \text{pH} = 14 - (-0.78) = 14.78$

$$[\text{OH}^-] = \frac{K_w}{[\text{H}^+]} = \frac{1 \times 10^{-14}}{6.0} = 1.7 \times 10^{-15}$$

58. 1.0 M NaOH yields $[OH^-] = 1.0\,M$ (100% ionized)

$pOH = -\log 1.0 = 0.00$

$pH = 14 - pOH = 14.00$

$$[H^+] = \frac{K_w}{[OH^-]} = \frac{1.0 \times 10^{-14}}{1.0} = 1 \times 10^{-14}$$

59. $pH + pOH = 14.0 \qquad pOH = 14.0 - pH$

(a) 0.00010 M HCl $[H^+] = 0.00010\,M = 1.0 \times 10^{-4}\,M$

$pH = -\log(1.0 \times 10^{-4}) = 4.00$

$pOH = 14.0 - 4.00 = 10.0$

(b) 0.010 M NaOH $[OH^-] = 0.010\,M = 1.0 \times 10^{-2}\,M$

$pOH = -\log(1.0 \times 10^{-2}) = 2.00$

$pH = 14.0 - 2.00 = 12.0$

60. $pH + pOH = 14.0$

(a) 0.00250 M NaOH $[OH^-] = 2.5 \times 10^{-3}\,M$

$pOH = -\log(2.5 \times 10^{-3}) = 2.60$

$pH = 14.0 - 2.60 = 11.4$

(b) $HClO \rightleftharpoons H^+ + ClO^-$

0.10 M $\quad x \quad x$

$$K_a = \frac{[H^+][ClO^-]}{[HClO]} = 3.5 \times 10^{-8}$$

$$\frac{(x)(x)}{0.10} = 3.5 \times 10^{-8}$$

$x^2 = (0.10)(3.5 \times 10^{-8}) \qquad x = \sqrt{3.5 \times 10^{-9}}$

$x = 5.9 \times 10^{-5} = [H^+]$

$pH = -\log(5.9 \times 10^{-5}) = 4.23$

$pOH = 14.0 - 4.23 = 9.8$

(c) $Fe(OH)_2(s) \rightleftharpoons Fe^{2+} + 2\,OH^-$

$x \quad x \quad 2x$

$K_{sp} = [Fe^{2+}][OH^-]^2 = (x)(2x)^2 = 8.0 \times 10^{-16}$

$4x^3 = 8.0 \times 10^{-16}$

$$x = \sqrt[3]{\frac{8.0 \times 10^{-16}}{4}} = 5.8 \times 10^{-6}$$

$$[OH^-] = 2x = 2(5.8 \times 10^{-6}) = 1.2 \times 10^{-5}$$

$$pOH = -\log(1.2 \times 10^{-5}) = 4.92$$

$$pH = 14.0 - 4.92 = 9.1$$

61. Calculate the $[OH^-]$. $[OH^-] = \frac{K_w}{[H^+]}$

(a) $[H^+] = 1.0 \times 10^{-4}$ $[OH^-] = \frac{1.0 \times 10^{-14}}{1.0 \times 10^{-4}} = 1.0 \times 10^{-10}$

(b) $[H^+] = 2.8 \times 10^{-6}$ $[OH^-] = \frac{1.0 \times 10^{-14}}{2.8 \times 10^{-6}} = 3.6 \times 10^{-9}$

62. Calculate the $[OH^-]$. $[OH^-] = \frac{K_w}{[H^+]}$

(a) $[H^+] = 4.0 \times 10^{-9}$ $[OH^-] = \frac{1.0 \times 10^{-14}}{4.0 \times 10^{-9}} = 2.5 \times 10^{-6}$

(b) $[H^+] = 8.9 \times 10^{-2}$ $[OH^-] = \frac{1.0 \times 10^{-14}}{8.9 \times 10^{-2}} = 1.1 \times 10^{-13}$

63. Calculate the $[H^+]$. $[H^+] = \frac{K_w}{[OH^-]}$

(a) $[OH^-] = 6.0 \times 10^{-7}$ $[H^+] = \frac{1.0 \times 10^{-14}}{6.0 \times 10^{-7}} = 1.7 \times 10^{-8}$

(b) $[OH^-] = 1 \times 10^{-8}$ $[H^+] = \frac{1 \times 10^{-14}}{1 \times 10^{-8}} = 1 \times 10^{-6}$

64. Calculate the $[H^+]$. $[H^+] = \frac{K_w}{[OH^-]}$

(a) $[OH^-] = 4.5 \times 10^{-6}$ $[H^+] = \frac{1.0 \times 10^{-14}}{4.5 \times 10^{-6}} = 2.2 \times 10^{-9}$

(b) $[OH^-] = 7.3 \times 10^{-4}$ $[H^+] = \frac{1 \times 10^{-14}}{7.3 \times 10^{-4}} = 1.4 \times 10^{-11}$

65. The molar solubilities of the salts and their ions are indicated below the formulas in the equilibrium equations.

(a) $BaSO_4(s) \rightleftharpoons \underset{3.9 \times 10^{-5}}{Ba^{2+}} + \underset{3.9 \times 10^{-5}}{SO_4^{2-}}$

$K_{sp} = [Ba^{2+}][SO_4^{2-}] = (3.9 \times 10^{-5})^2 = 1.5 \times 10^{-9}$

(b) $Ag_2CrO_4(s) \rightleftharpoons 2\,Ag^+ + CrO_4^{2-}$

$2(7.8 \times 10^{-5}) \quad 7.8 \times 10^{-5}$

$K_{sp} = [Ag^+]^2[CrO_4^{2-}] = (15.6 \times 10^{-5})^2(7.8 \times 10^{-5}) = 1.9 \times 10^{-12}$

(c) First change g/L $\longrightarrow$ mol/L

$$\left(\frac{0.67\text{ g }CaSO_4}{L}\right)\left(\frac{1\text{ mol}}{136.1\text{ g}}\right) = 4.9 \times 10^{-3}\text{ M }CaSO_4$$

$CaSO_4(s) \rightleftharpoons Ca^{2+} + SO_4^{2-}$

$4.9 \times 10^{-3} \quad 4.9 \times 10^{-3}$

$K_{sp} = [Ca^{2+}][SO_4^{2-}] = (4.9 \times 10^{-3})^2 = 2.4 \times 10^{-5}$

(d) First change g/L $\longrightarrow$ mol/L

$$\left(\frac{0.0019\text{ g AgCl}}{L}\right)\left(\frac{1\text{ mol}}{143.4\text{ g}}\right) = 1.3 \times 10^{-5}\text{ M AgCl}$$

$AgCl(s) \rightleftharpoons Ag^+ + Cl^-$

$1.3 \times 10^{-5} \quad 1.3 \times 10^{-5}$

$K_{sp} = [Ag^+][Cl^-] = (1.3 \times 10^{-5})^2 = 1.7 \times 10^{-10}$

66. The molar solubilities of the salts and their ions are indicated below the formulas in the equilibrium equations.

(a) $ZnS(s) \rightleftharpoons Zn^{2+} + S^{2-}$

$3.5 \times 10^{-12} \quad 3.5 \times 10^{-12}$

$K_{sp} = [Zn^{2+}][S^{2-}] = (3.5 \times 10^{-12})^2 = 1.2 \times 10^{-23}$

(b) $Pb(IO_3)_2(s) \rightleftharpoons Pb^{2+} + 2\,IO_3^-$

$4.0 \times 10^{-5} \quad 2(4.0 \times 10^{-5})$

$K_{sp} = [Pb^{2+}][IO_3^-]^2 = (4.0 \times 10^{-5})(8.0 \times 10^{-5})^2 = 2.6 \times 10^{-13}$

(c) First change g/L $\longrightarrow$ mol/L

$$\left(\frac{6.73 \times 10^{-3}\text{ g }Ag_3PO_4}{L}\right)\left(\frac{1\text{ mol}}{418.7\text{ g}}\right) = 1.61 \times 10^{-5}\text{ M }Ag_3PO_4$$

$Ag_3PO_4(s) \rightleftharpoons 3\,Ag^+ + PO_4^{3-}$

$3(1.61 \times 10^{-5}) \quad 1.61 \times 10^{-5}$

$K_{sp} = [Ag^+]^3[PO_4^{3-}] = (4.83 \times 10^{-5})^3(1.61 \times 10^{-5}) = 1.81 \times 10^{-18}$

(d) First change g/L $\longrightarrow$ mol/L

$$\left(\frac{2.33 \times 10^{-4}\ \text{g}\ Zn(OH)_2}{\text{L}}\right)\left(\frac{1\ \text{mol}}{99.41\ \text{g}}\right) = 2.34 \times 10^{-6}\ \text{M}\ Zn(OH)_2$$

$$Zn(OH)_2(s) \rightleftharpoons Zn^{2+} + 2\,OH^-$$

$$2.34 \times 10^{-6} \quad 2(2.34 \times 10^{-6})$$

$$K_{sp} = [Zn^{2+}][OH^-]^2 = (2.34 \times 10^{-6})(4.68 \times 10^{-6})^2 = 5.13 \times 10^{-17}$$

67. (a) $Ag_2SO_4(s) \rightleftharpoons 2\,Ag^+ + SO_4^{2-}$

$$2x \qquad x$$

x = molar solubility

$$K_{sp} = [Ag^+]^2[SO_4^{2-}] = (2x)^2(x) = 4x^3 = 1.5 \times 10^{-5}$$

$$x = \sqrt[3]{\frac{1.5 \times 10^{-5}}{4}} = 1.6 \times 10^{-2}\ \text{M}$$

(b) $Mg(OH)_2(s) \rightleftharpoons Mg^{2+} + 2\,OH^-$

$$x \qquad 2x$$

x = molar solubility

$$K_{sp} = [Mg^{2+}][OH^-]^2 = (x)(2x)^2 = 4x^3 = 7.1 \times 10^{-12}$$

$$x = \sqrt[3]{\frac{7.1 \times 10^{-12}}{4}} = 1.2 \times 10^{-4}\ \text{M}$$

68. The molar solubilities of the salts and their ions will be represented in terms of x below their formulas in the equilibrium equations.

(a) $BaCO_3(s) \rightleftharpoons Ba^{2+} + CO_3^{2-}$

$$x \qquad x$$

x = molar solubility

$$K_{sp} = [Ba^{2+}][CO_3^{2-}] = x^2 = 2.0 \times 10^{-9}$$

$$x = \sqrt{2.0 \times 10^{-9}} = 4.5 \times 10^{-5}\ \text{M}$$

(b) $AlPO_4(s) \rightleftharpoons Al^{3+} + PO_4^{3-}$

$$x \qquad x$$

x = molar solubility

$$K_{sp} = [Al^{3+}][PO_4^{3-}] = x^2 = 5.8 \times 10^{-19}$$

$$x = \sqrt{5.8 \times 10^{-19}} = 7.6 \times 10^{-10}\ \text{M}$$

69. (a) $\left(\dfrac{1.6 \times 10^{-2}\text{ mol Ag}_2\text{SO}_4}{\text{L}}\right)(0.100\text{ L})\left(\dfrac{311.9\text{ g}}{\text{mol}}\right) = 0.50\text{ g Ag}_2\text{SO}_4$

(b) $\left(\dfrac{1.2 \times 10^{-4}\text{ mol Mg(OH)}_2}{\text{L}}\right)(0.100\text{ L})\left(\dfrac{58.33\text{ g}}{\text{mol}}\right) = 7.0 \times 10^{-4}\text{ g Mg(OH)}_2$

70. (a) $\left(\dfrac{4.5 \times 10^{-5}\text{ mol BaCO}_3}{\text{L}}\right)(0.100\text{ L})\left(\dfrac{197.3\text{ g}}{\text{mol}}\right) = 8.9 \times 10^{-4}\text{ g BaCO}_3$

(b) $\left(\dfrac{7.6 \times 10^{-10}\text{ mol AlPO}_4}{\text{L}}\right)(0.100\text{ L})\left(\dfrac{122.0\text{ g}}{\text{mol}}\right) = 9.3 \times 10^{-9}\text{ g AlPO}_4$

71. The molar concentrations of ions, after mixing, are calculated and these concentrations are substituted into the equilibrium expression. The value obtained is compared to the K_{sp} of the salt. If the value is greater than the K_{sp}, precipitation occurs. If the value is less than the K_{sp}, no precipitation occurs.

100. mL 0.010 M $Na_2SO_4 \longrightarrow$ 100. mL 0.010 M SO_4^{2-}

100. mL 0.001 M $Pb(NO_3)_2 \longrightarrow$ 100. mL 0.001 M Pb^{2+}

Volume after mixing = 200. mL

Concentrations after mixing: $SO_4^{2-} = 0.0050$ M $Pb^{2+} = 0.0005$ M

$[Pb^{2+}][SO_4^{2-}] = (5.0 \times 10^{-3})(5 \times 10^{-4}) = 3 \times 10^{-6}$

$K_{sp} = 1.3 \times 10^{-8}$ which is less than 3×10^{-6}, therefore, precipitation occurs.

72. The molar concentrations of ions, after mixing, are calculated and these concentrations are substituted into the equilibrium expression. The value obtained is compared to the K_{sp} of the salt. If the value is greater than the K_{sp}, precipitation occurs. If the value is less than the K_{sp}, no precipitation occurs.

50.0 mL 1.0×10^{-4} M $AgNO_3 \longrightarrow$ 50.0 mL 1.0×10^{-4} M Ag^+

100. mL 1.0×10^{-4} M NaCl $\longrightarrow$ 100. mL 1.0×10^{-4} M Cl^-

Volume after mixing = 150. mL

Concentrations after mixing:

$$(1.0 \times 10^{-4}\text{ M Ag}^+)\left(\frac{50.0\text{ mL}}{150.\text{ mL}}\right) = 3.3 \times 10^{-5}\text{ M Ag}^+$$

$$(1.0 \times 10^{-4}\text{ M Cl}^-)\left(\frac{100.\text{ mL}}{150.\text{ mL}}\right) = 6.7 \times 10^{-5}\text{ M Cl}^-$$

$[Ag^+][Cl^-] = (3.3 \times 10^{-5})(6.7 \times 10^{-5}) = 2.2 \times 10^{-9}$

$K_{sp} = 1.7 \times 10^{-10}$ which is less than 2.2×10^{-9}, therefore, precipitation occurs.

73. The concentration of Br^- = 0.10 M in 1.0 L of 0.10 M NaBr. Substitute this Br^- concentration in the K_{sp} expression and solve for the $[Ag^+]$ in equilibrium with 0.10 M Br^-.

$$K_{sp} = [Ag^+][Br^-] = 5.0 \times 10^{-13}$$

$$[Ag^+] = \frac{5.0 \times 10^{-13}}{[Br^-]} = \frac{5.0 \times 10^{-13}}{0.10} = 5.0 \times 10^{-12}\ M$$

$$\left(\frac{5.0 \times 10^{-12}\ \text{mol}\ Ag^+}{L}\right)\left(\frac{1\ \text{mol AgBr}}{1\ \text{mol}\ Ag^+}\right)(1.0\ L) = 5.0 \times 10^{-12}\ \text{mol AgBr will dissolve}$$

74. $$\left(\frac{0.10\ \text{mol}\ MgBr_2}{L}\right)\left(\frac{2\ \text{mol}\ Br^-}{1\ \text{mol}\ MgBr_2}\right) = \left(\frac{0.20\ \text{mol}\ Br^-}{L}\right) = 0.20\ M\ Br^-\ \text{in solution.}$$

Substitute the Br^- concentration in the K_{sp} expression and solve for $[Ag^+]$ in equilibrium with 0.20 M Br^-.

$$[Ag^+] = \frac{5.0 \times 10^{-13}}{[Br^-]} = \frac{5.0 \times 10^{-13}}{0.20} = 2.5 \times 10^{-12}\ M$$

$$\left(\frac{2.5 \times 10^{-12}\ \text{mol}\ Ag^+}{L}\right)\left(\frac{1\ \text{mol AgBr}}{1\ \text{mol}\ Ag^+}\right)(1.0\ L) = 2.5 \times 10^{-12}\ \text{mol AgBr will dissolve}$$

75. $HC_2H_3O_2 \rightleftharpoons H^+ + C_2H_3O_2^-$

$$K_a = \frac{[H^+][C_2H_3O_2^-]}{[HC_2H_3O_2]} = 1.8 \times 10^{-5}$$

$$[H^+] = K_a\left(\frac{[HC_2H_3O_2]}{[C_2H_3O_2^-]}\right) \qquad [HC_2H_3O_2] = 0.20\ M \qquad [C_2H_3O_2^-] = 0.10\ M$$

$$[H^+] = (1.8 \times 10^{-5})\left(\frac{0.20}{0.10}\right) = 3.6 \times 10^{-5}\ M$$

$$pH = -\log(3.6 \times 10^{-5}) = 4.44$$

76. $HC_2H_3O_2 \rightleftharpoons H^+ + C_2H_3O_2^-$

$$K_a = \frac{[H^+][C_2H_3O_2^-]}{[HC_2H_3O_2]} = 1.8 \times 10^{-5}$$

$$[H^+] = K_a\left(\frac{[HC_2H_3O_2]}{[C_2H_3O_2^-]}\right) \qquad [HC_2H_3O_2] = 0.20\ M \qquad [C_2H_3O_2^-] = 0.20\ M$$

$$[H^+] = (1.8 \times 10^{-5})\left(\frac{0.20}{0.20}\right) = 1.8 \times 10^{-5}\ M$$

$$pH = -\log(1.8 \times 10^{-5}) = 4.74$$

77. Initially, the solution of NaCl is neutral. $[H^+] = 1 \times 10^{-7}$
$pH = -\log(1 \times 10^{-7}) = 7.0$
Final $H^+ = 2.0 \times 10^{-2}$ M
$pH = -\log(2.0 \times 10^{-2}) = 1.70$
Change in pH $= 7.0 - 1.70 = 5.3$ units in the unbuffered solution

78. Initially, $[H^+] = 1.8 \times 10^{-5}$
$pH = -\log(1.8 \times 10^{-5}) = 4.74$
Final $[H^+] = 1.9 \times 10^{-5}$
$pH = -\log(1.9 \times 10^{-5}) = 4.72$
Change in pH $= 4.74 - 4.72 = 0.02$ units in the buffered solution

79. The concentration of solid salt is not included in the K_{sp} equilibrium constant because the concentration of solid does not change. It is constant and part of the K_{sp}.

80. The energy diagram represents an exothermic reaction because the energy of the products is lower than the energy of the reactants. This means that energy was given off during the reaction.

81.

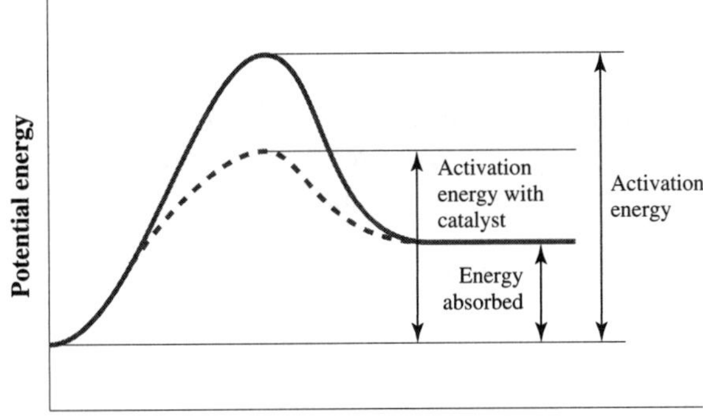

82. $H_2 + I_2 \rightleftharpoons 2\,HI$
The reaction is a 1 to 1 mole ratio of hydrogen to iodine. The data given indicates that hydrogen is the limiting reactant.

$$(2.10 \text{ mol } H_2)\left(\frac{2 \text{ mol HI}}{1 \text{ mol } H_2}\right) = 4.20 \text{ mol HI}$$

83. $H_2 + I_2 \rightleftharpoons 2\,HI$

(a) 2.00 mol H_2 and 2.00 mol I_2 will produce 4.00 mol HI assuming 100% yield. However, at 79% yield you get

$$4.00 \text{ mol HI} \times 0.79 = 3.16 \text{ mol HI}$$

(b) The addition of 0.27 mol I_2 makes the iodine present in excess and the 2.00 mol H_2 the limiting reactant. The yield increases to 85%.

$$(2.00 \text{ mol } H_2)\left(\frac{2 \text{ mol HI}}{1 \text{ mol } H_2}\right)(0.85) = 3.4 \text{ mol HI}$$

There will be 15% unreacted H_2 and I_2 plus the extra I_2 added.

$(0.15)(2.0 \text{ mol } H_2) = 0.30 \text{ mol } H_2$ present; also 0.30 mol I_2.

In addition to the 0.30 mol of unreacted I_2, will be the 0.27 mol I_2 added.

$0.27 \text{ mol} + 0.30 \text{ mol} = 0.57 \text{ mol } I_2$ present.

(c) $$K = \frac{[HI]^2}{[H_2][I_2]}$$

The formation of 3.16 mol HI required the reaction of 1.58 mol I_2 and 1.58 mol H_2. At equilibrium, the concentrations are:

3.16 mol HI; $\quad 2.00 - 1.58 = 0.42 \text{ mol } H_2 = 0.42 \text{ mol } I_2$

$$K_{eq} = \frac{(3.16)^2}{(0.42)(0.42)} = 57$$

In the calculation of the equilibrium constant, the actual number of moles of reactants and products present at equilibrium can be used in the calculation in place of molar concentrations. This occurs because the reaction is gaseous and the liters of HI produced equals the sum of the liters of H_2 and I_2 reacting. In the equilibrium expression, the volumes will cancel.

84. $H_2 + I_2 \rightleftharpoons 2\,HI$

$$(64.0 \text{ g HI})\left(\frac{1 \text{ mol}}{127.9 \text{ g}}\right) = 0.500 \text{ mol HI present}$$

$$(0.500 \text{ mol HI})\left(\frac{1 \text{ mol } I_2}{2 \text{ mol HI}}\right) = 0.250 \text{ mol } I_2 \text{ reacted}$$

$$(0.500 \text{ mol HI})\left(\frac{1 \text{ mol } H_2}{2 \text{ mol HI}}\right) = 0.250 \text{ mol } H_2 \text{ reacted}$$

$$(6.00 \text{ g } H_2)\left(\frac{1 \text{ mol}}{2.016 \text{ g}}\right) = 2.98 \text{ mol } H_2 \text{ initially present}$$

$$(200. \text{ g } I_2)\left(\frac{1 \text{ mol}}{253.8 \text{ g}}\right) = 0.788 \text{ mol } I_2 \text{ initially present}$$

At equilibrium, moles present are:

0.500 mol HI; $2.98 - 0.250 = 2.73$ mol H_2

$0.788 - 0.250 = 0.538$ mol I_2

85. $PCl_3(g) + Cl_2(g) \rightleftharpoons PCl_5(g)$

$$K_{eq} = \frac{[PCl_5]}{[PCl_3][Cl_2]}$$

The concentrations are:

$$PCl_5 = \frac{0.22\text{ mol}}{20.\text{ L}} = 0.011\text{ M}$$

$$PCl_3 = \frac{0.10\text{ mol}}{20.\text{ L}} = 0.0050\text{ M}$$

$$Cl_2 = \frac{1.50\text{ mol}}{20.\text{ L}} = 0.075\text{ M}$$

$$K_{eq} = \frac{0.011}{(0.0050)(0.075)} = 29$$

86. $100°C - 30°C = 70°C$ temperature increase. This increase is equal to seven 10°C increments. The reaction rate will be increased by $2^7 = 128$ times.

87. $NH_4^+ \rightleftharpoons NH_3 + H^+ \quad K_{eq} = 5.6 \times 10^{-10}$

0.30 M $\quad y \quad y$

$y = [H^+] = [NH_3]$

$$\frac{[NH_3][H^+]}{[NH_4^+]} = \frac{[y][y]}{[0.30]} = 5.6 \times 10^{-10}$$

$y^2 = (5.6 \times 10^{-10})(0.30) = 1.7 \times 10^{-10}$

$y = [H^+] = 1.3 \times 10^{-5}$

$pH = -\log(1.3 \times 10^{-5}) = 4.89$ (on acidic solution)

88. pH of an acetic acid-acetate buffer

$HC_2H_3O_2 \rightleftharpoons H^+ + C_2H_3O_2^-$

$$K_a = \frac{[H^+][C_2H_3O_2^-]}{[HC_2H_3O_2]} = \frac{[H^+][0.30]}{0.20} = 1.8 \times 10^{-5}$$

$$H^+ = \frac{(1.8 \times 10^{-5})(0.20)}{0.30} = 2.7 \times 10^{-5}$$

$pH = -\log(2.7 \times 10^{-5}) = 4.57$

89. Concentration of Ba^{2+} in solution

$BaCl_2(aq) + Na_2CrO_4(aq) \rightleftharpoons BaCrO_4(s) + NaCl(aq)$

$BaCrO_4 \rightleftharpoons Ba^{2+} + CrO_4^{2-} \qquad K_{sp} = 8.5 \times 10^{-11}$

Determine the moles of Ba^{2+} and CrO_4^{2-} in solution

$$\left(\frac{0.10\text{ mol}}{\text{L}}\right)(0.050\text{ L}) = 0.0050\text{ mol Ba}^{2+}$$

$$\left(\frac{0.15\text{ mol}}{\text{L}}\right)(0.050\text{ L}) = 0.0075\text{ mol CrO}_4^{2-}$$

Excess CrO_4^{2-} in solution $= 0.0025$ mol (2.5×10^{-3})

Concentration of CrO_4^{2-} in solution (total volume = 100 mL)

$$\frac{2.5 \times 10^{-3}\text{ mol CrO}_4^{2-}}{0.100\text{ L}} = 2.5 \times 10^{-2}\text{ M CrO}_4^{2-}$$

Now using the K_{sp}, calculate the Ba^{2+} remaining in solution.

$$[Ba^{2+}][CrO_4^{2-}] = [Ba^{2+}][2.5 \times 10^{-2}] = 8.5 \times 10^{-11}$$

$$[Ba^{2+}] = \frac{8.5 \times 10^{-11}}{2.5 \times 10^{-2}} = 3.4 \times 10^{-9}\text{ mol/L}$$

90. **Hypochlorous acid** $HOCl \rightleftharpoons H^+ + OCl^-$

Equilibrium concentrations:

$[H^+] = [OCl^-] = 5.9 \times 10^{-5}$ M

$[HOCl] = 0.1 - 5.9 \times 10^{-5} = 0.10$ M (neglecting 5.9×10^{-5})

$$K_a = \frac{[H^+][OCl^-]}{[HOCl]} = \frac{(5.9 \times 10^{-5})(5.9 \times 10^{-5})}{0.10} = 3.5 \times 10^{-8}$$

Propanoic acid $HC_3H_5O_2 \rightleftharpoons H^+ + C_3H_5O_2^-$

Equilibrium concentrations:

$[H^+] = [C_3H_5O_2^-] = 1.4 \times 10^{-3}$ M

$[HC_3H_5O_2] = 0.15 - 1.4 \times 10^{-3} = 0.15$ M (neglecting 1.4×10^{-3})

$$K_a = \frac{[H^+][C_3H_5O_2^-]}{[HC_3H_5O_2]} = \frac{(1.4 \times 10^{-3})(1.4 \times 10^{-3})}{0.15} = 1.3 \times 10^{-5}$$

Hydrocyanic acid $HCN \rightleftharpoons H^+ + CN^-$

Equilibrium concentrations:

$[H^+] = [CN^-] = 8.9 \times 10^{-6}$ M

$[HCN] = 0.20 - 8.9 \times 10^{-6} = 0.20$ M (neglecting 8.9×10^{-6})

$$K_a = \frac{[H^+][CN^-]}{[HCN]} = \frac{(8.9 \times 10^{-6})^2}{0.20} = 4.0 \times 10^{-10}$$

91. Let y = M CaF_2 dissolving

$CaF_2(s) \rightleftharpoons Ca^{2+} + 2\,F^-$

$\quad\quad\quad\quad y \quad\quad 2y$

y = molar solubility

(a) $K_{sp} = [Ca^{2+}][F^-]^2 = (y)(2y)^2 = 4y^3 = 3.9 \times 10^{-11}$

$$y = \sqrt[3]{\frac{3.9 \times 10^{-11}}{4}} = 2.1 \times 10^{-4}\text{ M } (CaF_2\text{ dissolved})$$

$$\left(\frac{2.1 \times 10^{-4}\text{ mol } CaF_2}{\text{L}}\right)\left(\frac{1\text{ mol } Ca^{2+}}{1\text{ mol } CaF_2}\right) = 2.1 \times 10^{-4}\text{ M } Ca^{2+}$$

$$\left(\frac{2.1 \times 10^{-4}\text{ mol } CaF_2}{\text{L}}\right)\left(\frac{2\text{ mol } F^-}{1\text{ mol } CaF_2}\right) = 4.2 \times 10^{-4}\text{ M } F^-$$

(b) $$\left(\frac{2.1 \times 10^{-4}\text{ mol } CaF_2}{\text{L}}\right)(0.500\text{ L})\left(\frac{78.08\text{ g}}{\text{mol}}\right) = 8.2 \times 10^{-3}\text{ g } CaF_2$$

92. The molar concentrations of ions, after mixing, are calculated and these concentrations are substituted into the equilibrium expression. The value obtained is compared to the K_{sp} of the salt. If the value is greater than the K_{sp}, precipitation occurs. If the value is less than the K_{sp}, no precipitation occurs.

(a) 100. mL 0.010 M $Na_2SO_4 \longrightarrow$ 100. mL 0.010 M SO_4^{2-}

100. mL 0.001 M $Pb(NO_3)_2 \longrightarrow$ 100. mL 0.001 M Pb^{2+}

Volume after mixing = 200. mL

Concentrations after mixing: $SO_4^{2-} = 0.0050$ M $\quad Pb^{2+} = 0.0005$ M

$[Pb^{2+}][SO_4^{2-}] = (5.0 \times 10^{-3})(5 \times 10^{-4}) = 3 \times 10^{-6}$

$K_{sp} = 1.3 \times 10^{-8}$ which is less than 3×10^{-6}, therefore, precipitation occurs.

(b) 50.0 mL 1.0×10^{-4} M $AgNO_3 \longrightarrow$ 50.0 mL 1.0×10^{-4} M Ag^+

100. mL 1.0×10^{-4} M NaCl $\longrightarrow$ 100. mL 1.0×10^{-4} M Cl^-

Volume after mixing = 150. mL

Concentrations after mixing:

$$(1.0 \times 10^{-4})\left(\frac{50.0\text{ mL}}{150.\text{ mL}}\right) = 3.3 \times 10^{-5}\text{ M } Ag^+$$

$$(1.0 \times 10^{-4})\left(\frac{100.\text{ mL}}{150.\text{ mL}}\right) = 6.7 \times 10^{-5}\text{ M } Cl^-$$

$[Ag^+][Cl^-] = (3.3 \times 10^{-5})(6.7 \times 10^{-5}) = 2.2 \times 10^{-9}$

$K_{sp} = 1.7 \times 10^{-10}$ which is less than 2.2×10^{-9}, therefore, precipitation occurs.

(c) Convert g $Ca(NO_3)_2$ to g Ca^{2+}

$$\left(\frac{1.0\text{ g Ca(NO}_3)_2}{0.150\text{ L}}\right)\left(\frac{1\text{ mol}}{164.1\text{ g}}\right)\left(\frac{1\text{ mol Ca}^{2+}}{1\text{ mol Ca(NO}_2)_2}\right) = 0.041\text{ M Ca}^{2+}$$

250 mL 0.01 M NaOH $\longrightarrow$ 250 mL 0.01 M OH^-

Final volume $= 4.0 \times 10^2$ mL

Concentration after mixing:

$$(0.041\text{ M Ca}^{2+})\left(\frac{150\text{ mL}}{4.0 \times 10^2\text{ mL}}\right) = 0.015\text{ M Ca}^{2+}$$

$$(0.01\text{ M OH}^-)\left(\frac{250\text{ mL}}{4.0 \times 10^2\text{ mL}}\right) = 0.0063\text{ M OH}^-$$

$[Ca^{2+}][OH^-]^2 = (0.015)(0.0063)^2 = 6.0 \times 10^{-7}$

$K_{sp} = 1.3 \times 10^{-6}$ which is greater than 6.0×10^{-7}, therefore, no precipitation occurs.

93. With a known Ba^{2+} concentration, the SO_4^{2-} concentration can be calculated using the K_{sp} value. $BaSO_4(s) \rightleftharpoons Ba^{2+} + SO_4^{2-}$

$K_{sp} = [Ba^{2+}][SO_4^{2-}] = 1.5 \times 10^{-9}$ $\quad Ba^{2+} = 0.050$ M

(a) $[SO_4^{2-}] = \dfrac{K_{sp}}{[Ba^{2+}]} = \dfrac{1.5 \times 10^{-9}}{0.050} = 3.0 \times 10^{-8}$ M SO_4^{2-} in solution

(b) M SO_4^{2-} = M $BaSO_4$ in solution

$$\left(\frac{3.0 \times 10^{-8}\text{ mol BaSO}_4}{\text{L}}\right)(0.100\text{ L})\left(\frac{233.4\text{ g}}{\text{mol}}\right)$$

$= 7.0 \times 10^{-7}$ g $BaSO_4$ remain in solution

94. If $[Pb^{2+}][Cl^-]^2$ exceeds the K_{sp}, precipitation will occur.

$K_{sp} = [Pb^{2+}][Cl^-]^2 = 2.0 \times 10^{-5}$

0.050 M $Pb(NO_3)_2$ $\longrightarrow$ 0.050 M Pb^{2+}

0.010 M NaCl $\longrightarrow$ 0.010 M Cl^-

$(0.050)(0.010)^2 = 5.0 \times 10^{-6}$

$[Pb^{2+}][Cl^-]^2$ is smaller than the K_{sp} value. Therefore, no precipitate of $PbCl_2$ will form.

95. $[Ba^{2+}][SO_4^{2-}] = 1.5 \times 10^{-9}$ $\quad [Sr^{2+}][SO_4^{2-}] = 3.5 \times 10^{-7}$

Both cations are present in equal concentrations (0.10 M). Therefore, as SO_4^{2-} is added, the K_{sp} of $BaSO_4$ will be exceeded before that of $SrSO_4$. $BaSO_4$ precipitates first.

96. $2\,SO_2(g) + O_2(g) \rightleftharpoons 2\,SO_3(g)$

$$K_{eq} = \frac{[SO_3]^2}{[SO_2]^2[O_2]} = \frac{(11.0)^2}{(4.20)^2(0.60 \times 10^{-3})} = 1.1 \times 10^4$$

97. $(0.048\text{ g BaF}_2)\left(\frac{1\text{ mol}}{175.3\text{ g}}\right) = 2.7 \times 10^{-4}\text{ mol BaF}_2$

$$\left(\frac{2.7 \times 10^{-4}\text{ mol}}{0.015\text{ L}}\right) = 1.8 \times 10^{-2}\text{ M BaF}_2\text{ dissolved}$$

$BaF_2(s) \rightleftharpoons Ba^{2+} + 2\,F^-$

1.8×10^{-2} $\quad 2(1.8 \times 10^{-2})$ (molar concentration)

$$K_{sp} = [Ba^{2+}][F^-]^2 = (1.8 \times 10^{-2})(3.6 \times 10^{-2})^2 = 2.3 \times 10^{-5}$$

98. $N_2 + 3\,H_2 \rightleftharpoons 2\,NH_3$

$$K_{eq} = \frac{[NH_3]^2}{[N_2][H_2]^3} = 4.0 \qquad \text{Let } y = [NH_3]$$

$$4.0 = \frac{y^2}{(2.0)(2.0)^3} \qquad y^2 = 64 \qquad y = \sqrt{64}$$

$y = 8.0\text{ M} = [NH_3]$

99. Total volume of mixture = 40.0 mL (0.0400 L)

$$K_{sp} = [Sr^{2+}][SO_4^{2-}] = 7.6 \times 10^{-7}$$

$$[Sr^{2+}] = \frac{(1.0 \times 10^{-3}\text{ M})(0.0250\text{ L})}{0.0400\text{ L}} = 6.3 \times 10^{-4}\text{ M}$$

$$[SO_4^{2-}] = \frac{(2.0 \times 10^{-3}\text{ M})(0.0150\text{ L})}{0.0400\text{ L}} = 7.5 \times 10^{-4}\text{ M}$$

$$[Sr^{2+}][SO_4^{2-}] = (6.3 \times 10^{-4})(7.5 \times 10^{-4}) = 4.7 \times 10^{-7}$$

$4.7 \times 10^{-7} < 7.6 \times 10^{-7}$ no precipitation should occur.

100. First change g $Hg_2I_2 \longrightarrow$ mol Hg_2I_2

$$\left(\frac{3.04 \times 10^{-7}\text{ g Hg}_2\text{I}_2}{\text{L}}\right)\left(\frac{1\text{ mol}}{655.0\text{ g}}\right) = 4.64 \times 10^{-10}\text{ M Hg}_2\text{I}_2 \quad \text{(molar solubility)}$$

$Hg_2I_2 \rightleftharpoons Hg_2^{2+} + 2\,I^-$

$4.64 \times 10^{-10}\text{ M}$ $\quad 2(4.64 \times 10^{-10}\text{ M})$

$$K_{sp} = [Hg_2^{2+}][I^-]^2 = (4.64 \times 10^{-10})(9.28 \times 10^{-10})^2 = 4.00 \times 10^{-28}$$

101. $3\,O_2(g) + \text{heat} \rightleftharpoons 2\,O_3(g)$

Three ways to increase ozone

(a) increase heat

(b) increase amount of O_2

(c) increase pressure

(d) remove O_3 as it is made

102. $H_2O(l) \rightleftharpoons H_2O(g)$

Conditions on the second day

(a) the temperature could have been cooler

(b) the humidity in the air could have been higher

(c) the air pressure could have been greater

103. $CO(g) + H_2O(g) \rightleftharpoons CO_2(g) + H_2(g)$

(c) is the correct answer

$$K_{eq} = \frac{[CO_2][H_2]}{[CO][H_2O]} = 1$$

With equal concentrations of products and reactants, the K_{eq} value will equal 1.

104. (a) $K_{eq} = \dfrac{[O_3]^2}{[O_2]^3}$

(b) $K_{eq} = \dfrac{[H_2O(l)]}{[H_2O(g)]}$

(c) $K_{eq} = \dfrac{[MgO][CO_2]}{[MgCO_3]}$

(d) $K_{eq} = \dfrac{[Bi_2S_3][H^+]^6}{[Bi^{3+}]^2[H_2S]^3}$

105.

2A	+	B	$\rightleftharpoons$	C	
1.0 M		1.0 M		0	Initial conditions
$1.0 - 2(0.30)$		$1.0 - 0.30$		0.30	Equilibrium concentrations
0.4 M		0.7 M		0.30 M	

$$K_{eq} = \frac{[C]}{[A]^2[B]} = \frac{0.30}{(0.4)^2(0.7)} = 3$$

106. Since the second reaction is the reverse of the first, the K_{eq} value of the second reaction will be the reciprocal of the K_{eq} value of the first reaction.

$$K_{eq} = \frac{[I_2][Cl_2]}{[ICl]^2} = 2.2 \times 10^{-3} \quad \text{(first reaction)}$$

$$K_{eq} = \frac{[ICl]^2}{[I_2][Cl_2]} \qquad K_{eq} = \frac{1}{2.2 \times 10^{-3}} = 450$$

107. $HNO_2(aq) \rightleftharpoons H^+(aq) + NO_2^-(aq)$

OH^- reacts with H^+ and equilibrium shifts to the right.

(a) After an initial increase, $[OH^-]$ will be neutralized and equilibrium shifts to the right.

(b) $[H^+]$ will be reduced (reacts with OH^-). Equilibrium shifts to the right.

(c) $[NO_2^-]$ increases as equilibrium shifts to the right.

(d) $[HNO_2]$ decreases and equilibrium shifts to the right

108. $CaSO_4(s) \rightleftharpoons Ca^{2+}(aq) + SO_4^{2-}(aq)$

$K_{sp} = [Ca^{2+}][SO_4^{2-}] = 2.0 \times 10^{-4}$

Let x = moles $CaSO_4$ that dissolve per L $= [Ca^{2+}] = [SO_4^{2-}]$

$(x)(x) = 2.0 \times 10^{-4} \qquad x = \sqrt{2.0 \times 10^{-4}}$

$x = 0.014\text{ M } CaSO_4$

$\text{M} \longrightarrow \text{moles} \longrightarrow \text{grams}$

$$\left(\frac{0.014\text{ mol } CaSO_4}{\text{L}}\right)(0.600\text{ L})\left(\frac{136.2\text{ g}}{\text{mol}}\right) = 1.1\text{ g } CaSO_4$$

109. $PbF_2(s) \rightleftharpoons Pb^{2+} + 2\,F^-$

change g $PbF_2 \longrightarrow$ mol PbF_2

$$\left(\frac{0.098\text{ g } PbF_2}{0.400\text{ L}}\right)\left(\frac{1\text{ mol}}{245.2\text{ g}}\right) = 1.0 \times 10^{-3}\text{ mol/L} = 1.0 \times 10^{-3}\text{ M } PbF_2$$

$K_{sp} = (Pb^{2+})(F^-)^2$

$[Pb^{2+}] = 1.0 \times 10^{-3}$; $[F^-] = 2(1.0 \times 10^{-3}) = 2.0 \times 10^{-3}$

$K_{sp} = (1.0 \times 10^{-3})(2.0 \times 10^{-3})^2 = 4.0 \times 10^{-9}$

110. Treat this as an equilibrium where W = whole nuts, S = shell halves, and K = kernels

W	$\rightleftharpoons$	2 S	+	K	
144		0		0	amount before cracking
$144 - x$		$2x$		x	x = number of kernels after cracking

$144 - x + 2x + x = 194$ total pieces

$144 + 2x = 194; \quad 2x = 50$

$x = 25$ kernels; 50 shell halves; 119 whole nuts left

$$K_{eq} = \frac{(2x)^2(x)}{144 - x} = \frac{(50)^2(25)}{119} = 5.3 \times 10^2$$

111. $SO_2(g) + NO_2(g) \rightleftharpoons SO_3(g) + NO(g)$

$SO_2(g)$	$NO_2(g)$	$SO_3(g)$	$NO(g)$	
0.50 M	0.50 M	0	0	Initial conditions
$0.50 - x$	$0.50 - x$	x	x	Equilibrium concentrations

$$K_{eq} = \frac{[SO_3][NO]}{[SO_2][NO_2]} = \frac{x^2}{(0.50 - x)^2} = 81$$

Take the square root of both sides

$$\frac{x}{0.50 - x} = 9.0 \qquad x = 0.45\text{ M}$$

$[SO_3] = [NO] = 0.45\text{ M}$

$[SO_2] = [NO_2] = 0.05\text{ M}$

CHAPTER 17

OXIDATION–REDUCTION

1. Oxidation of a metal occurs when the metal loses electrons. The easier it is for a metal to lose electrons, the more active the metal is.

2. (a) Iodine is oxidized. Its oxidation number increases from 0 to +5.
   (b) Chlorine is reduced. Its oxidation number decreases from 0 to −1.

3. The higher metal on the list is more reactive.
   (a) Al (b) Ba (c) Ni

4. If the free element is higher on the list than the ion with which it is paired, the reaction occurs.
   (a) Yes. $Zn(s) + Cu^{2+}(aq) \longrightarrow Zn^{2+}(aq) + Cu(s)$
   (b) No reaction
   (c) Yes. $Sn(s) + 2\,Ag^{+}(aq) \longrightarrow Sn^{2+}(aq) + 2\,Ag(s)$
   (d) No reaction
   (e) Yes. $Ba(s) + FeCl_2(aq) \longrightarrow BaCl_2(aq) + Fe(s)$
   (f) No reaction
   (g) Yes. $Ni(s) + Hg(NO_3)_2(aq) \longrightarrow Ni(NO_3)_2(aq) + Hg(l)$
   (h) Yes. $2\,Al(s) + 3\,CuSO_4(aq) \longrightarrow Al_2(SO_4)_3(aq) + 3\,Cu(s)$

5. Copper is more active than silver. Therefore, copper undergoes oxidation more easily than silver. Accordingly, it is more difficult for copper ion to undergo reduction than it is for silver ion. When a silver wire is placed in a solution of copper (II) nitrate one might predict that copper crystals would form on the silver wire. However for copper to go from an oxidation state of +2 to an oxidation state of 0 it would have to gain electrons (reduction) and silver would have to lose electrons (oxidation). This will not happen because copper is more active than silver.

6. (a) $2\,Al + Fe_2O_3 \longrightarrow Al_2O_3 + 2\,Fe + \text{Heat}$
   (b) Al is above Fe in the activity series, which indicates Al is more active than Fe.
   (c) No. Iron is less active than aluminum and will not displace aluminum from its compounds.
   (d) Yes. Aluminum is above chromium in the activity series and will displace Cr^{3+} from its compounds.

7. (a) $2\,Al(s) + 6\,HCl(aq) \longrightarrow 2\,AlCl_3(aq) + 3\,H_2(g)$

$2\,Al(s) + 3\,H_2SO_4(aq) \longrightarrow Al_2(SO_4)_3(aq) + 3\,H_2(g)$

(b) $2\,Cr(s) + 6\,HCl(aq) \longrightarrow 2\,CrCl_3(aq) + 3\,H_2(g)$

$2\,Cr(s) + 3\,H_2SO_4(aq) \longrightarrow Cr_2(SO_4)_3(aq) + 3\,H_2(g)$

(c) $Au(s) + HCl(aq) \longrightarrow$ no reaction

$Au(s) + H_2SO_4(aq) \longrightarrow$ no reaction

(d) $Fe(s) + 2\,HCl(aq) \longrightarrow FeCl_2(aq) + H_2(g)$

$Fe(s) + H_2SO_4(aq) \longrightarrow FeSO_4(aq) + H_2(g)$

(e) $Cu(s) + HCl(aq) \longrightarrow$ no reaction

$Cu(s) + H_2SO_4(aq) \longrightarrow$ no reaction

(f) $Mg(s) + 2\,HCl(aq) \longrightarrow MgCl_2(aq) + H_2(g)$

$Mg(s) + H_2SO_4(aq) \longrightarrow MgSO_4(aq) + H_2(g)$

(g) $Hg(l) + HCl(aq) \longrightarrow$ no reaction

$Hg(l) + H_2SO_4(aq) \longrightarrow$ no reaction

(h) $Zn(s) + 2\,HCl(aq) \longrightarrow ZnCl_2(aq) + H_2(g)$

$Zn(s) + H_2SO_4(aq) \longrightarrow ZnSO_4(aq) + H_2(g)$

8. The oxidation number for an atom in an ionic compound is the same as the charge of the ion that resulted when that atom lost or gained electrons to form an ionic bond. In a covalently bonded compound electrons are shared between the two atoms making up the bond. Those shared electrons are assigned to the atom in the bond with a higher electronegativity giving it a negative oxidation number.

9. The anode is the positively charged electrode and attracts negatively charged ions (anions). The cathode is the negatively charged electrode and attracts positively charge ions (cations).

10. (a) Oxidation occurs at the anode. The reaction is

$2\,Cl^-(aq) \longrightarrow Cl_2(g) + 2\,e^-$

(b) Reduction occurs at the cathode. The reaction is

$Ni^{2+}(aq) + 2\,e^- \longrightarrow Ni(s)$

(c) The net chemical reaction is

$Ni^{2+}(aq) + 2\,Cl^-(aq) \xrightarrow[\text{energy}]{\text{electrical}} Ni(s) + Cl_2(g)$

11. In Figure 17.3, electrical energy is causing chemical reactions to occur. In Figure 17.4, chemical reactions are used to produce electrical energy.

12. (a) It would not be possible to monitor the voltage produced, but the reactions in the cell would still occur.

    (b) If the salt bridge were removed, the reaction would stop. Ions must be mobile to maintain an electrical neutrality of ions in solution. The two solutions would be isolated with no complete electrical circuit.

13. Oxidation and reduction are complementary processes because one does not occur without the other. The loss of e^- in oxidation is accompanied by a gain of e^- in reduction.

14. $Ca^{2+} + 2\,e^- \longrightarrow Ca$ cathode reaction, reduction

    $2\,Br^- \longrightarrow Br_2 + 2\,e^-$ anode reaction, oxidation

15. During electroplating of metals, the metal is plated by reducing the positive ions of the metal in the solution. The plating will occur at the cathode, the source of the electrons. With an alternating current, the polarity of the electrode would be constantly changing, so at one instant the metal would be plating and the next instant the metal would be dissolving.

16. Since lead dioxide and lead(II) sulfate are insoluble, it is unnecessary to have salt bridges in the cells of a lead storage battery.

17. The electrolyte in a lead storage battery is dilute sulfuric acid. In the discharge cycle, ${SO_4}^{2-}$, is removed from solution as it reacts with PbO_2 and H^+ to form $PbSO_4(s)$ and H_2O. Therefore, the electrolyte solution contains less H_2SO_4 and becomes less dense.

18. If Hg^{2+} ions are reduced to metallic mercury, this would occur at the cathode, because reduction takes place at the cathode.

19. In both electrolytic and voltaic cells, oxidation and reduction reactions occur. In an electrolytic cell an electric current is forced through the cell causing a chemical change to occur. In voltaic cells, spontaneous chemical changes occur, generating an electric current.

20. In some voltaic cells, the reactants at the electrodes are in solution. For the cell to function, these reactants must be kept separated. A salt bridge permits movement of ions in the cell. This keeps the solution neutral with respect to the charged particles (ions) in the solution.

21. The oxidation number of the underlined element is indicated by the number following the formula.

(a)	$\underline{Na}Cl$	+1	(c)	$\underline{Pb}O_2$	+4	(e)	$H_2\underline{S}O_3$	+4
(b)	$Fe\underline{Cl}_3$	−1	(d)	$Na\underline{N}O_3$	+5	(f)	$\underline{N}H_4Cl$	−3

22. The oxidation number of the underlined element is indicated by the number following the formula.

(a)	$K\underline{Mn}O_4$	$+7$	(c)	$\underline{N}H_3$	-3	(e)	$K_2\underline{Cr}O_4$	$+6$
(b)	$\underline{I}_2$	0	(d)	$K\underline{Cl}O_3$	$+5$	(f)	$K_2\underline{Cr}_2O_7$	$+6$

23. The oxidation number of the underlined element is indicated by the number following the formula.

(a)	$\underline{S}^{2-}$	-2	(c)	$Na_2\underline{O}_2$	-1
(b)	$\underline{N}O_2^-$	$+3$	(d)	$\underline{Bi}^{3+}$	$+3$

24. The oxidation number of the underlined element is indicated by the number following the formula.

(a)	$\underline{O}_2$	0	(c)	$Fe(\underline{O}H)_3$	-2
(b)	$\underline{As}O_4^{3-}$	$+5$	(d)	$\underline{I}O_3^-$	$+5$

25.

Balanced half-reaction	Changing Element	Type of reaction
(a) $Zn^{2+} + 2\,e^- \longrightarrow Zn$	Zn	reduction
(b) $2\,Br^- \longrightarrow Br_2 + 2\,e^-$	Br	oxidation
(c) $MnO_4^- + 8\,H^+ + 5\,e^- \longrightarrow Mn^{2+} + 4\,H_2O$	Mn	reduction
(d) $Ni \longrightarrow Ni^{2+} + 2\,e^-$	Ni	oxidation

26.

Balanced half-reactions	Changing Element	Type of reaction
(a) $SO_3^{2-} + H_2O \longrightarrow SO_4^{2-} + 2\,H^+ + 2\,e^-$	S	oxidation
(b) $NO_3^- + 4\,H^+ + 3\,e^- \longrightarrow NO + 2\,H_2O$	N	reduction
(c) $S_2O_4^{2-} + 2\,H_2O \longrightarrow 2\,SO_3^{2-} + 4\,H^+ + 2\,e^-$	S	oxidation
(d) $Fe^{2+} \longrightarrow Fe^{3+} + 1\,e^-$	Fe	oxidation

27. (1) $Cr + HCl \longrightarrow CrCl_3 + H_2$

    (a) Cr is oxidized, H is reduced

    (b) HCl is the oxidizing agent, Cr the reducing agent

    (2) $SO_4^{2-} + I^- + H^+ \longrightarrow H_2S + I_2 + H_2O$

    (a) I is oxidized, S is reduced

    (b) SO_4^{2-} is the oxidizing agent, I^- the reducing agent

28. (1) $AsH_3 + Ag^+ + H_2O \longrightarrow H_3AsO_4 + Ag + H^+$

   (a) As is oxidized, Ag is reduced

   (b) Ag^+ is the oxidizing agent, AsH_3 the reducing agent

   (2) $Cl_2 + NaBr \longrightarrow NaCl + Br_2$

   (a) Br is oxidized, Cl is reduced

   (b) Cl_2 is the oxidizing agent, NaBr the reducing agent

29. (a) correctly balanced

   (b) correctly balanced

   (c) incorrectly balanced

   $Mg(s) + 2\,HCl(aq) \longrightarrow Mg^{2+}(aq) + 2Cl^-(aq) + H_2(g)$

   (d) incorrectly balanced

   $3\,CH_3OH(aq) + Cr_2O_7{}^{2-}(aq) + 8\,H^+(aq) \longrightarrow 2\,Cr^{3+}(aq) + 3\,CH_2O(aq) + 7\,H_2O(l)$

30. (a) incorrectly balanced

   $3\,MnO_2(s) + 4\,Al(s) \longrightarrow 3\,Mn(s) + 2\,Al_2O_3(s)$

   (b) correctly balanced

   (c) correctly balanced

   (d) incorrectly balanced

   $8\,H_2O(l) + 2\,MnO_4{}^-(aq) + 7\,S^{2-}(aq) \longrightarrow 2\,MnS(s) + 16\,OH^-(aq) + 5\,S(s)$

31. Balancing oxidation-reduction equations

(a)

$Zn + S \longrightarrow ZnS$

ox $Zn^0 \longrightarrow Zn^{2+} + 2\,e^-$

red $S^0 + 2\,e^- \longrightarrow S^{2-}$ Add half-reactions

$Zn + S \longrightarrow ZnS$ the 2 e⁻ cancel

(b)

$AgNO_3 + Pb \longrightarrow Pb(NO_3)_2 + Ag$

ox $Pb^0 \longrightarrow Pb^{2+} + 2\,e^-$

red $Ag^+ + 1\,e^- \longrightarrow Ag^0$ Multiply by 2, add the half-reactions

$Pb + 2\,Ag^+ \longrightarrow Pb^{2+} + 2\,Ag$ the 2 e^- cancel

Transfer the coefficients to the original equation and complete the balancing by inspection.

$2\,AgNO_3 + Pb \longrightarrow Pb(NO_3)_2 + 2\,Ag$

(c) $Fe_2O_3 + CO \longrightarrow Fe + CO_2$

ox $C^{2+} \longrightarrow C^{4+} + 2\,e^-$ Multiply by 3

red $Fe^{3+} + 3\,e^- \longrightarrow Fe^0$ Multiply by 2, add, the 6 e^- cancel

$3\,C^{2+} + 2\,Fe^{3+} \longrightarrow 3\,C^{4+} + 2\,Fe$

Transfer the coefficients to the original equation (the coefficient 2 in front of the Fe^{3+} becomes the subscript 2 in Fe_2O_3). Complete the balancing by inspection.

$Fe_2O_3 + 3\,CO \longrightarrow 2\,Fe + 3\,CO_2$

(d) $H_2S + HNO_3 \longrightarrow S + NO + H_2O$

$S^{2-} \longrightarrow S^0 + 2\,e^-$ Multiply by 3

$N^{5+} + 3\,e^- \longrightarrow N^{2+}$ Multiply by 2, add, the 6 e^-

$3\,S^{2-} + 2\,N^{5+} \longrightarrow 3\,S + 2\,N^{2+}$

Transfer the coefficients to the original equations and complete the balancing by inspection.

$3\,H_2S + 2\,HNO_3 \longrightarrow 3\,S + 2\,NO + 4\,H_2O$

(e) $MnO_2 + HBr \longrightarrow MnBr_2 + Br_2 + H_2O$

$Br^- \longrightarrow Br^0 + 1\,e^-$ Multiply by 2

$Mn^{4+} + 2\,e^- \longrightarrow Mn^{2+}$ Add equations and the 2 e^- cancel

$Mn^{4+} + 2\,Br^- \longrightarrow Mn^{2+} + 2\,Br^0$

Transfer the coefficients to the original equation. The coefficient 2 in front of the Br^- becomes the subscript 2 in the Br_2. Also, 2 more Br^- ions are required to account for the 2 Br^- ions that do not change oxidation numbers. These 2 are part of the compound $MnBr_2$.

$MnO_2 + 4\,HBr \longrightarrow MnBr_2 + Br_2 + 2\,H_2O$

32. (a) Balancing oxidation-reduction equations

$Cl_2 + KOH \longrightarrow KCl + KClO_3 + H_2O$

$Cl^0 \longrightarrow Cl^{5+} + 5\,e^-$

$Cl^0 + e^- \longrightarrow Cl^-$ Multiply by 5, add, the 5 e^- cancel

$3\,Cl_2 \longrightarrow Cl^{5+} + 5\,Cl^-$ 6 Cl^0 becomes 3 Cl_2

Transfer the coefficients to the original equations and complete the balancing by inspection.

$3\,Cl_2 + 6\,KOH \longrightarrow KClO_3 + 5\,KCl + 3\,H_2O$

(b) $Ag + HNO_3 \longrightarrow AgNO_3 + NO + H_2O$

$Ag^0 \longrightarrow Ag^+ + e^-$ Multiply by 3, add,

$N^{5+} + 3\,e^- \longrightarrow N^{2+}$ the 3 e^- cancel

$3\,Ag + N^{5+} \longrightarrow 3\,Ag^+ + N^{2+}$

Transfer the coefficients to the original equations and complete the balancing by inspection.

$3\,Ag + 4\,HNO_3 \longrightarrow 3\,AgNO_3 + NO + 2\,H_2O$

(c) $CuO + NH_3 \longrightarrow N_2 + Cu + H_2O$

$N^{3-} \longrightarrow N^0 + 3\,e^-$ Multiply by 2

$Cu^{2+} + 2\,e^- \longrightarrow Cu^0$ Multiply by 3, add, the 6 e^- cancel

$2\,N^{3-} + 3\,Cu^{2+} \longrightarrow N_2 + 3\,Cu$

Transfer the coefficients to the original equations and complete the balancing by inspection.

$3\,CuO + 2\,NH_3 \longrightarrow N_2 + 3\,Cu + 3\,H_2O$

(d) $PbO_2 + Sb + NaOH \longrightarrow PbO + NaSbO_2 + H_2O$

$Sb^0 \longrightarrow Sb^{3+} + 3\,e^-$ Multiply by 2

$Pb^{4+} + 2\,e^- \longrightarrow Pb^{2+}$ Multiply by 3, add, the 6 e^- cancel

$2\,Sb + 3\,Pb^{4+} \longrightarrow 2\,Sb^{3+} + 3\,Pb^2$

Transfer the coefficients to the original equations and complete the balancing by inspection.

$3\,PbO_2 + 2\,Sb + 2\,NaOH \longrightarrow 3\,PbO + 2\,NaSbO_2 + H_2O$

(e) $H_2O_2 + KMnO_4 + H_2SO_4 \longrightarrow O_2 + MnSO_4 + K_2SO_4 + H_2O$

${O_2}^{2-} \longrightarrow {O_2}^0 + 2\,e^-$ Multiply by 5

$Mn^{7+} + 5\,e^- \longrightarrow Mn^{2+}$ Multiply by 2, add, the 10 e^- cancel

$5\,{O_2}^{2-} + 2\,Mn^{7+} \longrightarrow 5\,O_2 + 2\,Mn^{2+}$

Transfer the coefficients to the original equations and complete the balancing by inspection.

$5\,H_2O_2 + 2\,KMnO_4 + 3\,H_2SO_4 \longrightarrow 5\,O_2 + 2\,MnSO_4 + K_2SO_4 + 8\,H_2O$

33. (a) $Zn + {NO_3}^- \longrightarrow Zn^{2+} + {NH_4}^+$ (acidic solution)

Step 1 Write half-reaction equations. Balance except H and O.

$Zn \longrightarrow Zn^{2+}$

${NO_3}^- \longrightarrow {NH_4}^+$

Step 2 Balance H and O using H_2O and H^+

$Zn \longrightarrow Zn^{2+}$

$10\,H^+ + NO_3^- \longrightarrow NH_4^+ + 3\,H_2O$

Step 3 Balance electrically with electrons

$Zn \longrightarrow Zn^{2+} + 2\,e^-$

$10\,H^+ + NO_3^- + 8\,e^- \longrightarrow NH_4^+ + 3\,H_2O$

Step 4 Equalize the loss and gain of electrons

$4\,(Zn \longrightarrow Zn^{2+} + 2\,e^-)$

$10\,H^+ + NO_3^- + 8\,e^- \longrightarrow NH_4^+ + 3\,H_2O$

Step 5 Add the half-reactions–electrons cancel

$10\,H^+ + 4\,Zn + NO_3^- \longrightarrow 4\,Zn^{2+} + NH_4^+ + 3\,H_2O$

(b) $NO_3^- + S \longrightarrow NO_2 + SO_4^{2-}$ (acidic solution)

Step 1 Write half-reaction equations. Balance except H and O.

$S \longrightarrow SO_4^{2-}$

$NO_3^- \longrightarrow NO_2$

Step 2 Balance H and O using H_2O and H^+

$4\,H_2O + S \longrightarrow SO_4^{2-} + 8\,H^+$

$2\,H^+ + NO_3^- \longrightarrow NO_2 + H_2O$

Step 3 Balance electrically with electrons

$4\,H_2O + S \longrightarrow SO_4^{2-} + 8\,H^+ + 6\,e^-$

$2\,H^+ + NO_3^- + e^- \longrightarrow NO_2 + H_2O$

Step 4 and 5 Equalize the loss and gain of electrons; add the half-reactions

$4\,H_2O + S \longrightarrow SO_4^{2-} + 8\,H^+ + 6\,e^-$

$6\,(2\,H^+ + NO_3^- + e^- \longrightarrow NO_2 + H_2O)$

$4\,H^+ + S + 6\,NO_3^- \longrightarrow 6\,NO_2 + SO_4^{2-} + 2\,H_2O$

$4\,H_2O$, $8\,H^+$ and $6\,e^-$ canceled from each side

(c) $PH_3 + I_2 \longrightarrow H_3PO_2 + I^-$ (acidic solution)

Step 1 Write half-reaction equations. Balance except H and O.

$PH_3 \longrightarrow H_3PO_2$

$I_2 \longrightarrow 2I^-$

Step 2 Balance H and O using H_2O and H^+

$2\,H_2O + PH_3 \longrightarrow H_3PO_2 + 4\,H^+$

$I_2 \longrightarrow 2\,I^-$

Step 3 Balance electrically with electrons

$2\,H_2O + PH_3 \longrightarrow H_3PO_2 + 4\,H^+ + 4\,e^-$

$I_2 + 2\,e^- \longrightarrow 2\,I^-$

Step 4 and 5 Equalize the loss and gain of electrons; add the half-reactions

$$2\,H_2O + PH_3 \longrightarrow H_3PO_2 + 4\,H^+ + 4\,e^-$$
$$\underline{2\,(I_2 + 2\,e^- \longrightarrow 2\,I^-)}$$
$$PH_3 + 2\,H_2O + 2\,I_2 \longrightarrow H_3PO_2 + 4\,I^- + 4\,H$$

(d) $Cu + NO_3^- \longrightarrow Cu^{2+} + NO$ (acidic solution)

Step 1 Write half-reaction equations. Balance except H and O.

$Cu \longrightarrow Cu^{2+}$

$NO_3^- \longrightarrow NO$

Step 2 Balance H and O using H_2O and H^+

$Cu \longrightarrow Cu^{2+}$

$4H^+ + NO_3^- \longrightarrow NO + 2\,H_2O$

Step 3 Balance electrically with electrons

$Cu \longrightarrow Cu^{2+} + 2\,e^-$

$4H^+ + NO_3^- + 3\,e^- \longrightarrow NO + 2\,H_2O$

Step 4 and 5 Equalize the loss and gain of electrons; add the half-reactions

$$3\,(Cu \longrightarrow Cu^{2+} + 2\,e^-)$$
$$\underline{2\,(4\,H^+ + NO_3^- + 3\,e^- \longrightarrow NO + 2\,H_2O)}$$
$$3\,Cu + 8\,H^+ + 2\,NO_3^- \longrightarrow 3\,Cu^{2+} + 2\,NO + 4\,H_2O$$

(e) $ClO_3^- + Cl^- \longrightarrow Cl_2$ (acidic solution)

Step 1 Write half-reaction equations. Balance except H and O.

$Cl^- \longrightarrow Cl^0$

$ClO_3^- \longrightarrow Cl^0$

Step 2 Balance H and O using H_2O and H^+

$Cl^- \longrightarrow Cl^0$

$6\,H^+ + ClO_3^- \longrightarrow Cl^0 + 3\,H_2O$

Step 3 Balance electrically with electrons

$Cl^- \longrightarrow Cl^0 + e^-$

$6\,H^+ + ClO_3^- + 5\,e^- \longrightarrow Cl^0 + 3\,H_2O$

Step 4 and 5 Equalize the loss and gain of electrons; add the half-reactions

$5\,(Cl^- \longrightarrow Cl^0 + e^-)$

$6\,H^+ + ClO_3^- + 5\,e^- \longrightarrow Cl^0 + 3\,H_2O$

$6\,H^+ + ClO_3^- + 5\,Cl^- \longrightarrow 3\,Cl_2 + 3\,H_2O$

34. (a) $ClO_3^- + I^- \longrightarrow I_2 + Cl^-$ (acidic solution)

Step 1 Write half-reaction equations. Balance except H and O.

$2\,I^- \longrightarrow I_2$

$ClO_3^- \longrightarrow Cl^-$

Step 2 Balance H and O using H_2O and H^+

$2\,I^- \longrightarrow I_2$

$6\,H^+ + ClO_3^- \longrightarrow Cl^- + 3\,H_2O$

Step 3 Balance electrically with electrons

$2\,I^- \longrightarrow I_2 + 2\,e^-$

$6\,H^+ + ClO_3^- + 6\,e^- \longrightarrow Cl^- + 3\,H_2O$

Step 4 and 5 Equalize the loss and gain of electrons; add the half-reactions

$3\,(2\,I^- \longrightarrow I_2 + 2\,e^-)$

$6\,H^+ + ClO_3^- + 6\,e^- \longrightarrow Cl^- + 3\,H_2O$

$6\,H^+ + ClO_3^- + 6\,I^- \longrightarrow 3\,I_2 + Cl^- + 3\,H_2O$

(b) $Cr_2O_7^{2-} + Fe^{2+} \longrightarrow Cr^{3+} + Fe^{3+}$ (acidic solution)

Step 1 Write half-reaction equations. Balance except H and O.

$Fe^{2+} \longrightarrow Fe^{3+}$

$Cr_2O_7^{2-} \longrightarrow 2\,Cr^{3+}$

Step 2 Balance H and O using H_2O and H^+

$Fe^{2+} \longrightarrow Fe^{3+}$

$14\,H^+ + Cr_2O_7^{2-} \longrightarrow 2\,Cr^{3+} + 7\,H_2O$

Step 3 Balance electrically with electrons

$Fe^{2+} \longrightarrow Fe^{3+} + e^-$

$14\,H^+ + Cr_2O_7^{2-} + 6\,e^- \longrightarrow 2\,Cr^{3+} + 7\,H_2O$

Step 4 and 5 Equalize the loss and gain of electrons; add the half-reactions

$6\,(Fe^{2+} \longrightarrow Fe^{3+} + e^-)$

$14\,H^+ + Cr_2O_7^{2-} + 6\,e^- \longrightarrow 2\,Cr^{3+} + 7\,H_2O$

$14\,H^+ + Cr_2O_7^{2-} + 6\,Fe^{2+} \longrightarrow 2\,Cr^{3+} + 6\,Fe^{3+} + 7\,H_2O$

(c) $MnO_4^- + SO_2 \longrightarrow Mn^{2+} + SO_4^{2-}$ (acidic solution)

Step 1 Write half-reaction equations. Balance except H and O.

$SO_2 \longrightarrow SO_4^{2-}$

$MnO_4^- \longrightarrow Mn^{2+}$

Step 2 Balance H and O using H_2O and H^+

$2\,H_2O + SO_2 \longrightarrow SO_4^{2-} + 4\,H^+$

$8\,H^+ + MnO_4^- \longrightarrow Mn^{2+} + 4\,H_2O$

Step 3 Balance electrically with electrons

$2\,H_2O + SO_2 \longrightarrow SO_4^{2-} + 4\,H^+ + 2\,e^-$

$8\,H^+ + MnO_4^- + 5\,e^- \longrightarrow Mn^{2+} + 4\,H_2O$

Step 4 and 5 Equalize the loss and gain of electrons; add the half-reactions

$5\,(2\,H_2O + SO_2 \longrightarrow SO_4^{2-} + 4\,H^+ + 2\,e^-)$

$2\,(8\,H^+ + MnO_4^- + 5\,e^- \longrightarrow Mn^{2+} + 4\,H_2O)$

$2\,H_2O + 2\,MnO_4^- + 5\,SO_2 \longrightarrow 4\,H^+ + 2\,Mn^{2+} + 5\,SO_4^{2-}$

$8\,H_2O$, $16\,H^+$, and $10\,e^-$ canceled from each side

(d) $H_3AsO_3 + MnO_4^- \longrightarrow H_3AsO_4 + Mn^{2+}$ (acidic solution)

Step 1 Write half-reaction equations. Balance except H and O.

$H_3AsO_3 \longrightarrow H_3AsO_4$

$MnO_4^- \longrightarrow Mn^{2+}$

Step 2 — Balance H and O using H_2O and H^+

$H_2O + H_3AsO_3 \longrightarrow 2\,H^+ + H_3AsO_4$

$8\,H^+ + MnO_4^- \longrightarrow Mn^{2+} + 4\,H_2O$

Step 3 — Balance electrically with electrons

$H_2O + H_3AsO_3 \longrightarrow 2\,H^+ + H_3AsO_4 + 2\,e^-$

$8\,H^+ + MnO_4^- + 5\,e^- \longrightarrow Mn^{2+} + 4\,H_2O$

Step 4 and 5 — Equalize the loss and gain of electrons; add the half-reactions

$5\,(H_2O + H_3AsO_3 \longrightarrow 2\,H^+ + H_3AsO_4 + 2\,e^-)$

$2\,(8\,H^+ + MnO_4^- + 5\,e^- \longrightarrow Mn^{2+} + 4\,H_2O)$

$6\,H^+ + 5\,H_3AsO_3 + 2\,MnO_4^- \longrightarrow 5\,H_3AsO_4 + 2\,Mn^{2+} + 3\,H_2O$

$5\,H_2O$, $10\,H^+$, and $10\,e^-$ canceled from each side

(e) $Cr_2O_7^{2-} + H_3AsO_3 \longrightarrow Cr^{3+} + H_3AsO_4$ (acidic solution)

Step 1 — Write half-reaction equations. Balance except H and O.

$H_3AsO_3 \longrightarrow H_3AsO_4$

$Cr_2O_7^{2-} \longrightarrow 2\,Cr^{3+}$

Step 2 — Balance H and O using H_2O and H^+

$H_2O + H_3AsO_3 \longrightarrow 2\,H^+ + H_3AsO_4$

$14\,H^+ + Cr_2O_7^{2-} \longrightarrow 2\,Cr^{3+} + 7\,H_2O$

Step 3 — Balance electrically with electrons

$H_2O + H_3AsO_3 \longrightarrow 2\,H^+ + H_3AsO_4 + 2\,e^-$

$14\,H^+ + Cr_2O_7^{2-} + 6\,e^- \longrightarrow 2\,Cr^{3+} + 7\,H_2O$

Step 4 and 5 — Equalize the loss and gain of electrons; add the half-reactions

$3\,(H_2O + H_3AsO_3 \longrightarrow 2\,H^+ + H_3AsO_4 + 2\,e^-)$

$14\,H^+ + Cr_2O_7^{2-} + 6\,e^- \longrightarrow 2\,Cr^{3+} + 7\,H_2O$

$8\,H^+ + Cr_2O_7^{2-} + 3\,H_3AsO_3 \longrightarrow 2\,Cr^{3+} + 3\,H_3AsO_4 + 4\,H_2O$

$3\,H_2O$, $6\,H^+$, and $6\,e^-$ canceled from each side

35. (a) $Cl_2 + IO_3^- \longrightarrow Cl^- + IO_4^-$ (basic solution)

Step 1 — Write half-reaction equations. Balance except H and O.

$IO_3^- \longrightarrow IO_4^-$

$Cl_2 \longrightarrow 2\,Cl^-$

Step 2 Balance H and O using H_2O and H^+

$H_2O + IO_3^- \longrightarrow IO_4^- + 2\,H^+$

$Cl_2 \longrightarrow 2\,Cl^-$

Step 3 Add OH^- ions to both sides (same number as H^+ ions)

$2\,OH^- + H_2O + IO_3^- \longrightarrow IO_4^- + 2\,H^+ + 2\,OH^-$

$Cl_2 \longrightarrow 2\,Cl^-$

Step 4 Combine H^+ and OH^- to form H_2O; cancel H_2O where possible

$2\,OH^- + H_2O + IO_3^- \longrightarrow IO_4^- + 2\,H_2O$

$Cl_2 \longrightarrow 2\,Cl^-$

$2\,OH^- + IO_3^- \longrightarrow IO_4^- + H_2O$ (1 H_2O cancelled)

$Cl_2 \longrightarrow 2\,Cl^-$

Step 5 Balance electrically with electrons

$2\,OH^- + IO_3^- \longrightarrow IO_4^- + H_2O + 2\,e^-$

$Cl_2 + 2\,e^- \longrightarrow 2\,Cl^-$

Step 6 Electron loss and gain is balanced

Step 7 Add half-reactions

$2\,OH^- + IO_3^- + Cl_2 \longrightarrow IO_4^- + 2\,Cl^- + H_2O$

(b) $MnO_4^- + ClO_2^- \longrightarrow MnO_2 + ClO_4^-$ (basic solution)

Step 1 Write half-reaction equations. Balance except H and O.

$ClO_2^- \longrightarrow ClO_4^-$

$MnO_4^- \longrightarrow MnO_2$

Step 2 Balance H and O using H_2O and H^+

$2\,H_2O + ClO_2^- \longrightarrow ClO_4^- + 4\,H^+$

$MnO_4^- + 4\,H^+ \longrightarrow MnO_2 + 2\,H_2O$

Step 3 Add OH^- ions to both sides (same number as H^+ ions)

$4\,OH^- + 2\,H_2O + ClO_2^- \longrightarrow ClO_4^- + 4\,H^+ + 4\,OH^-$

$4\,OH^- + MnO_4^- + 4\,H^+ \longrightarrow MnO_2 + 2\,H_2O + 4\,OH^-$

Step 4 Combine H^+ and OH^- to form H_2O; cancel H_2O where possible

$$4\,OH^- + 2\,H_2O + ClO_2^- \longrightarrow ClO_4^- + 4\,H_2O$$

$$4\,H_2O + MnO_4^- \longrightarrow MnO_2 + 2\,H_2O + 4\,OH^-$$

$$4\,OH^- + ClO_2^- \longrightarrow ClO_4^- + 2\,H_2O \qquad (2\,H_2O \text{ cancelled})$$

$$2\,H_2O + MnO_4^- \longrightarrow MnO_2 + 4\,OH^- \qquad (2\,H_2O \text{ cancelled})$$

Step 5 Balance electrically with electrons

$$4\,OH^- + ClO_2^- \longrightarrow ClO_4^- + 2\,H_2O + 4\,e^-$$

$$2\,H_2O + MnO_4^- + 3\,e^- \longrightarrow MnO_2 + 4\,OH^-$$

Step 6 and 7 Equalize gain and loss of electrons; add half-reactions

$$3\,(4\,OH^- + ClO_2^- \longrightarrow ClO_4^- + 2\,H_2O + 4\,e^-)$$

$$4\,(2\,H_2O + MnO_4^- + 3\,e^- \longrightarrow MnO_2 + 4\,OH^-)$$

$$\overline{2\,H_2O + 4\,MnO_4^- + 3\,ClO_2^- \longrightarrow 4\,MnO_2 + 3\,ClO_4^- + 4\,OH^-}$$

$6\,H_2O$, $12\,OH^-$, and $12\,e^-$ canceled from each side

(c) $Se \longrightarrow SeO_3^{2-} + Se^{2-}$ (basic solution)

Step 1 Write half-reaction equations. Balance except H and O.

$$Se \longrightarrow SeO_3^{2-}$$

$$Se \longrightarrow Se^{2-}$$

Step 2 Balance H and O using H_2O and H^+

$$3\,H_2O + Se \longrightarrow SeO_3^{2-} + 6\,H^+$$

$$Se \longrightarrow Se^{2-}$$

Step 3 Add OH^- ions to both sides (same number as H^+ ions)

$$6\,OH^- + 3\,H_2O + Se \longrightarrow SeO_3^{2-} + 6\,H^+ + 6\,OH^-$$

$$Se \longrightarrow Se^{2-}$$

Step 4 Combine H^+ and OH^- to form H_2O; cancel H_2O where possible

$$6\,OH^- + 3\,H_2O + Se \longrightarrow SeO_3^{2-} + 6\,H_2O$$

$$Se \longrightarrow Se^{2-}$$

$$6\,OH^- + Se \longrightarrow SeO_3^{2-} + 3\,H_2O \qquad (3\,H_2O \text{ cancelled})$$

Step 5 Balance electrically with electrons

$$6\,OH^- + Se \longrightarrow SeO_3^{2-} + 3\,H_2O + 4\,e^-$$

$$Se + 2\,e^- \longrightarrow Se^{2-}$$

Step 6 and 7 Equalize gain and loss of electrons; add half-reactions

$$6\,OH^- + Se \longrightarrow SeO_3^{2-} + 3\,H_2O + 4\,e^-$$
$$2\,(Se + 2\,e^- \longrightarrow Se^{2-})$$
$$6\,OH^- + 3\,Se \longrightarrow SeO_3^{2-} + 2\,Se^{2-} + 3\,H_2O$$

(d) $Fe_3O_4 + MnO_4^- \longrightarrow Fe_2O_3 + MnO_2$ (basic solution)

Step 1 Write half-reaction equations. Balance except H and O.

$$2\,Fe_3O_4 \longrightarrow 3\,Fe_2O_3$$

$$MnO_4^- \longrightarrow MnO_2$$

Step 2 Balance H and O using H_2O and H^+

$$H_2O + 2\,Fe_3O_4 \longrightarrow 3\,Fe_2O_3 + 2\,H^+$$

$$4\,H^+ + MnO_4^- \longrightarrow MnO_2 + 2\,H_2O$$

Step 3 Add OH^- ions to both sides (same number as H^+ ions)

$$2\,OH^- + H_2O + 2\,Fe_3O_4 \longrightarrow 3\,Fe_2O_3 + 2\,H^+ + 2\,OH^-$$

$$4\,OH^- + 4\,H^+ + MnO_4^- \longrightarrow MnO_2 + 2\,H_2O + 4\,OH^-$$

Step 4 Combine H^+ and OH^- to form H_2O; cancel H_2O where possible

$$2\,OH^- + H_2O + 2\,Fe_3O_4 \longrightarrow 3\,Fe_2O_3 + 2\,H_2O$$

$$4\,H_2O + MnO_4^- \longrightarrow MnO_2 + 2\,H_2O + 4\,OH^-$$

$2\,OH^- + 2\,Fe_3O_4 \longrightarrow 3\,Fe_2O_3 + H_2O$ (1 H_2O cancelled)

$2\,H_2O + MnO_4^- \longrightarrow MnO_2 + 4\,OH^-$ (2 H_2O cancelled)

Step 5 Balance electrically with electrons

$$2\,OH^- + 2\,Fe_3O_4 \longrightarrow 3\,Fe_2O_3 + H_2O + 2\,e^-$$

$$2\,H_2O + MnO_4^- + 3\,e^- \longrightarrow MnO_2 + 4\,OH^-$$

Step 6 and 7 Equalize gain and loss of electrons; add half-reactions

$$3\,(2\,OH^- + 2\,Fe_3O_4 \longrightarrow 3\,Fe_2O_3 + H_2O + 2\,e^-)$$
$$2\,(2\,H_2O + MnO_4^- + 3\,e^- \longrightarrow MnO_2 + 4\,OH^-)$$
$$H_2O + 6\,Fe_3O_4 + 2\,MnO_4^- \longrightarrow 9\,Fe_2O_3 + 2\,MnO_2 + 2\,OH^-$$

$3\,H_2O$, $6\,OH^-$, and $6\,e^-$ canceled from each side

(e) $BrO^- + Cr(OH)_4^- \longrightarrow Br^- + CrO_4^{2-}$ (basic solution)

Step 1 Write half-reaction equations. Balance except H and O.

$Cr(OH)_4^- \longrightarrow CrO_4^{2-}$

$BrO^- \longrightarrow Br^-$

Step 2 Balance H and O using H_2O and H^+

$Cr(OH)_4^- \longrightarrow CrO_4^{2-} + 4\,H^+$

$2\,H^+ + BrO^- \longrightarrow Br^- + H_2O$

Step 3 Add OH^- ions to both sides (same number as H^+ ions)

$4\,OH^- + Cr(OH)_4^- \longrightarrow CrO_4^{2-} + 4\,H^+ + 4\,OH^-$

$2\,OH^- + 2\,H^+ + BrO^- \longrightarrow Br^- + H_2O + 2\,OH^-$

Step 4 Combine H^+ and OH^- to form H_2O; cancel H_2O where possible

$4\,OH^- + Cr(OH)_4^- \longrightarrow CrO_4^{2-} + 4\,H_2O$

$2\,H_2O + BrO^- \longrightarrow Br^- + H_2O + 2\,OH^-$

$H_2O + BrO^- \longrightarrow Br^- + 2\,OH^-$ (1 H_2O cancelled)

Step 5 Balance electrically with electrons

$4\,OH^- + Cr(OH)_4^- \longrightarrow CrO_4^{2-} + 4\,H_2O + 3\,e^-$

$H_2O + BrO^- + 2\,e^- \longrightarrow Br^- + 2\,OH^-$

Step 6 and 7 Equalize gain and loss of electrons; add half-reactions

$2\,(4\,OH^- + Cr(OH)_4^- \longrightarrow CrO_4^{2-} + 4\,H_2O + 3\,e^-)$

$3\,(H_2O + BrO^- + 2\,e^- \longrightarrow Br^- + 2\,OH^-)$

$2\,OH^- + 3\,BrO^- + 2\,Cr(OH)_4^- \longrightarrow 3\,Br^- + 2\,CrO_4^{2-} + 5\,H_2O$

3 H_2O, 6 OH^- and 6 e^- canceled from each side

36. (a) $MnO_4^- + SO_3^{2-} \longrightarrow MnO_2 + SO_4^{2-}$ (basic solution)

Step 1 Write half-reaction equations. Balance except H and O.

$SO_3^{2-} \longrightarrow SO_4^{2-}$

$MnO_4^- \longrightarrow MnO_2$

Step 2 Balance H and O using H_2O and H^+

$H_2O + SO_3^{2-} \longrightarrow SO_4^{2-} + 2\,H^+$

$MnO_4^- + 4\,H^+ \longrightarrow MnO_2 + 2\,H_2O$

Step 3 Add OH^- ions to both sides (same number as H^+ ions)

$$2\,OH^- + H_2O + SO_3^{2-} \longrightarrow SO_4^{2-} + 2\,H^+ + 2\,OH^-$$

$$4\,OH^- + MnO_4^- + 4\,H^+ \longrightarrow MnO_2 + 2\,H_2O + 4\,OH^-$$

Step 4 Combine H^+ and OH^- to form H_2O; cancel H_2O where possible

$$2\,OH^- + H_2O + SO_3^{2-} \longrightarrow SO_4^{2-} + 2\,H_2O$$

$$MnO_4^- + 4\,H_2O \longrightarrow MnO_2 + 2\,H_2O + 4\,OH^-$$

$$2\,OH^- + SO_3^{2-} \longrightarrow SO_4^{2-} + H_2O \qquad \text{(1 } H_2O \text{ cancelled)}$$

$$MnO_4^- + 2\,H_2O \longrightarrow MnO_2 + 4\,OH^- \qquad \text{(2 } H_2O \text{ cancelled)}$$

Step 5 Balance electrically with electrons

$$2\,OH^- + SO_3^{2-} \longrightarrow SO_4^{2-} + H_2O + 2\,e^-$$

$$3\,e^- + MnO_4^- + 2\,H_2O \longrightarrow MnO_2 + 4\,OH^-$$

Step 6 and 7 Equalize gain and loss of electrons; add half-reactions

$$3\,(2\,OH^- + SO_3^{2-} \longrightarrow SO_4^{2-} + H_2O + 2\,e^-)$$

$$2\,(MnO_4^- + 2\,H_2O + 3\,e^- \longrightarrow MnO_2 + 4\,OH^-)$$

$$H_2O + 2\,MnO_4^- + 3\,SO_3^{2-} \longrightarrow 2\,MnO_2 + 3\,SO_4^{2-} + 2\,OH^-$$

$3\,H_2O$, $4\,OH^-$, and $6\,e^-$ canceled from each side

(b) $ClO_2 + SbO_2^- \longrightarrow ClO_2^- + Sb(OH)_6^-$ (basic solution)

Step 1 Write half-reaction equations. Balance except H and O.

$$SbO_2^- \longrightarrow Sb(OH)_6^-$$

$$ClO_2 \longrightarrow ClO_2^-$$

Step 2 Balance H and O using H_2O and H^+

$$4\,H_2O + SbO_2^- \longrightarrow Sb(OH)_6^- + 2\,H^+$$

$$ClO_2 \longrightarrow ClO_2^-$$

Step 3 Add OH^- ions to both sides (same number as H^+ ions)

$$2\,OH^- + 4\,H_2O + SbO_2^- \longrightarrow Sb(OH)_6^- + 2\,H^+ + 2\,OH^-$$

$$ClO_2 \longrightarrow ClO_2^-$$

Step 4 Combine H^+ and OH^- to form H_2O; cancel H_2O where possible

$$2\,OH^- + 4\,H_2O + SbO_2^- \longrightarrow Sb(OH)_6^- + 2\,H_2O$$

$$ClO_2 \longrightarrow ClO_2^-$$

$$2\,OH^- + 2\,H_2O + SbO_2^- \longrightarrow Sb(OH)_6^- \qquad \text{(2 } H_2O \text{ cancelled)}$$

Step 5 — Balance electrically with electrons

$$2\,OH^- + 2\,H_2O + SbO_2^- \longrightarrow Sb(OH)_6^- + 2\,e^-$$

$$ClO_2 + e^- \longrightarrow ClO_2^-$$

Step 6 and 7 — Equalize gain and loss of electrons; add half-reactions

$$2\,H_2O + 2\,OH^- + SbO_2^- \longrightarrow Sb(OH)_6^- + 2\,e^-$$

$$2\,(ClO_2 + e^- \longrightarrow ClO_2^-)$$

$$2\,H_2O + 2\,ClO_2 + 2\,OH^- + SbO_2^- \longrightarrow 2\,ClO_2^- + Sb(OH)_6^-$$

(c) $Al + NO_3^- \longrightarrow NH_3 + Al(OH)_4^-$ (basic solution)

Step 1 — Write half-reaction equations. Balance except H and O.

$$Al \longrightarrow Al(OH)_4^-$$

$$NO_3^- \longrightarrow NH_3$$

Step 2 — Balance H and O using H_2O and H^+

$$4\,H_2O + Al \longrightarrow Al(OH)_4^- + 4\,H^+$$

$$9\,H^+ + NO_3^- \longrightarrow NH_3 + 3\,H_2O$$

Step 3 — Add OH^- ions to both sides (same number as H^+ ions)

$$4\,OH^- + 4\,H_2O + Al \longrightarrow Al(OH)_4^- + 4\,H^+ + 4\,OH^-$$

$$9\,OH^- + 9\,H^+ + NO_3^- \longrightarrow NH_3 + 3\,H_2O + 9\,OH^-$$

Step 4 — Combine H^+ and OH^- to form H_2O; cancel H_2O where possible

$$4\,OH^- + 4\,H_2O + Al \longrightarrow Al(OH)_4^- + 4\,H_2O$$

$$9\,H_2O + NO_3^- \longrightarrow NH_3 + 3\,H_2O + 9\,OH^-$$

$$4\,OH^- + Al \longrightarrow Al(OH)_4^- \qquad (4\,H_2O \text{ cancelled})$$

$$6\,H_2O + NO_3^- \longrightarrow NH_3 + 9\,OH^- \qquad (3\,H_2O \text{ cancelled})$$

Step 5 — Balance electrically with electrons

$$4\,OH^- + Al \longrightarrow Al(OH)_4^- + 3\,e^-$$

$$6\,H_2O + NO_3^- + 8\,e^- \longrightarrow NH_3 + 9\,OH^-$$

Step 6 and 7 — Equalize gain and loss of electrons; add half-reactions

$$8\,(4\,OH^- + Al \longrightarrow Al(OH)_4^- + 3\,e^-)$$

$$3\,(6\,H_2O + NO_3^- + 8\,e^- \longrightarrow NH_3 + 9\,OH^-)$$

$$8\,Al + 3\,NO_3^- + 18\,H_2O + 5\,OH^- \longrightarrow 3\,NH_3 + 8\,Al(OH)_4^-$$

27 OH^- and 24 e^- canceled from each side

(d) $P_4 \longrightarrow HPO_3^{2-} + PH_3$ (basic solution)

Step 1 Write half-reaction equations. Balance except H and O.

$P_4 \longrightarrow 4\,HPO_3^{2-}$

$P_4 \longrightarrow 4\,PH_3$

Step 2 Balance H and O using H_2O and H^+

$12\,H_2O + P_4 \longrightarrow 4\,HPO_3^{2-} + 20\,H^+$

$12\,H^+ + P_4 \longrightarrow 4\,PH_3$

Step 3 Add OH^- ions to both sides (same number as H^+ ions)

$20\,OH^- + 12\,H_2O + P_4 \longrightarrow 4\,HPO_3^{2-} + 20\,H^+ + 20\,OH^-$

$12\,OH^- + 12\,H^+ + P_4 \longrightarrow 4\,PH_3 + 12\,OH^-$

Step 4 Combine H^+ and OH^- to form H_2O; cancel H_2O where possible

$20\,OH^- + 12\,H_2O + P_4 \longrightarrow 4\,HPO_3^{2-} + 20\,H_2O$

$12\,H_2O + P_4 \longrightarrow 4\,PH_3 + 12\,OH^-$

$20\,OH^- + P_4 \longrightarrow 4\,HPO_3^{2-} + 8\,H_2O$ (12 H_2O cancelled)

Step 5 Balance electrically with electrons

$20\,OH^- + P_4 \longrightarrow 4\,HPO_3^{2-} + 8\,H_2O + 12\,e^-$

$12\,H_2O + P_4 + 12\,e^- \longrightarrow 4\,PH_3 + 12\,OH^-$

Step 6 and 7 Loss and gain of electrons are equal; add half-reactions

$8\,OH^- + 4\,H_2O + 2\,P_4 \longrightarrow 4\,HPO_3^{2-} + 4\,PH_3$

Divide equation by 2

$4\,OH^- + 2\,H_2O + P_4 \longrightarrow 2\,HPO_3^{2-} + 2\,PH_3$

(e) $Al + OH^- \longrightarrow Al(OH)_4^- + H_2$ (basic solution)

Step 1 Write half-reaction equations. Balance except H and O.

$Al \longrightarrow Al(OH)_4^-$

$OH^- \longrightarrow H_2$

Step 2 Balance H and O using H_2O and H^+

$4\,H_2O + Al \longrightarrow Al(OH)_4^- + 4\,H^+$

$3\,H^+ + OH^- \longrightarrow H_2 + H_2O$

Step 3 Add OH^- ions to both sides (same number as H^+ ions)

$4\,OH^- + 4\,H_2O + Al \longrightarrow Al(OH)_4^- + 4\,H^+ + 4\,OH^-$

$3\,OH^- + 3\,H^+ + OH^- \longrightarrow H_2 + H_2O + 3\,OH^-$

Step 4 Combine H^+ and OH^- to form H_2O; cancel H_2O where possible

$4\,OH^- + 4\,H_2O + Al \longrightarrow Al(OH)_4^- + 4\,H_2O$

$3\,H_2O + OH^- \longrightarrow H_2 + H_2O + 3\,OH^-$

$4\,OH^- + Al \longrightarrow Al(OH)_4^-$ (4 H_2O cancelled)

$2\,H_2O + OH^- \longrightarrow H_2 + 3\,OH^-$ (1 H_2O cancelled)

Step 5 Balance electrically with electrons

$4\,OH^- + Al \longrightarrow Al(OH)_4^- + 3\,e^-$

$2\,H_2O + OH^- + 2\,e^- \longrightarrow H_2 + 3\,OH^-$

Step 6 and 7 Equalize gain and loss of electrons; add half-reactions

$2\,(4\,OH^- + Al \longrightarrow Al(OH)_4^- + 3\,e^-)$

$3\,(2\,H_2O + OH^- + 2\,e^- \longrightarrow H_2 + 3\,OH^-)$

$2\,Al + 6\,H_2O + 2\,OH^- \longrightarrow 2\,Al(OH)_4^- + 3\,H_2$

$9\,OH^-$ and $6\,e^-$ canceled on each side

37. (a) $IO_3^- + I^- \longrightarrow I_2$ (acidic solution)

Step 1 Write half-reaction equations. Balance except H and O.

$2\,IO_3^- \longrightarrow I_2$

$2\,I^- \longrightarrow I_2$

Step 2 Balance H and O using H_2O and H^+

$12\,H^+ + 2\,IO_3^- \longrightarrow I_2 + 6\,H_2O$

$2\,I^- \longrightarrow I_2$

Step 3 Balance electrically with electrons

$12\,H^+ + 2\,IO_3^- + 10\,e^- \longrightarrow I_2 + 6\,H_2O$

$2\,I^- \longrightarrow I_2 + 2\,e^-$

Step 4 and 5 Equalize the loss, and gain of electrons; add the half-reaction.

$12\,H^+ + 2\,IO_3^- + 10\,e^- \longrightarrow I_2 + 6\,H_2O$

$5\,(2\,I^- \longrightarrow I_2 + 2\,e^-)$

$12\,H^+ + 2\,IO_3^- + 10\,I^- \longrightarrow 6\,I_2 + 6\,H_2O$

(b) $Mn^{2+} + S_2O_8^{2-} \longrightarrow MnO_4^- + SO_4^{2-}$ (acid solution)

Step 1 Write half-reaction equations. Balance except H and O

$$Mn^{2+} \longrightarrow MnO_4^-$$
$$S_2O_8^{2-} \longrightarrow 2\,SO_4^{2-}$$

Step 2 Balance H and O using H_2O and H^+

$$4\,H_2O + Mn^{2+} \longrightarrow MnO_4^- + 8\,H^+$$
$$S_2O_8^{2-} \longrightarrow 2\,SO_4^{2-}$$

Step 3 Balance electrically with electrons

$$4\,H_2O + Mn^{2+} \longrightarrow MnO_4^- + 8\,H^+ + 5\,e^-$$
$$2\,e^- + S_2O_8^{2-} \longrightarrow 2\,SO_4^{2-}$$

Step 4 and 5 Equalize the loss and gain of electrons; add the half-reactions

$$2\,(4\,H_2O + Mn^{2+} \longrightarrow MnO_4^- + 8\,H^+ + 5\,e^-)$$
$$5\,(2\,e^- + S_2O_8^{2-} \longrightarrow 2\,SO_4^{2-})$$
$$\overline{2\,Mn^{2+} + 5\,S_2O_8^{2-} + 8\,H_2O \longrightarrow 2\,MnO_4^- + 10\,SO_4^{2-} + 16\,H^+}$$

Each side has 2 Mn, 10 S, 16 H, and 48 O and a −6 charge.

(c) $Co(NO_2)_6^{3-} + MnO_4^- \longrightarrow Co^{2+} + Mn^{2+} + NO_3^-$ (acidic solution)

Step 1 Write half-reaction equations. Balance except H and O.

$$Co(NO_2)_6^{3-} \longrightarrow Co^{2+} + 6\,NO_3^-$$
$$MnO_4^- \longrightarrow Mn^{2+}$$

Step 2 Balance H and O using H_2O and H^+

$$6\,H_2O + Co(NO_2)_6^{3-} \longrightarrow Co^{2+} + 6\,NO_3^- + 12\,H^+$$
$$8\,H^+ + MnO_4^- \longrightarrow Mn^{2+} + 4\,H_2O$$

Step 3 Balance electrically with e^-

$$6\,H_2O + Co(NO_2)_6^{3-} \longrightarrow Co^{2+} + 6\,NO_3^- + 12\,H^+ + 11\,e^-$$
$$5\,e^- + 8\,H^+ + MnO_4^- \longrightarrow Mn^{2+} + 4\,H_2O$$

Step 4 Equalize the loss and gain of electrons.

$$5\,(6\,H_2O + Co(NO_2)_6^{3-} \longrightarrow Co^{2+} + 6\,NO_3^- + 12\,H^+ + 11\,e^-)$$
$$11\,(5\,e^- + 8\,H^+ + MnO_4^- \longrightarrow Mn^{2+} + 4\,H_2O)$$

Step 5 Add the half-reactions

$$5\,Co(NO_2)_6^{3-} + 11\,MnO_4^- + 28\,H^+ \longrightarrow 5\,Co^{2+} + 30\,NO_3^- + 11\,Mn^{2+} + 14\,H_2O$$

Each side has 5 Co, 30 N, 11 Mn, 28 H, 104 O and a + 2 charge.

38. (a) $Mo_2O_3 + MnO_4^- \longrightarrow MoO_3 + Mn^{2+}$ (acid solution)

Step 1 Write half-reactions equations. Balance except H and O

$Mo_2O_3 \longrightarrow 2\,MoO_3$

$MnO_4^- \longrightarrow Mn^{2+}$

Step 2 Balance H and O using H_2O and H^+

$3\,H_2O + Mo_2O_3 \longrightarrow 2\,MoO_3 + 6\,H^+$

$8\,H^+ + MnO_4^- \longrightarrow Mn^{2+} + 4\,H_2O$

Step 3 Balance electrically with electrons

$3\,H_2O + Mo_2O_3 \longrightarrow 2\,MoO_3 + 6\,H^+ + 6\,e^-$

$5\,e^- + 8\,H^+ + MnO_4^- \longrightarrow Mn^{2+} + 4\,H_2O$

Steps 4 and 5 Equalize the loss and gain of electrons; add the half-reactions.

$5\,(3\,H_2O + Mo_2O_3 \longrightarrow 2\,MoO_3 + 6\,H^+ + 6\,e^-)$

$6\,(5\,e^- + 8\,H^+ + MnO_4^- \longrightarrow Mn^{2+} + 4\,H_2O)$

$5\,Mo_2O_3 + 6\,MnO_4^- + 18\,H^+ \longrightarrow 10\,MoO_3 + 6\,Mn^{2+} + 9\,H_2O$

(b) $BrO^- + Cr(OH)_4^- \longrightarrow Br^- + CrO_4^{2-}$ (basic solution)

Step 1 Write half-reaction equation. Balance except H and O

$BrO^- \longrightarrow Br^-$

$Cr(OH)_4^- \longrightarrow CrO_4^{2-}$

Step 2 Balance H and O using H_2O and H^+

$2\,H^+ + BrO^- \longrightarrow Br^- + H_2O$

$Cr(OH)_4^- \longrightarrow CrO_4^{2-} + 4\,H^+$

Step 3 Add OH^- ions to both sides (same number as H^+)

$2\,OH^- + 2\,H^+ + BrO^- \longrightarrow Br^- + H_2O + 2\,OH^-$

$4\,OH^- + Cr(OH)_4^- \longrightarrow CrO_4^{2-} + 4\,H^+ + 4\,OH^-$

Step 4 — Combine H^+ and OH^- to form H_2O; cancel H_2O where possible

$$2\,H_2O + BrO^- \longrightarrow Br^- + H_2O + 2\,OH^-$$

$$4\,OH^- + Cr(OH)_4^- \longrightarrow CrO_4^{2-} + 4\,H_2O$$

$$H_2O + BrO^- \longrightarrow Br^- + 2\,OH^- \qquad \text{(1 } H_2O \text{ cancelled)}$$

Step 5 — Balance electrically with electrons

$$2\,e^- + H_2O + BrO^- \longrightarrow Br^- + 2\,OH^-$$

$$4\,OH^- + Cr(OH)_4^- \longrightarrow CrO_4^{2-} + 4\,H_2O + 3\,e^-$$

Steps 6 and 7 — Equalize loss and gain of electrons; add the half-reactions

$$3\,(2\,e^- + H_2O + BrO^- \longrightarrow Br^- + 2\,OH^-)$$

$$2\,(4\,OH^- + Cr(OH)_4^- \longrightarrow CrO_4^{2-} + 4\,H_2O + 3\,e^-)$$

$$3\,BrO^- + 2\,Cr(OH)_4^- + 2\,OH^- \longrightarrow 3\,Br^- + 2\,CrO_4^{2-} + 5\,H_2O$$

(c) $S_2O_3^{2-} + MnO_4^- \longrightarrow SO_4^{2-} + Mn^{2+}$ (basic solution)

Step 1 — Write half-reaction equations. Balance except H and O.

$$S_2O_3^{2-} \longrightarrow 2\,SO_4^{2-}$$

$$MnO_4^- \longrightarrow Mn^{2+}$$

Step 2 — Balance H and O using H_2O and H^+

$$5\,H_2O + S_2O_3^{2-} \longrightarrow 2\,SO_4^{2-} + 10\,H^+$$

$$8\,H^+ + MnO_4^- \longrightarrow Mn^{2+} + 4\,H_2O$$

Step 3 — Add OH^- ions to both sides (same number as H^+)

$$10\,OH^- + 5\,H_2O + S_2O_3^{2-} \longrightarrow 2\,SO_4^{2-} + 10\,H^+ + 10\,OH^-$$

$$8\,OH^- + 8\,H^+ + MnO_4^- \longrightarrow Mn^{2+} + 4\,H_2O + 8\,OH^-$$

Step 4 — Combine H^+ and OH^- to form H_2O; cancel H_2O where possible

$$10\,OH^- + 5\,H_2O + S_2O_3^{2-} \longrightarrow 2\,SO_4^{2-} + 10\,H_2O$$

$$8\,H_2O + MnO_4^- \longrightarrow Mn^{2+} + 4\,H_2O + 8\,OH^-$$

$$10\,OH^- + S_2O_3^{2-} \longrightarrow 2\,SO_4^{2-} + 5\,H_2O \qquad \text{(5 } H_2O \text{ cancelled)}$$

$$4\,H_2O + MnO_4^- \longrightarrow Mn^{2+} + 8\,OH^- \qquad \text{(4 } H_2O \text{ cancelled)}$$

Step 5 — Balance electrically with electrons

$$10\,OH^- + S_2O_3^{2-} \longrightarrow 2\,SO_4^{2-} + 5\,H_2O + 8\,e^-$$

$$5\,e^- + 4\,H_2O + MnO_4^- \longrightarrow Mn^{2+} + 8\,OH^-$$

Step 6 and 7 Equalize loss and gain of electrons; add half-reactions.

$$5\,(10\,OH^- + S_2O_3^{2-} \longrightarrow 2\,SO_4^{2-} + 5\,H_2O + 8\,e)$$
$$8\,(5\,e^- + 4\,H_2O + MnO_4^- \longrightarrow Mn^{2+} + 8\,OH^-)$$
$$\overline{5\,S_2O_3^{2-} + 8\,MnO_4^- + 7\,H_2O \longrightarrow 10\,SO_4^{2-} + 8\,Mn^{2+} + 14\,OH^-}$$

Each side has 10 S, 8 Mn, 14 H, 54 O and a −18 charge.

39.

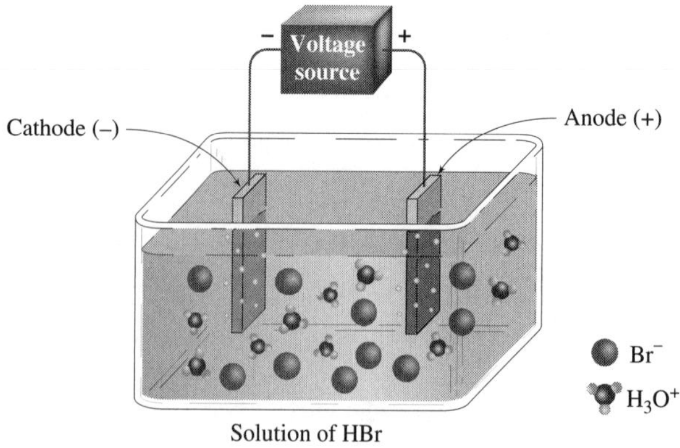

40. (a) $Pb + SO_4^{2-} \longrightarrow PbSO_4 + 2\,e^-$

$PbO_2 + SO_4^{2-} + 4\,H^+ + 2\,e^- \longrightarrow PbSO_4 + 2\,H_2O$

(b) The first reaction is oxidation (Pb^0 is oxidized to Pb^{2+}).

The second reaction is reduction (Pb^{4+} is reduced to Pb^{2+}).

(c) The first reaction (oxidation) occurs at the anode of the battery.

41. (a) The oxidizing agent is $KMnO_4$.

(b) The reducing agent is HCl.

(c) 5 moles of electrons $5\,e^- + Mn^{7+} \longrightarrow Mn^{2+}$

$$\left(\frac{5\text{ mol e}^-}{\text{mol KMnO}_4}\right)\left(\frac{6.022 \times 10^{23}\text{ e}^-}{\text{mol e}^-}\right) = 3.011 \times 10^{24}\,\frac{\text{electrons}}{\text{mol KMnO}_4}$$

42. Zinc is a more reactive metal than copper so when corrosion occurs the zinc preferentially reacts. Zinc is above hydrogen in the Activity series of metals; copper is below hydrogen.

43. $3\,Ag + 4\,HNO_3 \longrightarrow 3\,AgNO_3 + NO + 2\,H_2O$ (balanced)

g Ag $\longrightarrow$ mol Ag $\longrightarrow$ mol NO

$$(25.0\text{ g Ag})\left(\frac{1\text{ mol}}{107.9\text{ g}}\right)\left(\frac{1\text{ mol NO}}{3\text{ mol Ag}}\right) = 0.0772\text{ mol NO}$$

44. $3\,Cl_2 + 6\,KOH \longrightarrow KClO_3 + 5\,KCl + 3\,H_2O$

mol $KClO_3 \longrightarrow$ mol $Cl_2 \longrightarrow$ L Cl_2

$$(0.300\text{ g }KClO_3)\left(\frac{3\text{ mol }Cl_2}{1\text{ mol }KClO_3}\right)\left(\frac{22.4\text{ L}}{1\text{ mol}}\right) = 20.2\text{ L }Cl_2$$

45. $5\,H_2O_2 + 2\,KMnO_4 + 3\,H_2SO_4 \longrightarrow 5\,O_2 + 2\,MnSO_4 + K_2SO_4 + 8\,H_2O$

mL $H_2O_2 \longrightarrow$ g $H_2O_2 \longrightarrow$ mol $H_2O_2 \longrightarrow$ mol $KMnO_4 \longrightarrow$ g $KMnO_4$

$$(100.\text{ mL }H_2O_2\text{ solution})\left(\frac{1.031\text{ g}}{\text{mL}}\right)\left(\frac{9.0\text{ g }H_2O_2}{100.\text{ g }H_2O_2\text{ solution}}\right)\left(\frac{1\text{ mol}}{34.02\text{ g}}\right)$$

$$\left(\frac{2\text{ mol }KMnO_4}{5\text{ mol }H_2O_2}\right)\left(\frac{158.0\text{ g}}{\text{mol}}\right) = 17\text{ g }KMnO_4$$

46. $Cr_2O_7^{2-} + 3\,H_3AsO_3 + 8\,H^+ \longrightarrow 2\,Cr^{3+} + 3\,H_3AsO_4 + 4\,H_2O$

g $H_3AsO_3 \longrightarrow$ mol $H_3AsO_3 \longrightarrow$ mol $Cr_2O_7^{2-} \longrightarrow$ mL $Cr_2O_7^{2-}$

$$(5.00\text{ g }H_3AsO_3)\left(\frac{1\text{ mol}}{125.9\text{ g}}\right)\left(\frac{1\text{ mol }Cr_2O_7^{2-}}{3\text{ mol }H_3AsO_3}\right)\left(\frac{1000\text{ mL}}{0.200\text{ mol}}\right) = 66.2\text{ mL of }0.200\text{ M }K_2Cr_2O_7$$

47. $Cr_2O_7^{2-} + 6\,Fe^{2+} + 14\,H^+ \longrightarrow 2\,Cr^{3+} + 6\,Fe^{3+} + 7\,H_2O$

mL $FeSO_4 \longrightarrow$ mol $FeSO_4 \longrightarrow$ mol $Cr_2O_7^{2-} \longrightarrow$ mL $Cr_2O_7^{2-}$

$$(60.0\text{ mL }FeSO_4)\left(\frac{0.200\text{ mol}}{1000\text{ mL}}\right)\left(\frac{1\text{ mol }Cr_2O_7^{2-}}{6\text{ mol }FeSO_4}\right)\left(\frac{1000\text{ mL}}{0.200\text{ mol}}\right)$$

$$= 10.0\text{ mL of }0.200\text{ M }K_2Cr_2O_7$$

48. $2\,Al + 2\,OH^- + 6\,H_2O \longrightarrow 2\,Al(OH)_4^- + 3\,H_2$

g Al $\longrightarrow$ mol Al $\longrightarrow$ mol H_2

$$(100.0\text{ g Al})\left(\frac{1\text{ mol Al}}{26.98\text{ g}}\right)\left(\frac{3\text{ mol }H_2}{2\text{ mol Al}}\right) = 5.560\text{ mol }H_2$$

49. (a) $Cu^+ \longrightarrow Cu^{2+}$ is an oxidation, but when electrons are gained reduction should occur.
$Cu^+ + e^- \longrightarrow Cu^0$ or $Cu^+ \longrightarrow Cu^{2+} + e^-$

(b) When Pb^{2+} is reduced, it requires two individual electrons. $Pb^{2+} + 2\,e^- \longrightarrow Pb^0$. An electron has only a single negative charge (e^-).

50. The electrons lost by the species undergoing oxidation must be gained (or attracted) by another species which then undergoes reduction.

51. $A(s) + B^{2+}(aq) \longrightarrow NR$ — B^{2+} cannot take e^- from A

$A(s) + C^{+}(aq) \longrightarrow NR$ — C^{+} cannot take e^- from A

$D(s) + 2\,C^{+}(aq) \longrightarrow 2C(s) + D^{2+}(aq)$ — C^{+} takes e^- from D

$B(s) + D^{2+}(aq) \longrightarrow D(s) + B^{2+}(aq)$ — D^{2+} takes 2 e^- from B

Therefore, B^{2+} is least able to attract e^-, then D^{2+}, then C^+, then A^+

52. Sn^{4+} can only be an oxidizing agent.

$Sn^{4+} + 2\,e^- \longrightarrow Sn^{2+}$

$Sn^{4+} + 4\,e^- \longrightarrow Sn^{0}$

Sn^{0} can only be a reducing agent.

$Sn^{0} \longrightarrow Sn^{2+} + 2\,e^-$

$Sn^{0} \longrightarrow Sn^{4+} + 4\,e^-$

Sn^{2+} can be both oxidizing and reducing.

$Sn^{2+} + 2\,e^- \longrightarrow Sn^{0}$ (oxidizing)

$Sn^{2+} \longrightarrow Sn^{4+} + 2\,e^-$ (reducing)

53.

$Mn(OH)_2$	+2
MnF_3	+3
MnO_2	+4
K_2MnO_4	+6
$KMnO_4$	+7

$KMnO_4$ is the best oxidizing agent of the group, since its greater oxidation number (+7) makes it very attractive to electrons.

54. Equations (a) and (b) represent oxidation

(a) $Mg \longrightarrow Mg^{2+} + 2\,e^-$

(b) $SO_2 \longrightarrow SO_3; (S^{4+} \longrightarrow S^{6+} + 2\,e^-)$

55. (a) $MnO_2 + 2\,Br^- + 4\,H^+ \longrightarrow Mn^{2+} + Br_2 + 2\,H_2O$

(b) mL $Mn^{2+} \longrightarrow$ mol $Mn^{2+} \longrightarrow$ mol $MnO_2 \longrightarrow$ g MnO_2

$$(100.0\text{ mL }Mn^{2+})\left(\frac{0.05\text{ mol}}{1000\text{ mL}}\right)\left(\frac{1\text{ mol }MnO_2}{1\text{ mol }Mn^{2+}}\right)\left(\frac{86.94\text{ g}}{\text{mol}}\right) = 0.4\text{ g }MnO_2$$

(c) $$(100.0\text{ mL }Mn^{2+})\left(\frac{0.05\text{ mol}}{1000\text{ mL}}\right)\left(\frac{1\text{ mol }Br_2}{1\text{ mol }Mn^{2+}}\right) = 0.005\text{ mol }Br_2$$

$$PV = nRT \qquad V = \frac{nRT}{P}$$

$$V = \left(\frac{0.005\text{ mol}}{1.4\text{ atm}}\right)\left(\frac{0.0821\text{ L atm}}{\text{mol K}}\right)(323\text{ K}) = 0.09\text{ L }Br_2\text{ vapor}$$

56. (a) $F_2 + 2\,Cl^- \longrightarrow 2\,F^- + Cl_2$
   (b) $Br_2 + Cl^- \longrightarrow NR$
   (c) $I_2 + Cl^- \longrightarrow NR$
   (d) $Br_2 + 2\,I^- \longrightarrow 2\,Br^- + I_2$

57. $Mn(s) + 2\,HCl(aq) \longrightarrow Mn^{2+}(aq) + H_2(g) + 2\,Cl^-(aq)$

58. $4\,Zn + NO_3^- + 10\,H^+ \longrightarrow 4\,Zn^{2+} + NH_4^+ + 3\,H_2O$ See Exercise 33(a).

59.

	Equation 1	2	3	4	5
a	C oxidized	S oxidized	N oxidized	S oxidized	O_2^{2-} oxidized
b	O_2 reduced	N reduced	Cu reduced	O reduced	O_2^{2-} reduced
c	O_2, O.A.	HNO_3, O.A.	CuO, O.A.	H_2O_2, O.A.	H_2O_2, O.A.
d	C_3H_8, R.A.	H_2S, R.A.	NH_3, R.A.	Na_2SO_3, R.A.	H_2O_2, R.A.
e	$C^{2\frac{2}{3}+} \longrightarrow C^{4+}$	$S^{2-} \longrightarrow S^0$	$N^{3-} \longrightarrow N_2^{\,0}$	$S^{4+} \longrightarrow S^{6+}$	$O_2^{2-} \longrightarrow O_2^{\,0}$
f	$O^0 \longrightarrow O^{2-}$	$N^{5+} \longrightarrow N^{2+}$	$Cu^{2+} \longrightarrow Cu^0$	$O_2^{2-} \longrightarrow O^{2-}$	$O_2^{2-} \longrightarrow O^{2-}$

O.A. = Oxidizing agent
R.A. = Reducing agent

60. $Pb + 2\,Ag^+ \longrightarrow 2\,Ag + Pb^{2+}$
   (a) Pb is the anode
   (b) Ag is the cathode
   (c) Oxidation occurs at Pb (anode)
   (d) Reduction occurs at Ag (cathode)
   (e) Electrons flow from the lead electrode through the wire to the silver electrode.
   (f) Positive ions flow through the salt bridge towards the negatively charged strip of silver; negative ions flow toward the positively charged strip of lead.

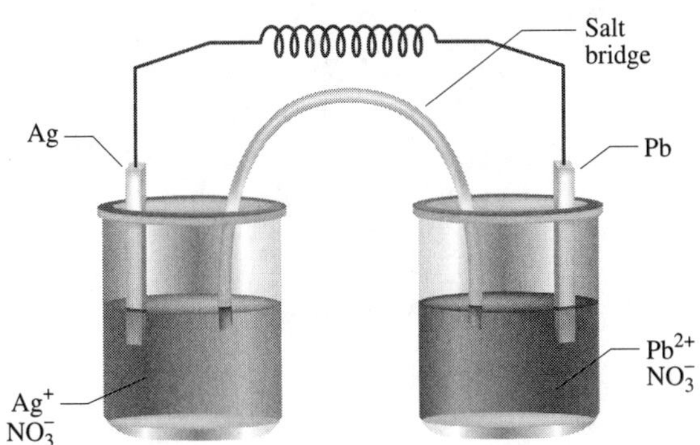

61. $8\,KI + 5\,H_2SO_4 \longrightarrow 4\,I_2 + H_2S + 4\,K_2SO_4 + 4\,H_2O$

start with grams I_2 and work towards g KI

$g\ I_2 \longrightarrow mol\ I_2 \longrightarrow mol\ KI \longrightarrow g\ KI$

$$(2.79\ g\ I_2)\left(\frac{1\ mol}{253.8\ g}\right)\left(\frac{8\ mol\ KI}{4\ mol\ I_2}\right)\left(\frac{166.0\ g}{mol}\right) = 3.65\ g\ KI\ in\ sample$$

$$\left(\frac{3.65\ g\ KI}{4.00\ g\ sample}\right)(100) = 91.3\%\ KI$$

62. $3\,Ag + 4\,HNO_3 \longrightarrow 3\,AgNO_3 + NO + 2\,H_2O$

$mol\ Ag \longrightarrow mol\ NO$

$$(0.500\ mol\ Ag)\left(\frac{1\ mol\ NO}{3\ mol\ Ag}\right) = 0.167\ mol\ NO$$

$$PV = nRT \qquad V = \frac{nRT}{P}$$

$$P = (744\ torr)\left(\frac{1\ atm}{760.\ torr}\right) = 0.979\ atm$$

$T = 301\ K$

$$V = \frac{(0.167\ mol\ NO)(0.0821\ L\ atm/mol\ K)(301\ K)}{(0.979\ atm)} = 4.22\ L\ NO$$

CHAPTER 18

NUCLEAR CHEMISTRY

1. (a) Gamma radiation requires the most shielding.

   (b) Alpha radiation requires the least shielding.

2. Alpha particles are deflected less than beta particles while passing through a magnetic field, because they are much heavier (more than 7,000 times heavier) than beta particles.

3. Pairs of nuclides that would be found in the fission reaction of U-235. Any two nuclides, whose atomic numbers add up to 92 and mass numbers (in the range of 70-160) add up to 230-234. Examples include:

   $^{90}_{38}Sr$ and $^{141}_{54}Xe$ $\quad$ $^{139}_{56}Ba$ and $^{94}_{36}Kr$ $\quad$ $^{101}_{42}Mo$ and $^{131}_{50}Sn$

4. Contributions to the early history of radioactivity include:

   (a) Henri Becquerel: He discovered radioactivity.

   (b) Marie and Pierre Curie: They discovered the elements polonium and radium.

   (c) Wilhelm Roentgen: He discovered X rays and developed the technique of producing them. While this was not a radioactive phenomenon, it triggered Becquerel's discovery of radioactivity.

   (d) Ernest Rutherford: He discovered alpha and beta particles, established the link between radioactivity and transmutation, and produced the first successful man-made transmutation.

   (e) Otto Hahn and Fritz Strassmann: They were first to produce nuclear fission.

5. Chemical reactions are caused by atoms or ions coming together, so are greatly influenced by temperature and concentration, which affect the number of collisions. Radioactivity is a spontaneous reaction of an individual nucleus, and is independent of such influences.

6. The term isotope is used with reference to atoms of the same element that contains different masses. For example, $^{12}_{6}C$ and $^{14}_{6}C$. The term nuclide is used in nuclear chemistry to infer any isotope of any atom.

7. $(5 \times 10^9 \text{ years})\left(\dfrac{1 \text{ half-life}}{7.6 \times 10^7 \text{ years}}\right) = 70 \text{ half-lives}$

   Even if plutonium-244 had been present in large quantities five billion years ago, no measureable amount would survive after 70 half-lives.

8.

	charge	mass	nature of particles	penetrating power
Alpha	+2	4 amu	He nucleus	low
Beta	−1	$\frac{1}{1837 \text{ amu}}$	electron	moderate
Gamma	0	0	electromagnetic radiation	high

9. Natural radioactivity is the spontaneous disintegration of those radioactive nuclides found in nature. Artificial radioactivity is the spontaneous disintegration of radioactive isotopes produced synthetically.

10. A radioactive disintegration series starts with a particular radionuclide and progresses stepwise by alpha and beta emissions to other radionuclides, ending at a stable nuclide. For example:

$$^{238}_{92}\text{U} \xrightarrow{\text{14 steps}} {}^{206}_{82}\text{Pb(stable)}$$

11. Transmutation is the conversion of one element into another by natural or artificial means. The nucleus of an atom is bombarded by various particles (alpha, beta, protons, etc.). The fast moving particles are captured by the nucleus, forming an unstable nucleus, which decays to another kind of atom. For example:

$$^{9}_{4}\text{Be} + {}^{4}_{2}\text{He} \longrightarrow {}^{12}_{6}\text{C} + {}^{1}_{0}\text{n}$$

12. $^{232}_{90}\text{Th} \xrightarrow{-\alpha} {}^{228}_{88}\text{Ra} \xrightarrow{-\beta} {}^{228}_{89}\text{Ac} \xrightarrow{-\beta} {}^{228}_{90}\text{Th} \xrightarrow{-\alpha} {}^{224}_{88}\text{Ra} \xrightarrow{-\alpha} {}^{220}_{86}\text{Rn} \xrightarrow{-\alpha}$

$^{216}_{84}\text{Po} \xrightarrow{-\alpha} {}^{212}_{82}\text{Pb} \xrightarrow{-\beta} {}^{212}_{83}\text{Bi} \xrightarrow{-\beta} {}^{212}_{84}\text{Po} \xrightarrow{-\alpha} {}^{208}_{82}\text{Pb}$

13. $^{237}_{93}\text{Np}$ loses seven alpha particles and four beta particles.

Determination of the final product: $^{209}_{83}\text{Bi}$

nuclear charge $= 93 - 7(2) + 4(1) = 83$

mass $= 237 - 7(4) = 209$

14. Decay of bismuth-211

$^{211}_{83}\text{Bi} \longrightarrow {}^{4}_{2}\text{He} + {}^{207}_{81}\text{Tl} \qquad {}^{207}_{81}\text{Tl} \longrightarrow {}^{0}_{-1}\text{e} + {}^{207}_{82}\text{Pb}$

15. Two Germans, Otto Hahn and Fritz Strassmann, were the first scientists to report nuclear fission. The fission resulted from bombarding uranium nuclei with neutrons.

16. Natural uranium is 99+%U-238. Commercial nuclear reactors use U-235 enriched uranium as a fuel. Slow neutrons will cause the fission of U-235, but not U-238. Fast

neutrons are capable of a nuclear reaction with U-238 to produce fissionable Pu-239. A breeder reactor converts nonfissionable U-238 to fissionable Pu-239, and in the process, manufactures more fuel than it consumes.

17. The fission reaction in a nuclear reactor and in an atomic bomb are essentially the same. The difference is that the fissioning is "wild" or uncontrolled in the bomb. In a nuclear reactor, the fissioning rate is controlled by means of moderators, such as graphite, to slow the neutrons and control rods of cadmium or boron to absorb some of the neutrons.

18. A certain amount of fissionable material (a critical mass) must be present before a self-sustaining chain reaction can occur. Without a critical mass, too many neutrons from fissions will escape, and the reaction cannot reach a chain reaction status, unless at least one neutron is captured for every fission that occurs.

19. The mass defect is the difference between the mass of an atom and the sum of the masses of the number of protons, neutrons, and electrons in that atom. The energy equivalent of this mass defect is known as the nuclear binding energy.

20. When radioactive rays pass through normal matter, they cause that matter to become ionized (usually by knocking out electrons). Therefore, the radioactive rays are classified as ionizing radiation.

21. Some biological hazards associated with radioactivity are:
    (a) High levels of radiation can cause nausea, vomiting, diarrhea, and death. The radiation produces ionization in the cells, particularly in the nucleus of the cells.
    (b) Long-term exposure to low levels of radiation can weaken the body and cause malignant tumors.
    (c) Radiation can damage DNA molecules in the body causing mutations, which by reproduction, can be passed on to succeeding generations.

22. Strontium-90 has two characteristics that create concern. Its half-life is 28 years, so it remains active for a long period of time (disintegrating by emitting β radiation). The other characteristic is that Sr-90 is chemically similar to calcium, so when it is present in milk Sr-90 is deposited in bone tissue along with calcium. Red blood cells are produced in the bone marrow. If the marrow is subjected to beta radiation from strontium-90, the red blood cells will be destroyed, increasing the incidence of leukemia and bone cancer.

23. A radioactive "tracer" is a radioactive material, whose presence is traced by a Geiger counter or some other detecting device. Tracers are often injected into the human body, animals, and plants to determine chemical pathways, rates of circulation, etc. For example, use of a tracer could determine the length of time for material to travel from the root system to the leaves in a tree.

24. In living species, the ratio of carbon-14 to carbon-12 is constant due to the constant C-14/C-12 ratio in the atmosphere and food sources. When a species dies, life processes stop. The C-14/C-12 ratio decreases with time because C-14 is radioactive and decays according to its half-life, while the amount of C-12 in the species remains constant. Thus, the age of an archaeological artifact containing carbon can be calculated by comparing the C-14/C-12 ratio in the artifact with the C-14/C-12 ratio in the living species.

25. Radioactivity could be used to locate a leak in an underground pipe by using a water soluble tracer element. Dissolve the tracer in water and pass the water through the pipe. Test the ground along the path of the pipe with a Geiger counter until radioactivity from the leak is detected. Then dig.

26. The half-life of carbon-14 is 5730 years.

$$(4 \times 10^6 \text{ years})\left(\frac{1 \text{ half-life}}{5730 \text{ years}}\right) = 7 \times 10^2 \text{ half-lives}$$

700 half-lives would pass in 4 million years. Not enough C-14 would remain to allow detection with any degree of reliability. C-14 dating would not prove useful in this case.

27. The disadvantages of nuclear power include the danger of contamination from radioactive material and the radioactive waste products that accumulate, some having half-lives of thousands of years.

28. The hazards associated with an atomic bomb explosion include shock waves, heat and radiation from alpha particles, beta particles, gamma rays and ultraviolet rays. Gamma rays and X-rays cause burns, sterilization and gene mutation. If the bomb explodes close to the ground radioactive material is carried by dust particles and is spread over wide areas.

29. Heavy elements undergo fission and lighter elements undergo fusion.

30. In a nuclear power plant, a controlled nuclear fission reaction provides heat energy that is used to produce steam. The steam turns a turbine that generates electricity.

31.

	Protons	Neutrons	Nucleons
(a) $^{35}_{17}\text{Cl}$	17	18	35
(b) $^{226}_{88}\text{Ra}$	88	138	226

32.

	Protons	Neutrons	Nucleons
(a) $^{235}_{92}\text{U}$	92	143	235
(b) $^{82}_{35}\text{Br}$	35	47	82

33. When a nucleus loses an alpha particle, its atomic number decreases by two, and its mass number decreases by four.

34. When a nucleus loses a beta particle, its atomic number increases by one, and its mass number remains unchanged.

35. Equations for alpha decay:
   (a) $^{218}_{85}\text{At} \longrightarrow {}^{4}_{2}\text{He} + {}^{214}_{83}\text{Bi}$
   (b) $^{221}_{87}\text{Fr} \longrightarrow {}^{4}_{2}\text{He} + {}^{217}_{85}\text{At}$

36. Equations for alpha decay:
   (a) $^{192}_{78}\text{Pt} \longrightarrow {}^{4}_{2}\text{He} + {}^{188}_{76}\text{Os}$
   (b) $^{210}_{84}\text{Po} \longrightarrow {}^{4}_{2}\text{He} + {}^{206}_{82}\text{Pb}$

37. Equations for beta decay:
   (a) $^{14}_{6}\text{C} \longrightarrow {}^{0}_{-1}\text{e} + {}^{14}_{7}\text{N}$
   (b) $^{137}_{55}\text{Cs} \longrightarrow {}^{0}_{-1}\text{e} + {}^{137}_{56}\text{Ba}$

38. Equations for beta decay:
   (a) $^{239}_{93}\text{Np} \longrightarrow {}^{0}_{-1}\text{e} + {}^{239}_{94}\text{Pu}$
   (b) $^{90}_{38}\text{Sr} \longrightarrow {}^{0}_{-1}\text{e} + {}^{90}_{39}\text{Y}$

Paired Exercises

39. (a) alpha-emission (b) beta-emission then gamma-emission (c) beta-emission

40. (a) gamma-emission (b) alpha-emission then beta-emission
   (c) alpha-emission then gamma-emission

41. $^{13}_{6}\text{C} + {}^{1}_{0}\text{n} \longrightarrow {}^{14}_{6}\text{C}$

42. $^{30}_{15}\text{P} \longrightarrow {}^{30}_{14}\text{Si} + {}^{0}_{+1}\text{e}$

43. (a) $^{66}_{29}\text{Cu} \longrightarrow {}^{66}_{30}\text{Zn} + {}^{0}_{-1}\text{e}$
   (b) $^{0}_{-1}\text{e} + {}^{7}_{4}\text{Be} \longrightarrow {}^{7}_{3}\text{Li}$
   (c) $^{27}_{13}\text{Al} + {}^{4}_{2}\text{He} \longrightarrow {}^{30}_{14}\text{Si} + {}^{1}_{1}\text{H}$
   (d) $^{85}_{37}\text{Rb} + {}^{1}_{0}\text{n} \longrightarrow {}^{82}_{35}\text{Br} + {}^{4}_{2}\text{He}$

44. (a) $^{27}_{13}Al + ^{4}_{2}He \longrightarrow ^{30}_{15}P + ^{1}_{0}n$

(b) $^{27}_{14}Si \longrightarrow ^{0}_{+1}e + ^{27}_{13}Al$

(c) $^{12}_{6}C + ^{2}_{1}H \longrightarrow ^{13}_{7}N + ^{1}_{0}n$

(d) $^{82}_{35}Br \longrightarrow ^{82}_{36}Kr + ^{0}_{-1}e$

45. $(112\ \text{years})\left(\dfrac{1\ \text{half-life}}{28\ \text{years}}\right) = 4\ \text{half-lives}$

In 4 half-lives 1/16th or $(^1/_2)^4$ of the starting amount would remain.

$\dfrac{1.00\ \text{mg Sr-90}}{16} = 0.0625$ mg Sr-90 remains after 112 years.

46. $\dfrac{240\ \text{Cts/min}}{2} = 120\ \text{Cts/min};\quad \dfrac{120\ \text{Cts/min}}{2} = 60\ \text{Cts/min};\quad \dfrac{60\ \text{Cts/min}}{2} = 30\ \text{Cts/min};$

3 half-lives are required to reduce the count from 240 to 30 counts/min.

$1980 + (3 \times 28) = 2064$. One eighth of the original amount Sr-90 remains. $\left[\left(\dfrac{1}{2}\right)^3 = \dfrac{1}{8}\right]$

47. $^{234}_{90}Th \xrightarrow{-\beta} ^{234}_{91}Pa \xrightarrow{-\beta} ^{234}_{92}U \xrightarrow{-\alpha} ^{230}_{90}Th \xrightarrow{-\alpha} ^{226}_{88}Ra \xrightarrow{-\alpha} ^{222}_{86}Rn \xrightarrow{-\alpha} ^{218}_{84}Po$

48. $^{226}_{88}Ra \xrightarrow{-\alpha} ^{222}_{86}Rn \xrightarrow{-\alpha} ^{218}_{84}Po \xrightarrow{-\alpha} ^{214}_{82}Pb \xrightarrow{-\beta} ^{214}_{83}Bi \xrightarrow{-\beta} ^{214}_{84}Po \xrightarrow{-\alpha} ^{210}_{82}Pb$
$\xrightarrow{-\beta} ^{210}_{83}Bi$

49. (a) $^{235}_{92}U + ^{1}_{0}n \longrightarrow ^{94}_{38}Sr + ^{139}_{54}Xe + 3\ ^{1}_{0}n + \text{energy}$

Mass loss = mass of reactants − mass of products

Mass of reactants = 235.0439 amu + 1.0087 amu = 236.0526 amu

Mass of products = 93.9154 amu + 138.9179 amu + 3(1.0087 amu) = 235.8594 amu

Mass lost = 236.0526 amu − 235.8594 amu = 0.1932 amu

$$(0.1932\ \text{amu})\left(\frac{1.000\ \text{g}}{6.022 \times 10^{23}\ \text{amu}}\right)\left(\frac{9.0 \times 10^{13}\ \text{J}}{1.00\ \text{g}}\right) = 2.9 \times 10^{-11}\ \text{J/atom U-235}$$

(b) $$\left(\frac{2.9 \times 10^{-11}\ \text{J}}{\text{atom}}\right)\left(\frac{6.022 \times 10^{23}\ \text{atoms}}{\text{mol}}\right) = 1.7 \times 10^{13}\ \text{J/mol}$$

(c) $$\left(\frac{0.1932\ \text{amu}}{236.0526\ \text{amu}}\right)(100) = 0.08185\%\ \text{mass loss}$$

50. (a) $^{1}_{1}H + ^{2}_{1}H \longrightarrow ^{3}_{2}He + \text{energy}$

Mass loss = mass of reactants − mass of products

Mass of reactants = 1.00794 g/mol + 2.01410 g/mol = 3.02204 g/mol

Mass of products = 3.01603 g/mol

Mass lost = 3.02204 g/mol − 3.01603 g/mol = 0.00601 g/mol

$$\left(\frac{0.00601\text{ g}}{\text{mol}}\right)\left(\frac{9.0 \times 10^{13}\text{ J}}{\text{g}}\right) = 5.4 \times 10^{11}\text{ J/mol}$$

(b) $\left(\frac{0.00601\text{ g}}{3.02204\text{ g}}\right)(100) = 0.199\%$ mass loss

51. $^{232}_{90}Th \longrightarrow ^{208}_{82}Pb$

$$^{232}_{90}Th \xrightarrow{-\alpha} ^{228}_{88}Ra \xrightarrow{-\beta} ^{228}_{89}Ac \xrightarrow{-\beta} ^{228}_{90}Th \xrightarrow{-\alpha} ^{224}_{88}Ra \xrightarrow{-\alpha} ^{220}_{86}Rn$$

$$\xrightarrow{-\alpha} ^{216}_{84}Po \xrightarrow{-\beta} ^{216}_{85}At \xrightarrow{-\alpha} ^{212}_{83}Bi \xrightarrow{-\beta} ^{212}_{84}Po \xrightarrow{-\alpha} ^{208}_{82}Pb$$

52. $\frac{1}{8} \times 8.0 = 1.0$ g left after 7/8 disappears.

$$8.0\text{ g} \xrightarrow{\text{half-life}} 4.0\text{ g} \xrightarrow{\text{half-life}} 2.0\text{ g} \xrightarrow{\text{half-life}} 1.0\text{ g}$$

It takes 3 half-lives

3.64 days × 3 = 10.9 days to disintegrate from 8.0 g to 1.0 g

53. $^{249}_{98}Cf + ^{15}_{7}N \longrightarrow + 4\,^{1}_{0}N + ^{260}_{105}Db$

54. $^{226}_{88}Ra$ contains 138 neutrons and 88 electrons

mass of neutron = 1.0087 amu   mass of electron = 0.00055 amu

$$\frac{(138)(1.0087\text{ amu})}{226\text{ amu}}(100) = 61.59\%\text{ neutrons by mass}$$

$$\frac{(88)(0.00055\text{ amu})}{226\text{ amu}}(100) = 0.021\%\text{ electrons by mass}$$

55. $(0.0100\text{ g RaCl}_2)\left(\frac{226.0\text{ g Ra}}{296.9\text{ g RaCl}_2}\right)\left(\frac{\$50{,}000}{1\text{ g Ra}}\right) = \$381$

56. 100% to 25% requires 2 half-lives. The half-life of C-14 is 5730 years. The specimen will be the age of two half-lives:

(2)(5730 years) = 11,460 years old.

57. $16.0\text{ g} \longrightarrow 8.0\text{ g} \longrightarrow 4.0\text{ g} \longrightarrow 2.0\text{ g} \longrightarrow 1.0\text{ g} \longrightarrow 0.50\text{ g}$

16.0 g to 0.50 g requires five half-lives.

$$\frac{90\text{ minutes}}{5\text{ half-lives}} = 18\text{ minutes/half-life}$$

58. (a) $^{7}_{3}Li$ is made up of 3 protons, 4 neutrons, and 3 electrons.

Calculated mass

3 protons	3(1.0073 g)	=	3.0219 g
4 neutrons	4(1.0087 g)	=	4.0348 g
3 electrons	3(0.00055 g)	=	0.0017 g
calculated mass			7.0584 g

Mass defect = calculated mass − actual mass

Mass defect = 7.0584 g − 7.0160 g = 0.0424 g/mol

(b) Binding energy

$$\left(\frac{0.0424\text{ g}}{\text{mol}}\right)\left(\frac{9.0 \times 10^{13}\text{ J}}{\text{g}}\right) = 3.8 \times 10^{12}\text{ J/mol}$$

59. $^{235}_{92}U \longrightarrow {}^{207}_{82}Pb$

Mass loss: 235 − 207 = 28

Net proton loss (atomic number): 92 p − 82 p = 10 p

The mass loss is equivalent to 7 alpha particles (28/4). A loss of 7 alpha particles gives a loss of 14 protons. A decrease in the atomic number to 78 (14 protons) is due to the loss of 7 alpha particles (92 − 14 = 78). Therefore, a loss of 4 beta particles is required to increase the atomic number from 78 to 82.

The total loss = 7 alpha particles and 4 beta particles.

60. (a) Geiger counter: Radiation passes through a thin glass window into a chamber filled with argon gas and containing two electrodes. Some of the argon ionizes, sending a momentary electrical impulse between the electrodes to the detector. This signal is amplified electronically and read out on a counter or as a series of clicks.

(b) Scintillation counter: Radiation strikes a scintillator, which is composed of molecules that emit light in the presence of ionizing radiation. A light sensitive detector counts the flashes and converts them into a digital readout.

(c) Film badge: Radiation penetrates a film holder. The silver grains in the film darken when exposed to radiation. The film is developed at regular intervals.

61. (3 days)(24 hours/day) = 72 hours

$72\text{ hr} + 6\text{ hr} = 78\text{ hr}$

$\dfrac{78\text{ hr}}{13\dfrac{\text{hr}}{t_{\frac{1}{2}}}} = 6\text{ half-lives}$ $\qquad (10\text{ mg})\left(\dfrac{1}{2}\right)^6 = 0.16\text{ mg remaining}$

62. Fission is the splitting of a heavy nuclide into two or more intermediate-sized fragments with the conversion of some mass into energy. Fission occurs in nuclear reactors, or atomic bombs.

Example: ${}^{235}_{92}\text{U} + {}^{1}_{0}\text{n} \longrightarrow {}^{143}_{54}\text{Xe} + {}^{90}_{38}\text{Sr} + 3\,{}^{1}_{0}\text{n}$

Fusion is the process of combining two relatively small nuclei to form a single larger nucleus. Fusion occurs on the sun, or in a hydrogen bomb.

Example: ${}^{3}_{1}\text{H} + {}^{2}_{1}\text{H} \longrightarrow {}^{4}_{2}\text{He} + {}^{1}_{0}\text{n} + \text{energy}$

63.

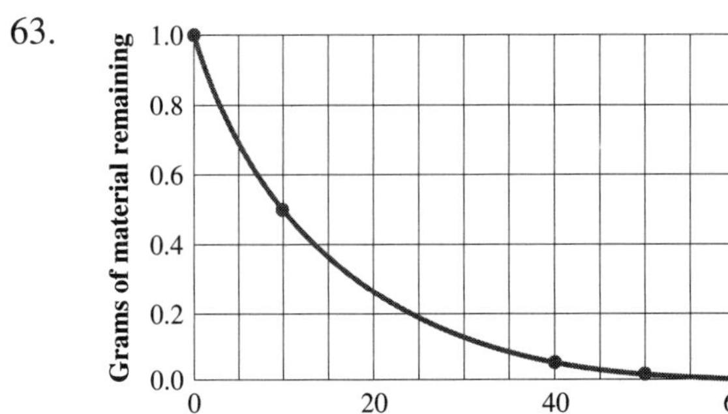

The graph produces a curve for radioactive decay which never actually crosses the *x*-axis (where mass = 0), it simply approaches that point.

64. (a) ${}^{235}_{92}\text{U} + {}^{1}_{0}\text{n} \longrightarrow {}^{143}_{54}\text{Xe} + 3\,{}^{1}_{0}\text{n} + {}^{90}_{38}\text{Sr}$

(b) ${}^{235}_{92}\text{U} + {}^{1}_{0}\text{n} \longrightarrow {}^{102}_{39}\text{Y} + 3\,{}^{1}_{0}\text{n} + {}^{131}_{53}\text{I}$

(c) ${}^{14}_{7}\text{N} + {}^{1}_{0}\text{n} \longrightarrow {}^{1}_{1}\text{H} + {}^{14}_{6}\text{C}$

65. (a) $H_2O(l) \longrightarrow H_2O(g)$

Energy$_2$: Weakest bond changes requires the least energy.

(b) $2H_2(g) + O_2(g) \longrightarrow 2H_2O(g)$

Energy$_1$: medium-sized value involved in interatomic bonds.

(c) ${}^{2}_{1}\text{H} + {}^{2}_{1}\text{H} \longrightarrow {}^{3}_{1}\text{H} + {}^{1}_{1}\text{H}$

Energy$_3$: Nuclear process; greatest amount of energy involved.

66. $^{236}_{92}U \longrightarrow\ ^{90}_{38}Sr + 3\ ^{1}_{0}n + \ ^{143}_{54}Xe$

67. (a) beta emission: $^{29}_{12}Mg \longrightarrow\ ^{0}_{-1}e + \ ^{29}_{13}Al$

(b) alpha emission: $^{150}_{60}Nd \longrightarrow\ ^{4}_{2}He + \ ^{146}_{58}Ce$

(c) positron emission: $^{72}_{33}As \longrightarrow\ ^{0}_{+1}e + \ ^{72}_{32}Ge$

68. (a) $^{87}_{37}Rb \longrightarrow\ ^{0}_{-1}e + \ ^{87}_{38}Sr$

(b) $^{87}_{38}Sr \longrightarrow\ ^{0}_{+1}e + \ ^{87}_{37}Rb$

69.

$t_{\frac{1}{2}}$, hours	0	12.5	25.0	37.5	50.0	62.5	75.0	87.5	100.
Amount, mg	15.4	7.7	3.85	1.93	0.965	0.483	0.242	0.121	0.0605

Fraction of K-42 remaining $\frac{0.0605\text{ mg}}{15.4\text{ mg}} = 0.00393(\text{or } 0.393\%)$

No. After an additional eight half-lives there would be less than one microgram (0.000001 g) remaining.

$$(200\text{ hrs})\left(\frac{1\text{ half-life}}{12.5\text{ hrs}}\right) = 16\text{ half-lives}$$

$$\text{Amount remaining} = (15.4\text{ mg})\left(\frac{1}{2}\right)^{16}\left(\frac{10^3\mu\text{g}}{\text{mg}}\right) = 0.235\ \mu\text{g}$$

70. $(270\text{ years})\left(\frac{1\text{ half-life}}{30\text{ years}}\right) = 9\text{ half-lives}$ work down from 270 years.

$t_{\frac{1}{2}}$, years	270	240	210	180	150	120	90	60	30	0
Amount, g	15.0	30.0	60.0	120.	240.	480.	960.	1920	3840	7680

There would have been 7680 g originally

71. Element 114 would fall below lead on the periodic table. It would be a metal and would most likely form +2 and +4 ions in solution (like lead).

72. 1.00 g Co-60

(a) one half-life: $\frac{1.00\text{ g}}{2} = 0.500\text{ g left}$

(b) two half-lives: $\frac{0.500\text{ g}}{2} = 0.250\text{ g left}$

(c) four half-lives: $2^4 = 16$; $\frac{1}{16}$ left $\frac{1.00\text{ g}}{16} = 0.0625\text{ g left}$

(d) ten half-lives: $2^{10} = 1024$; $\frac{1}{1024}$ left $\frac{1.00\text{ g}}{1024} = 9.77 \times 10^{4}\text{ g}$ left

73. (a) $^{11}_{5}\text{B} \longrightarrow {}^{4}_{2}\text{He} + {}^{7}_{3}\text{Li}$

(b) $^{88}_{38}\text{Sr} \longrightarrow {}^{0}_{-1}\text{e} + {}^{88}_{39}\text{Y}$

(c) $^{107}_{47}\text{Ag} + {}^{1}_{0}\text{n} \longrightarrow {}^{108}_{47}\text{Ag}$

(d) $^{41}_{19}\text{K} \longrightarrow {}^{1}_{1}\text{H} + {}^{40}_{18}\text{Ar}$

(e) $^{116}_{51}\text{Sb} + {}^{0}_{-1}\text{e} \longrightarrow {}^{116}_{50}\text{Sn}$

74. C-14 content is 1/16 of that in living plants. This means that four half-lives have passed. ^{14}C half-life is 5730 years.

$$\left(\frac{5730\text{ years}}{\text{half-life}}\right)(4\text{ half-lives}) = 22{,}920\text{ years } (2.29 \times 10^{4}\text{ years})$$

75. 1 Curie = 3.7×10^{10} disintegrations/sec

1 becquerel = 1 disintegration/sec

Therefore there are 3.7×10^{10} becquerels/1 Curie

$$\left(\frac{3.7 \times 10^{10}\text{ becquerel}}{1\text{ Curie}}\right)(1.24\text{ Curies}) = 4.6 \times 10^{10}\text{ becquerels}$$

CHAPTER 19

INTRODUCTION TO ORGANIC CHEMISTRY

1. Carbon atoms have the characteristic of bonding extensively with one another. They form organic compounds containing carbon chains of varying lengths and structure. Consequently, a great many compounds of carbon exist.

2. The most common geometric pattern of covalent bonds about carbon atoms is the tetrahedral arrangement of bonds. A simple example is the methane molecule, CH_4, with hydrogen atoms at the corners of the tetrahedron and the carbon atom at the center.

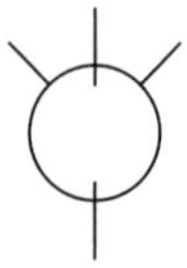

3. In addition to single bonds, carbon atoms can also form double and triple bonds. For examples, see Question 4.(b)

4. Lewis structures for:

   (a) a carbon atom

   (b) molecules of

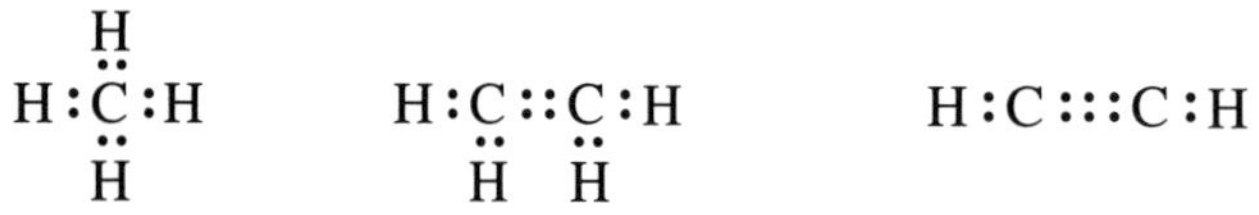

methane   ethene (ethylene)   ethyne (acetylene)

5. Names and formulas of the first ten normal alkanes

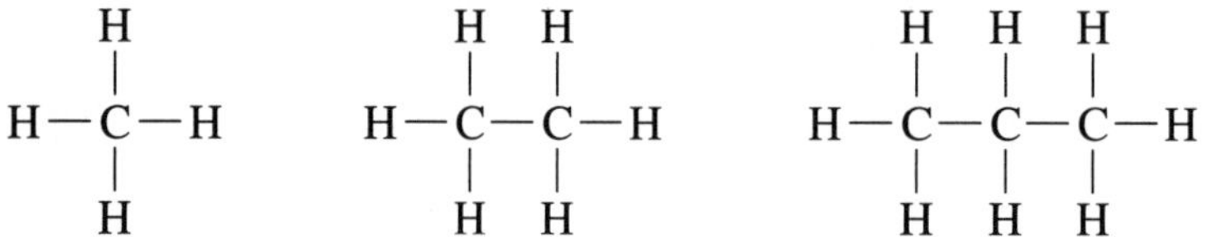

methane   ethane   propane

```
    H   H   H   H
    |   |   |   |
H - C - C - C - C - H
    |   |   |   |
    H   H   H   H
```

butane

```
    H   H   H   H   H   H   H   H
    |   |   |   |   |   |   |   |
H - C - C - C - C - C - C - C - C - H
    |   |   |   |   |   |   |   |
    H   H   H   H   H   H   H   H
```

octane

```
    H   H   H   H   H
    |   |   |   |   |
H - C - C - C - C - C - H
    |   |   |   |   |
    H   H   H   H   H
```

pentane

```
    H   H   H   H   H   H   H   H   H
    |   |   |   |   |   |   |   |   |
H - C - C - C - C - C - C - C - C - C - H
    |   |   |   |   |   |   |   |   |
    H   H   H   H   H   H   H   H   H
```

nonane

```
    H   H   H   H   H   H
    |   |   |   |   |   |
H - C - C - C - C - C - C - H
    |   |   |   |   |   |
    H   H   H   H   H   H
```

hexane

```
    H   H   H   H   H   H   H   H   H   H
    |   |   |   |   |   |   |   |   |   |
H - C - C - C - C - C - C - C - C - C - C - H
    |   |   |   |   |   |   |   |   |   |
    H   H   H   H   H   H   H   H   H   H
```

decane

```
    H   H   H   H   H   H   H
    |   |   |   |   |   |   |
H - C - C - C - C - C - C - C - H
    |   |   |   |   |   |   |
    H   H   H   H   H   H   H
```

heptane

6. Alkyl groups

```
    H
    |
H - C -
    |
    H
```

CH_3-

methyl

```
    H   H
    |   |
H - C - C -
    |   |
    H   H
```

CH_3CH_2-

ethyl

```
    H   H   H
    |   |   |
H - C - C - C -
    |   |   |
    H   H   H
```

$CH_3CH_2CH_2-$

propyl

```
    H   H   H
    |   |   |
H - C - C - C - H
    |   |   |
    H       H
```

$(CH_3)_2CH-$

isopropyl

```
    H   H   H   H
    |   |   |   |
H - C - C - C - C -
    |   |   |   |
    H   H   H   H
```

$CH_3CH_2CH_2CH_2-$

n-butyl

```
    H   H       H
    |   |   |   |
H - C - C - C - C - H
    |   |   |   |
    H   H   H   H
```

$CH_3CH_2\overset{|}{C}HCH_3$

s-butyl

```
    H         H         H
    |         |         |
H - C ------- C ------- C -
    |         |         |
    H     H - C - H     H
              |
              H
```

$(CH_3)_2CHCH_2-$

isobutyl

```
    H                   H
    |         |         |
H - C ------- C ------- C - H
    |         |         |
    H     H - C - H     H
              |
              H
```

$(CH_3)_3C-$

t-butyl

7. Compound (a) is an alkene and belongs to a homologous series of compounds represented by the formula C_nH_{2n}.
   Compounds (b) (c), and (d) are all alkanes, C_nH_{2n+2}

8. The word ethylene represents a compound containing a double bond. All the other words represent structures having no double bonds.

9. The single most important reaction of alkanes is combustion.

10. Structure of vinyl acetylene, C_4H_4, $CH_2{=}CH{-}C{\equiv}CH$

11. Formula of C_6H_8

    The formula would be C_6H_{14} if the compound were a saturated hydrocarbon. This formula is 6 H atoms short of being saturated. Removing two H atoms from a saturated hydrocarbon forms a carbon-carbon double bond. Removing four H atoms can form a carbon-carbon triple bond. Therefore, C_6H_8 can contain three carbon-carbon double bonds or one double bond and one triple bond.

12. Ethylene glycol is superior to methyl alcohol as an antifreeze because of its low volatility. Methyl alcohol is much more volatile than water. If the radiator leaks under pressure (normally steam), it would primarily leak methanol vapor, thus losing the antifreeze. Ethylene glycol has a lower volatility and a higher boiling point than water, so it does not present this problem.

13. (a) Methanol, taken internally, is poisonous and capable of causing blindness and death. Breathing methanol vapors is also very dangerous.

    (b) Physiologically, ethanol acts as a food, as a drug, and as a poison. It is a food, in a limited sense. The body is able to metabolize small amounts of it to carbon dioxide and water, resulting in the production of energy. As a drug, ethanol is often mistakenly considered to be a stimulant, but it is, in fact, a depressant. In moderate quantities, ethanol causes drowsiness and depresses brain functions. In larger quantities, ethanol causes nausea, vomiting, impaired perception, and a lack of coordination. Consumption of very large quantities may cause unconsciousness, and even death.

14. The structure for 1-methylpentane would be $CH_3CH_2CH_2CH_2CH_2CH_3$. The name 1-methylpentane is not based on the longest continuous carbon chain of 6 atoms. Therefore, the correct name is hexane.

15. (a) organic (b) inorganic (c) organic (d) organic (e) inorganic

16. (a) inorganic (b) organic (c) inorganic (d) organic (e) organic

17. Organic compounds are compounds in which carbon combines with hydrogen, oxygen, nitrogen and/or sulfur. Each compound identified as organic above contains some combination of carbon with hydrogen, oxygen, and/or sulfur.

18. Organic compounds are compounds in which carbon combines with hydrogen, oxygen, nitrogen and/or sulfur. Each compound identified as organic above contains some combination of carbon with hydrogen, oxygen, nitrogen and sulfur.

19. Names of alkyl groups

    (a) $C_5H_{11}-$ pentyl

    (b) $C_7H_{15}-$ heptyl

20. Names of alkyl groups
    (a) $C_8H_{17}-$ octyl
    (b) $C_{10}H_{21}-$ decyl

21. Hexanes

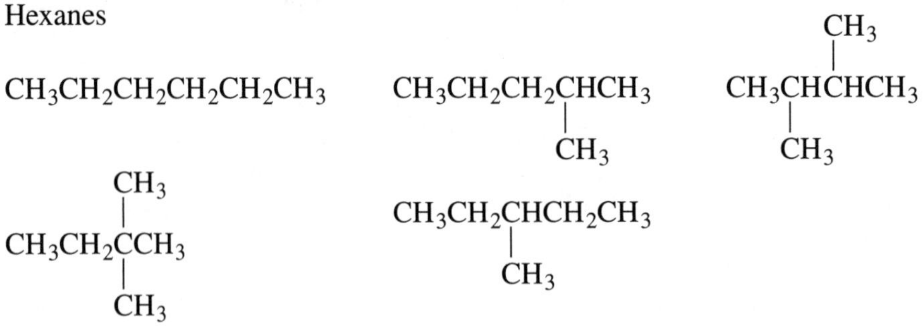

$CH_3CH_2CH_2CH_2CH_2CH_3$

$CH_3CH_2CH_2CHCH_3$
|
CH_3

CH_3
|
$CH_3CHCHCH_3$
|
CH_3

CH_3
|
$CH_3CH_2CCH_3$
|
CH_3

$CH_3CH_2CHCH_2CH_3$
|
CH_3

22. Heptanes

$CH_3CH_2CH_2CH_2CH_2CH_2CH_3$

$CH_3CH_2CH_2CH_2CHCH_3$
|
CH_3

CH_3
|
$CH_3CH_2CCH_2CH_3$
|
CH_3

CH_3
|
$CH_3CH_2CH_2CCH_3$
|
CH_3

$CH_3CH_2CH_2CHCH_2CH_3$
|
CH_3

$CH_3CHCH_2CHCH_3$
| |
CH_3 CH_3

CH_3
|
$CH_3CH_2CHCHCH_3$
|
CH_3

CH_3 CH_3
| |
$CH_3CH—CCH_3$
|
CH_3

$CH_3CH_2CHCH_2CH_3$
|
CH_2CH_3

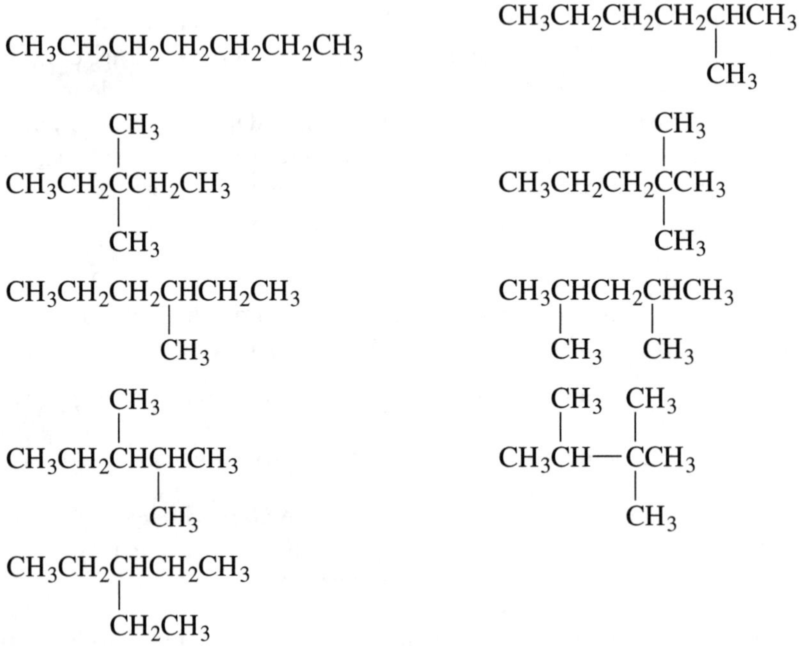

23. IUPAC names
    (a) CH_3CH_2Cl chloroethane
    (b) $CH_3CHClCH_3$ 2-chloropropane
    (c) $(CH_3)_2CHCH_2Cl$ 1-chloro-2-methylpropane

24. IUPAC names
    (a) $CH_3CH_2CH_2Cl$ 1-chloropropane

(b) $(CH_3)_3CCl$ 2-chloro-2-methylpropane

(c) $CH_3CHClCH_2CH_3$ 2-chlorobutane

25. (a) 4-ethyl-2-methylheptane
    (b) 4,6-diethyl-2-methyloctane

26. (a) 3,6-dimethyloctane
    (b) 3-ethyl-2-methylhexane (or 3-isopropylhexane)

27. Structural formulas

(a) 2,4-dimethylpentane

CH_3 CH_3
| |
$CH_3CHCH_2CHCH_3$

(b) 2,2-dimethylpentane

CH_3
|
$CH_3CCH_2CH_2CH_3$
|
CH_3

(c) 3-isopropyloctane

$CH_3CH_2CHCH_2CH_2CH_2CH_2CH_3$
|
$CH(CH_3)_2$

28. (a) 4-ethyl-2-methylhexane

$CH_3CHCH_2CHCH_2CH_3$
| |
CH_3 CH_2CH_3

(b) 4-*t*-butylheptane

$CH_3CH_2CH_2CHCH_2CH_2CH_3$
|
$C(CH_3)_3$

(c) 4-ethyl-7-isopropyl-2,4,8-trimethyldecane

CH_3 CH_3 CH_3
| | |
$CH_3CHCH_2CCH_2CH_2CHCHCH_2CH_3$
| |
CH_2CH_3 $CH(CH_3)_2$

29. (a)

$CH_3CHCH_2CH_3$
|
CH_3

1 2 3 4
$CH_3CH_2CHCH_3$
|
CH_3

3-methylbutane is an incorrect name because the carbon atoms were numbered from the wrong end of the chain. The correct name is 2-methylbutane.

(b)

$CH_3CHCH_2CH_3$
|
CH_2CH_3

2-ethylbutane

$CH_3CH_2CHCH_2CH_3$
|
CH_3

3-methylpentane

2-ethylbutane is incorrect. The longest carbon chain in the molecule was not used in determining the root name. The correct name is 3-methylpentane.

30. (a)

$$\begin{array}{l} \quad\;\; CH_3 \\ \quad\;\;\; | \\ CH_3CCH_2CH_3 \\ \quad\;\;\; | \\ \quad\;\; CH_3 \end{array}$$

2-dimethylbutane is incorrect. Each methyl group is attached to carbon 2 on the chain and requires a number to identify its location on the chain. The correct name is 2,2-dimethylbutane.

(b)

$$\begin{array}{l} CH_3CHCH_2CHCH_3 \\ \quad\;\;\; | \qquad\quad | \\ \quad\;\; CH_3 \quad CH_2CH_3 \end{array}$$

2-ethyl-4-methylpentane is incorrect. The longest continuous chain in the molecule was not used to name the compound. The correct name is 2,4-dimethylhexane.

31. (a) (1 isomer) CH_3Br

(b) (1 isomer) CH_2Cl_2

(c) (1 isomer) CH_3CH_2Cl

(d) (2 isomers) $CH_3CH_2CH_2Br$

$$\begin{array}{l} CH_3CHCH_3 \\ \quad\;\;\; | \\ \quad\;\;\, Br \end{array}$$

32. (a) (4 isomers)

$CH_3CH_2CH_2CH_2I$

$$\begin{array}{l} CH_3CHCH_2CH_3 \\ \quad\;\;\; | \\ \quad\;\;\;\, I \end{array}$$

$$\begin{array}{l} \quad\;\; CH_3 \\ \quad\;\;\; | \\ CH_3C{-}I \\ \quad\;\;\; | \\ \quad\;\; CH_3 \end{array}$$

$$\begin{array}{l} CH_3CHCH_2I \\ \quad\;\;\; | \\ \quad\;\; CH_3 \end{array}$$

(b) (4 isomers)

$CH_3CH_2CHCl_2$ $\quad$ $CH_3CCl_2CH_3$

$CH_3CHClCH_2Cl$ $\quad$ $CH_2ClCH_2CH_2Cl$

(c) (5 isomers)

$CH_3CH_2CHBrCl$ $\quad$ $CH_3CHClCH_2Br$

$CH_3CHBrCH_2Cl$ $\quad$ $CH_2ClCH_2CH_2Br$

$CH_3CClBrCH_3$

(d) (9 isomers)

$CH_3CH_2CH_2CHCl_2$

$CH_3CH_2CHClCH_2Cl$

$CH_3CHClCH_2CH_2Cl$

$CH_2ClCH_2CH_2CH_2Cl$

$CH_3CH_2CCl_2CH_3$

$CH_3CHClCHClCH_3$

$CH_3CH(CH_3)CHCl_2$ (CH$_3$ branch on the second carbon)

$CH_3CCl(CH_3)CH_2Cl$ (CH$_3$ branch on the second carbon)

$CH_3CH(CH_2Cl)CH_2Cl$ (CH$_2$Cl branch on the second carbon)

33. (a) $CH_3CH_2CH{=}CH_2 + Cl_2 \longrightarrow CH_3CH_2CHClCH_2Cl$
    (b) $CH_2{=}CH_2 + HBr \longrightarrow CH_3CH_2Br$

34. (a) $CH_3CH{=}CH_2 + Br_2 \longrightarrow CH_3CHBrCH_2Br$
    (b) $CH_3CH{=}CH_2 + HBr \longrightarrow CH_3CHBrCH_3$

35. (a) $CH_2{=}CH_2$ alkene
    (b) $CH{\equiv}CH$ alkyne
    (c) CH_3CH_2Cl alkyl halide
    (d) CH_3CH_2OH alcohol

36. (a) CH_3OCH_3 ether
    (b) CH_3CHO aldehyde
    (c) CH_3COOH carboxylic acid
    (d) $HCOOCH_3$ ester

37. (a) ethene
    (b) ethyne
    (c) chloroethane
    (d) ethanol

38. (a) methoxymethane
    (b) ethanal
    (c) ethanoic acid
    (d) methyl methanoate

39. (a) chloromethane CH_3Cl
    (b) vinyl chloride $CH_2{=}CHCl$
    (c) chloroform $CHCl_3$
    (d) 1,1-dibromoethene $CH_2{=}CBr_2$

40. (a) hexachloroethane CCl_3CCl_3
(b) iodoethyne $CH{\equiv}CI$
(c) 6-bromo-3-methyl-3-hexene-1-yne $BrCH_2CH_2CH{=}C(CH_3)C{\equiv}CH$
(d) 1,2-dibromoethene $CHBr{=}CHBr$

41. (a) 2,5-dimethyl-3-hexene
$CH_3CH(CH_3)CH{=}CHCH(CH_3)CH_3$
(b) 2-ethyl-3-methyl-1-pentene
$CH_2{=}C(CH_2CH_3)CH(CH_3)CH_2CH_3$
(c) 4-methyl-2-pentene
$CH_3CH{=}CHCH(CH_3)CH_3$

42. (a) 1,2-diphenylethene
$C_6H_5CH{=}CHC_6H_5$
(b) 3-pentene-1-yne
$CH{\equiv}CCH{=}CHCH_3$
(c) 3-phenyl-1-butyne
$CH{\equiv}CCH(C_6H_5)CH_3$

43. (a) $CH_3CH{=}C(CH_3)CH_2CH_2CH_3$ 3-methyl-2-hexene
(b) $CH_3C(CH_3){=}C(CH_3)CH_3$ 2,3-dimethyl-2-butene

44. (a) $CH_3CH_2CH(CH(CH_3)_2)CH{=}CH_2$ 3-isopropyl-1-pentene
(b) $CH_3CH_2CH{=}C(CH_3)CH_2CH_3$ 3-methyl-3-hexene

45. Complete the reactions and name the products.

(a) $CH_2{=}CHCH_3 + Br_2 \longrightarrow CH_2BrCHBrCH_3$ (1,2-dibromopropane)

(b) $CH_2{=}CH_2 + HBr \longrightarrow CH_3CH_2Br$ (bromoethane)

(c) $CH_3CH{=}CHCH_3 + H_2 \xrightarrow[\text{Pt, 25°C}]{\text{1 atm}} CH_3CH_2CH_2CH_3$ (butane)

46. (a) $CH_2{=}CH_2 + H_2O \xrightarrow{H^+} CH_3CH_2OH$ (ethanol)

(b) $CH{\equiv}CH + 2Br_2 \longrightarrow CHBr_2CHBr_2$ (1,1,2,2-tetrabromoethane)

(c) $CH_2{=}CH_2 + H_2 \xrightarrow[\text{Pt, 25°C}]{\text{1 atm}} CH_3CH_3$ (ethane)

47. Names of aromatic compounds

(a) OH

phenol

(b) CH_3

toulene

(c) COOH

benzoic acid

(d) NH_2

aniline

(e) Cl Cl

ortho-dichlorobenzene
(1,2-dichlorobenzene)

(f)

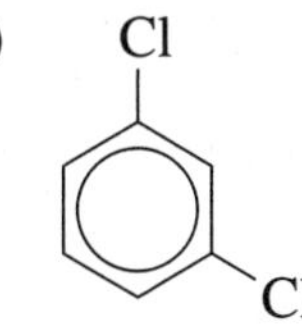

meta-dichlorobenzene
(1,3-dichlorobenzene)

48. (a)

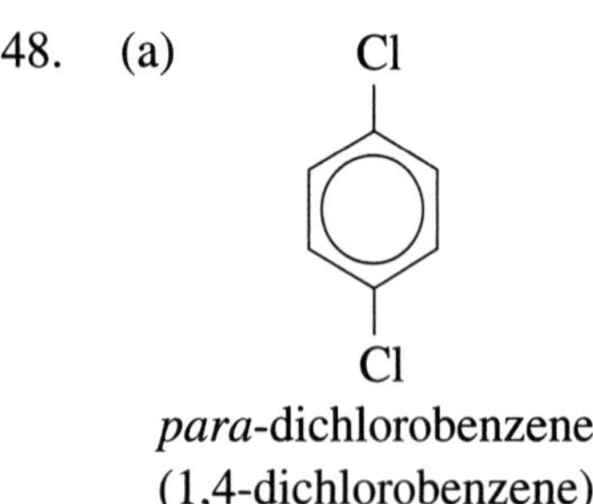

para-dichlorobenzene
(1,4-dichlorobenzene)

(b) OH

NO_2

para-nitrophenol

(c) CH_3

Br

Br

Br

2,4,5-tribromotoluene

(d) CH_2CH_3

ethylbenzene

(e) CH_3

OH

para-methylphenol

(f) OH

C=O

CH_3

NO_2

2-methyl-3-nitrobenzoic acid

49. (a)

benzene

(b) CH_3

toluene

(c) COOH

benzoic acid

(d) NH_2

aniline

50. (a)

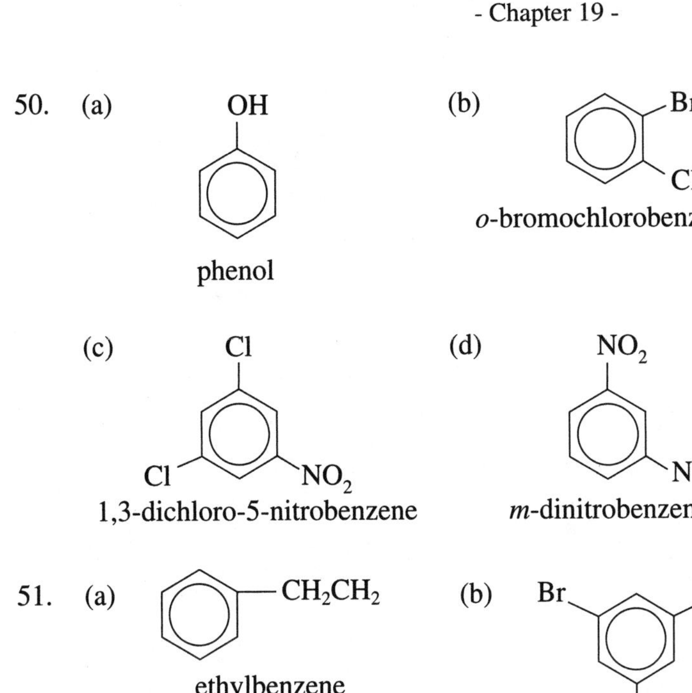

Br

1,3,5-tribromobenzene

52. (a) CH_3 — C — CH_3, CH_3

t-butylbenzene

(b) CH_3 — C — H

1,1-diphenylethane

53. Trichlorobenzene, $C_6H_3Cl_3$

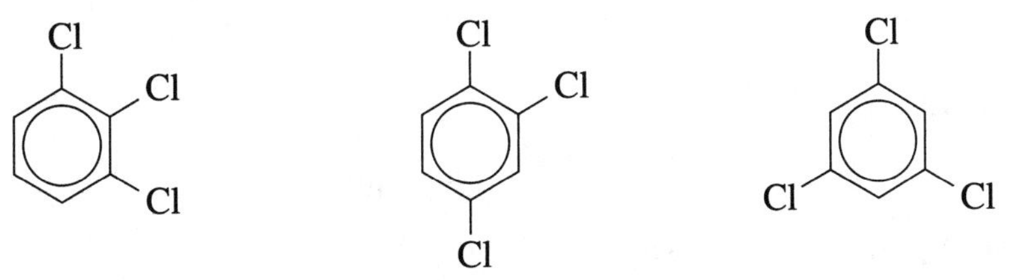

1,2,3-trichlorobenzene 1,2,4-trichlorobenzene 1,3,5-trichlorobenzene

54. Dichlorobromobenzene, $C_6H_3Cl_2Br$ (6 isomers)

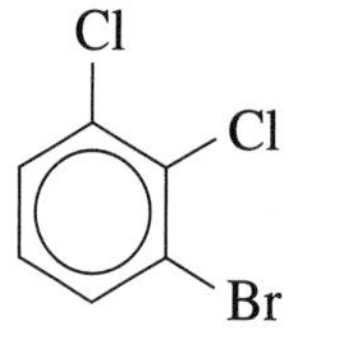

1,2-dichloro-3-bromobenzene

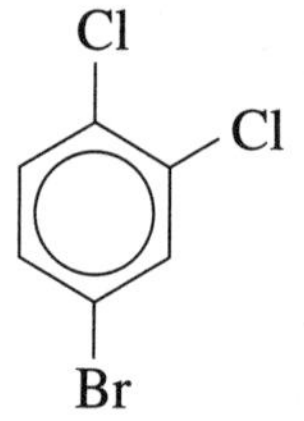

1,2-dichloro-4-bromobenzene

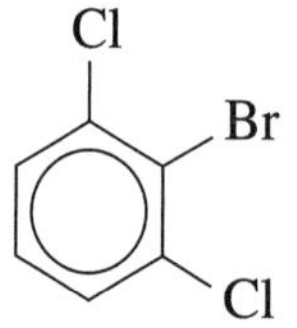

1,3-dichloro-2-bromobenzene

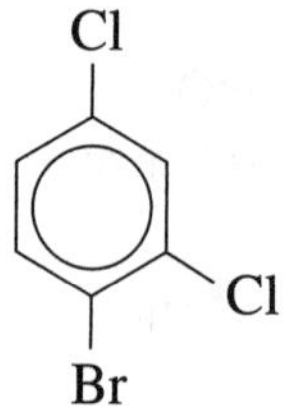

1,3-dichloro-4-bromobenzene

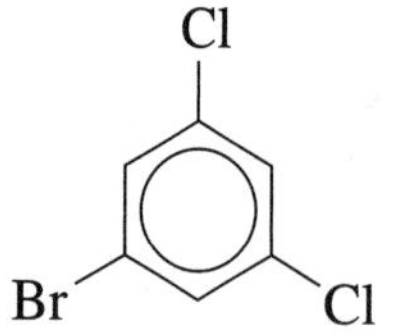

1,3-dichloro-5-bromobenzene

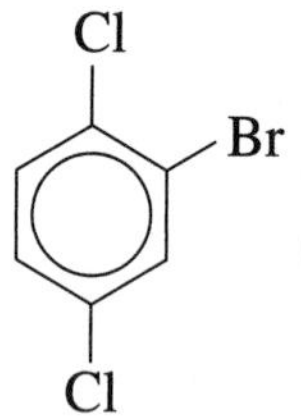

1,4-dichloro-2-bromobenzene

55. (a) $CH_3CH_2CH(OH)CH_3$ — 2-butanol — secondary

(b) $CH_3CH(CH_3)CH_2CH_2CH(CH_3)—OH$ — 5-methyl-2-hexanol — secondary

56. (a) $CH_3CH(CH_2CH_3)CH(OH)CH_2CH_2CH(CH_2CH_3)CH_3$ — 3,7-dimethyl-4-nonanol — secondary

(b) $HOCH_2CH_2CH(CH_3CHCH_3)CH_2CH_2CH_3$ — 3-isopropyl-1-hexanol — primary

57. Structural formulas

(a) 2-pentanol

$$CH_3CH_2CH_2\underset{\displaystyle OH}{\underset{|}{C}}HCH_3$$

(b) isopropyl alcohol

$$CH_3\underset{\displaystyle OH}{\underset{|}{C}}HCH_3$$

58. (a) 2,2-dimethyl-1-heptanol

$$HOCH_2\overset{\displaystyle CH_3}{\overset{|}{\underset{\displaystyle CH_3}{\underset{|}{C}}}}CH_2CH_2CH_2CH_2CH_3$$

(b) 1,3-propanediol

$$HOCH_2CH_2CH_2OH$$

59. Names of aldehydes

(a) $H_2C{=}O$ — methanal (formaldehyde)

(b) $CH_3CH_2CH_2\underset{\displaystyle H}{\underset{|}{C}}{=}O$ — butanal

(c) $CH_3\underset{\displaystyle CH_3}{\underset{|}{C}}HCH_2\underset{\displaystyle H}{\underset{|}{C}}{=}O$ — 3-methylbutanal

60. (a) $C_6H_5-\overset{\displaystyle H}{\overset{|}{C}}{=}O$ — benzaldehyde

(b) $O{=}\underset{\displaystyle H}{\underset{|}{C}}CH_2CH_2\underset{\displaystyle H}{\underset{|}{C}}{=}O$ — butanedial

(c) $CH_3\underset{\displaystyle OH}{\underset{|}{C}}HCH_2\overset{\displaystyle O}{\overset{\|}{C}}-H$ — 3-hydroxybutanal

61. Names of ketones

(a) $CH_3\underset{\displaystyle O}{\underset{\|}{C}}CH_3$ — propanone, acetone, dimethyl ketone

(b) $CH_3CH_2\overset{\overset{\displaystyle O}{\|}}{C}CH_3$ — 2-butanone, methyl ethyl ketone

(c) 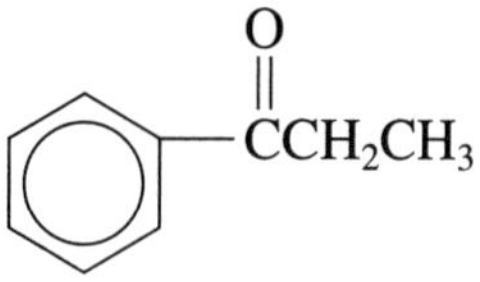 $C_6H_5\text{—}\overset{\overset{\displaystyle O}{\|}}{C}CH_2CH_3$ — 1-phenyl-1-propanone, phenyl ethyl ketone

62. (a) $CH_3\overset{\overset{\displaystyle O}{\|}}{C}\text{—}C(CH_3)_2CH_3$ — 3,3-dimethyl-2-butanone, methyl *t*-butyl ketone

(b) $CH_3\overset{\overset{\displaystyle O}{\|}}{C}CH_2CH_2\overset{\overset{\displaystyle O}{\|}}{C}CH_3$ — 2,5-hexanedione

(c) $CH_3C(CH_3)(OH)CH_3\overset{\overset{\displaystyle O}{\|}}{C}CH_3$ — 4-hydroxy-4-methyl-2-pentanone

63. (a) $CH_3CHBrCOOH$ — 2-bromopropanoic acid

(b) $CH_2{=}CHCH_2COOH$ — 3-butenoic acid

(c) $CH_3CH_2CH_2COOH$ — butanoic acid

(d) 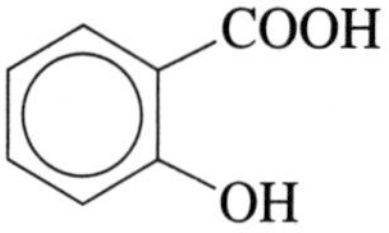 $o\text{-}HOC_6H_4COOH$ (benzene ring with COOH and OH) — salicylic acid
o-hydroxybenzoic acid

64. (a) $CH_3CH(CH_2CH_3)COOH$ — 2-methylbutanoic acid

(b) 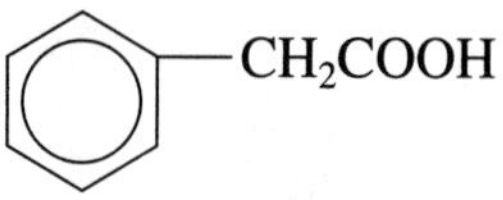 $C_6H_5\text{—}CH_2COOH$ — phenylacetic acid

(c) $CH_3CH_2CH_2CH_2COOH$ — pentanoic acid

(d) 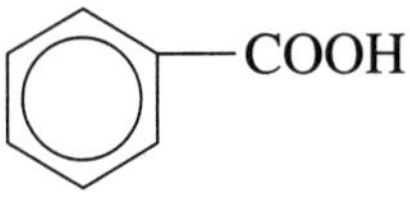 C_6H_5-COOH benzoic acid

65. Esters

(a) Ethyl formate $H-\overset{\overset{O}{\|}}{C}-O-CH_2CH_3$

(b) Methyl ethanoate $CH_3\overset{\overset{O}{\|}}{C}-O-CH_3$

(c) Isopropyl propanoate $CH_3CH_2\overset{\overset{O}{\|}}{C}-O-\underset{\underset{CH_3}{|}}{C}HCH_3$

66. Esters

(a) *n*-nonyl acetate $CH_3\overset{\overset{O}{\|}}{C}-O-CH_2(CH_2)_7CH_3$

(b) Ethyl benzoate $C_6H_5-\overset{\overset{O}{\|}}{C}-O-CH_2CH_3$

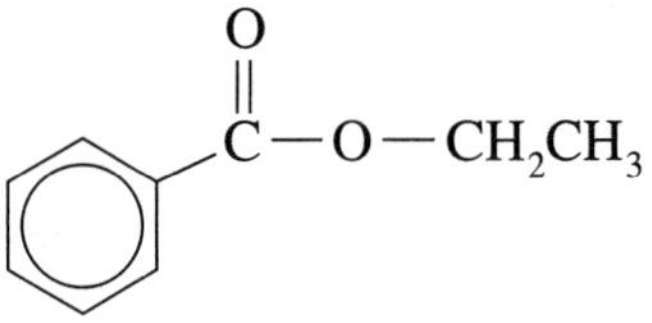

(c) Methyl salicylate $C_6H_4(OH)-\overset{\overset{O}{\|}}{C}-O-CH_3$ (OH ortho)

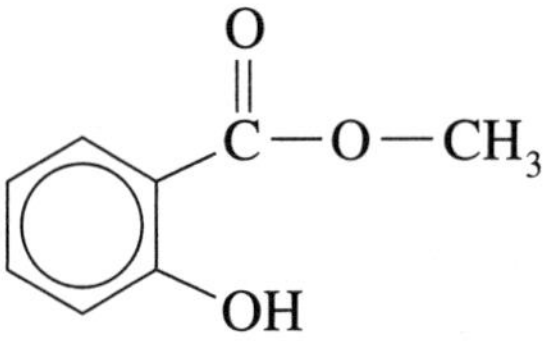

67. (a) $CH_3\overset{\overset{O}{\|}}{C}-O-CH_2CH_3$ ethyl ethanoate

(b) 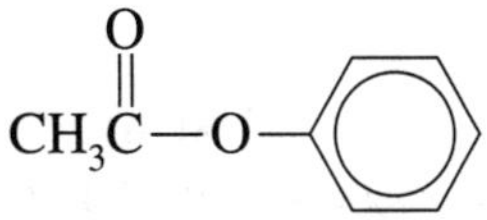 $CH_3\overset{\overset{O}{\|}}{C}-O-C_6H_5$ phenyl ethanoate

(c) $CH_3CH_2CH_2\overset{\overset{O}{\|}}{C}-O-\overset{\overset{CH_3}{|}}{C}HCH_3$ isopropyl butanoate

68. (a) $H-\underset{\underset{O}{\|}}{C}-O-CH(CH_3)_2$ isopropyl methanoate

(b) 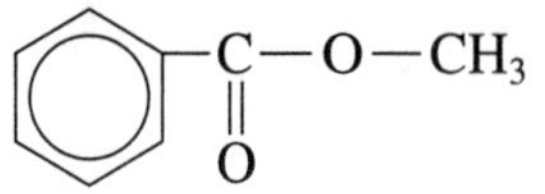 $C_6H_5-\underset{\underset{O}{\|}}{C}-O-CH_3$ methyl benzoate

(c) $CH_3\underset{\underset{CH_3}{|}}{CH}CH_2\overset{\overset{O}{\|}}{C}-O-CH_2CH_3$ ethyl-3-methylbutanoate

69. (a) $CH_3COOH + NaOH \longrightarrow CH_3COO^-Na^+ + H_2O$

(b) $CH_3\underset{\underset{OH}{|}}{CH}COOH + NH_3 \longrightarrow CH_3\underset{\underset{OH}{|}}{CH}COO^-NH_4^+$

70. (a) $CH_3COOH + KOH \longrightarrow CH_3COO^-K^+ + H_2O$

(b)

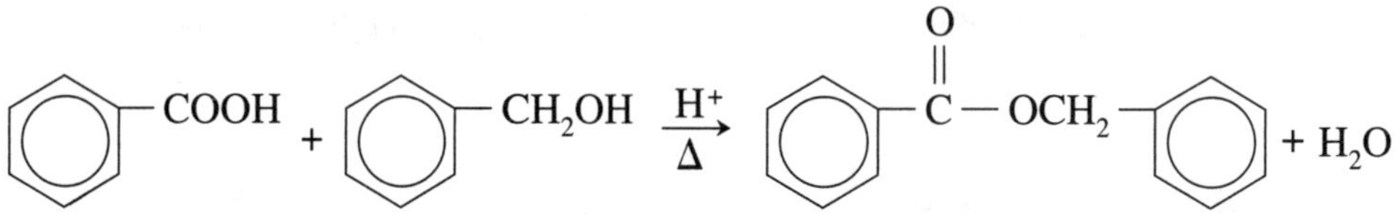

$$C_6H_5-COOH + C_6H_5-CH_2OH \xrightarrow[\Delta]{H^+} C_6H_5-\overset{\overset{O}{\|}}{C}-OCH_2-C_6H_5 + H_2O$$

71. (a) $+CH_2-CH_2+_n$ polyethylene

(b) $+CH_2-\underset{\underset{Cl}{|}}{CH}+_n$ polyvinyl chloride

72. (a) $+CH_2-\underset{\underset{CN}{|}}{CH}+_n$ polyacrylonitrile

(b) $+CF_2-CF_2+_n$ teflon

73. (a) incorrect, should be 2-methylbutane
(b) incorrect, should be *meta*-dibromobenzene
(c) incorrect, should be 3-methyl-1-butyne
(d) incorrect, should be 3-pentanol
(e) correct

74. (a) correct
(b) incorrect, should be 2-butanone
(c) incorrect, should be 4-methyl-2-pentene
(d) incorrect, should be methyl propanoate
(e) incorrect, should be 3-methylpentane

75. (a) hydrocarbon (b) aromatic (c) alkyne (d) alcohol (e) carboxylic acid

76. (a) alkyl halide (b) ketone (c) alkene (d) ester (e) hydrocarbon

77. Alkanes are nonpolar molecules. Therefore, there is little attraction between them because there are no partial positive and partial negative charges to attract one another. As a result, it takes very little energy to cause alkanes to boil because there are no intermolecular attractive forces to be overcome.

78. (a) C_6H_{14}, 5 isomers

$CH_3CH_2CH_2CH_2CH_2CH_3$

CH_3
|
$CH_3CHCH_2CH_2CH_3$

CH_3
|
$CH_3CH_2CHCH_2CH_3$

CH_3
|
$CH_3CHCHCH_3$
|
CH_3

CH_3
|
$CH_3CCH_2CH_3$
|
CH_3

b) C_5H_{10}, 5 isomers

$CH_2{=}CHCH_2CH_2CH_3$

$CH_3CH{=}CHCH_2CH_3$

$CH_2{=}CHCHCH_3$
|
CH_3

$CH_2{=}CCH_2CH_3$
|
CH_3

$CH_3CH{=}CCH_3$
|
CH_3

(c) C_7H_{16}, 9 isomers

$CH_3CH_2CH_2CH_2CH_2CH_2CH_3$

$CH_3CHCH_2CH_2CH_2CH_3$
|
CH_3

CH_3
|
$CH_3CH_2CHCH_2CH_2CH_3$

CH_3
|
$CH_3{-}C{-}CH_2CH_2CH_3$
|
CH_3

CH_3
|
$CH_3CH_2CCH_2CH_3$
|
CH_3

CH_3
|
$CH_3CHCHCH_2CH_3$
|
CH_3

CH_3
|
$CH_3CHCH_2CHCH_3$
|
CH_3

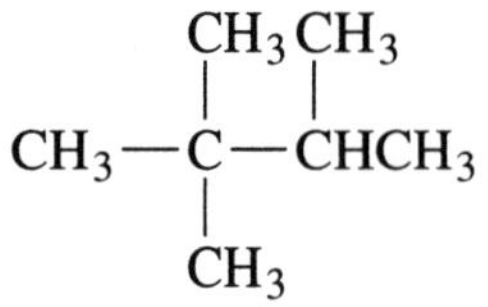

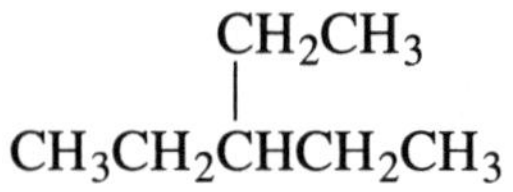

```
     CH3 CH3
      |   |
CH3—C—CHCH3
      |
     CH3
```

```
           CH2CH3
           |
CH3CH2CHCH2CH3
```

79. (a) Isomers of pentyne, C_5H_8 (3 isomers)

1-pentyne
$CH_3CH_2CH_2C{\equiv}CH$

2-pentyne
$CH_3CH_2C{\equiv}CCH_3$

3-methyl-1-butyne

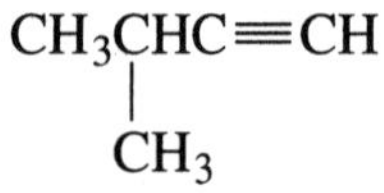

```
CH3CHC≡CH
    |
    CH3
```

(b) Isomers of hexyne, C_6H_{10} (7 isomers)

1-hexyne
$CH_3CH_2CH_2CH_2C{\equiv}CH$

2-hexyne
$CH_3CH_2CH_2C{\equiv}CCH_3$

3-hexyne
$CH_3CH_2C{\equiv}CCH_2CH_3$

3-methyl-1-pentyne

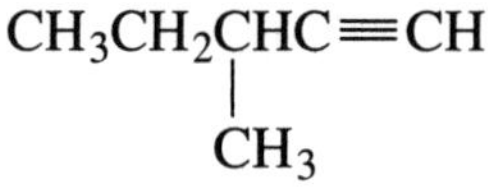

```
CH3CH2CHC≡CH
      |
      CH3
```

4-methyl-1-pentyne

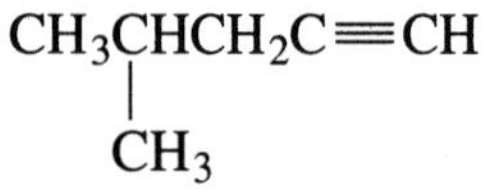

```
CH3CHCH2C≡CH
    |
    CH3
```

4-methyl-2-pentyne

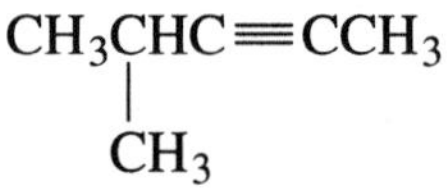

```
CH3CHC≡CCH3
    |
    CH3
```

3,3-dimethyl-1-butyne

```
    CH3
    |
CH3CC≡CH
    |
    CH3
```

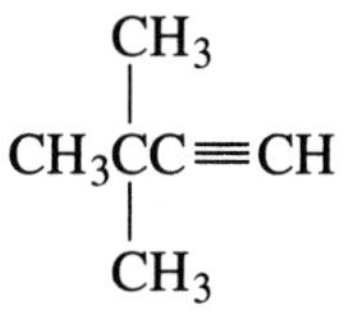

80. (8) Isomeric alcohols, formula $C_5H_{11}OH$

$CH_3CH_2CH_2CH_2CH_2OH$ 1-pentanol 1°

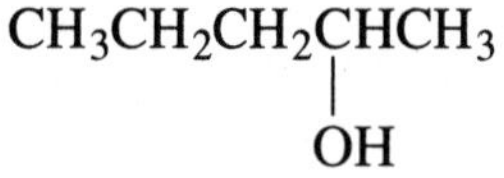

```
CH3CH2CH2CHCH3
          |
          OH
```

2-pentanol 2°

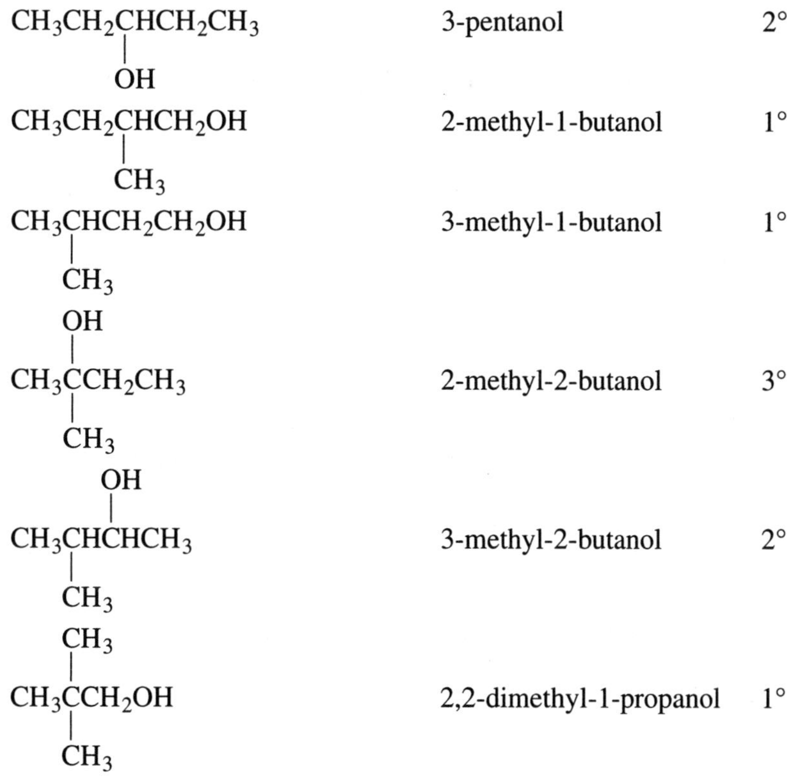

81. The molar mass of myricyl alcohol, an open chain saturated alcohol, contains 30 carbon atoms. The first three alcohols in the homologous series, CH_3OH, C_2H_5OH, and C_3H_7OH, lead us to the formula $C_{30}H_{61}OH$.

$$\text{molar mass} = (30)(12.01\text{ g/mol}) + (62)(1.008\text{ g/mol}) + (1)(16.00\text{ g/mol}) = 438.8\text{ g/mol}$$

82. (a) methyl ethyl ether $CH_3CH_2OCH_3$
(b) dimethyl ether CH_3OCH_3
(c) methyl ethyl ether $CH_3CH_2OCH_3$

83. Ethers having the formula $C_5H_{12}O$ are (IUPAC name in parentheses)

$CH_3OCH_2CH_2CH_2CH_3$ (1-methoxybutane) methyl *n*-butyl ether

$CH_3CH_2OCH_2CH_2CH_3$ (1-ethoxypropane) ethyl *n*-propyl ether

$CH_3CH_2\underset{\displaystyle CH_3}{\underset{|}{C}}HOCH_3$ (2-methoxybutane) methyl *s*-butyl ether

$CH_3OCH_2\underset{\displaystyle CH_3}{\underset{|}{C}}HCH_3$ (1-methoxy-2-methylpropane) methyl isobutyl ether

```
       CH3
        |
CH3OCCH3
        |
       CH3
```

(2-methoxy-2-methylpropane) methyl *t*-butyl ether

$CH_3CH_2OCH(CH_3)_2$ (2-ethoxypropane) ethyl isopropyl ether

84. Propanal

```
CH3CH2C=O
       |
       H
```

Propanone

```
CH3CCH3
    ||
    O
```

Propanal and propanone are isomers (C_3H_6O)

Butanal

```
CH3CH2CH2C=O
          |
          H
```

Butanone

```
CH3CH2CCH3
       ||
       O
```

Butanal and butanone are isomers (C_4H_8O). Both compounds have the same molecular formula.

85. Carboxylic acids, IUPAC name, common name

	IUPAC	common
HCOOH	methanoic acid	formic acid
CH_3COOH	ethanoic acid	acetic acid
CH_3CH_2COOH	propanoic acid	propionic acid
$CH_3CH_2CH_2COOH$	butanoic acid	butyric acid
$CH_3CH_2CH_2CH_2COOH$	pentanoic acid	valeric acid

86. Isomers of hexanoic acid, $CH_3(CH_2)_4COOH$ (8 isomers)

$CH_3(CH_2)_4COOH$ hexanoic acid

```
CH3CH2CH2CHCOOH
          |
          CH3
```

2-methylpentanoic acid

```
CH3CH2CHCH2COOH
       |
       CH3
```

3-methylpentanoic acid

```
CH3CHCH2CH2COOH
   |
   CH3
```

4-methylpentanoic acid

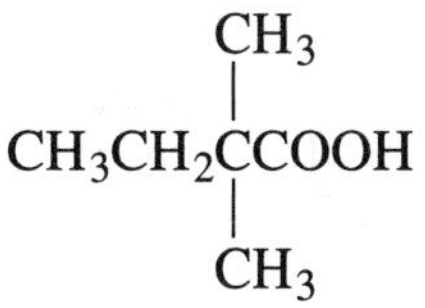

$$CH_3CH_2C(CH_3)_2COOH$$ 2,2-dimethylbutanoic acid

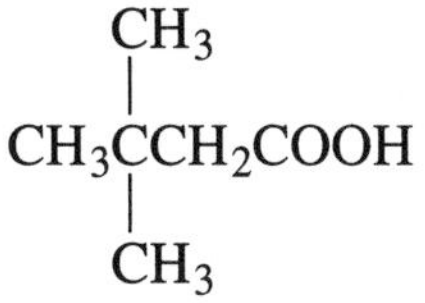

$$CH_3C(CH_3)_2CH_2COOH$$ 3,3-dimethylbutanoic acid

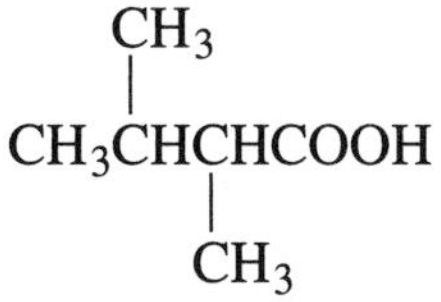

$$CH_3CH(CH_3)CH(CH_3)COOH$$ 2,3-dimethylbutanoic acid

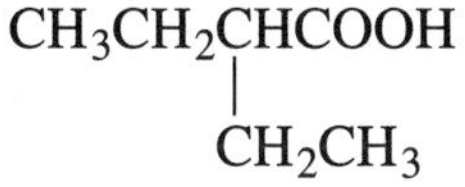

$$CH_3CH_2CH(CH_2CH_3)COOH$$ 2-ethylbutanoic acid

87. Preparation of esters

(a) $$HCOOH + CH_3CH_2OH \xrightarrow[\Delta]{H^+} HC(=O){-}OCH_2CH_3 + H_2O$$
ethyl formate

(b) $$CH_3CH_2COOH + CH_3OH \xrightarrow[\Delta]{H^+} CH_3CH_2C(=O){-}OCH_3 + H_2O$$
methyl propanoate

(c) $$C_6H_5{-}COOH + CH_3CH_2CH_2OH \xrightarrow[\Delta]{H^+} C_6H_5{-}C(=O){-}OCH_2CH_2CH_3 + H_2O$$
n-propyl benzoate

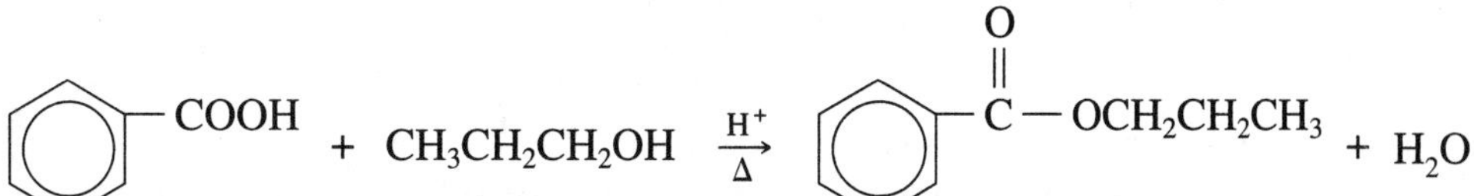

88. (a) $$\left(CH_2{-}CH(CH_3){-}CH_2{-}CH(CH_3){-}CH_2{-}CH(CH_3){-}CH_2{-}CH(CH_3)\right)_n$$ polypropylene

$$\left(CH_2{-}CH(CH_2CH_3){-}CH_2{-}CH(CH_2CH_3){-}CH_2{-}CH(CH_2CH_3){-}CH_2{-}CH(CH_2CH_3)\right)_n$$ poly-1-butene

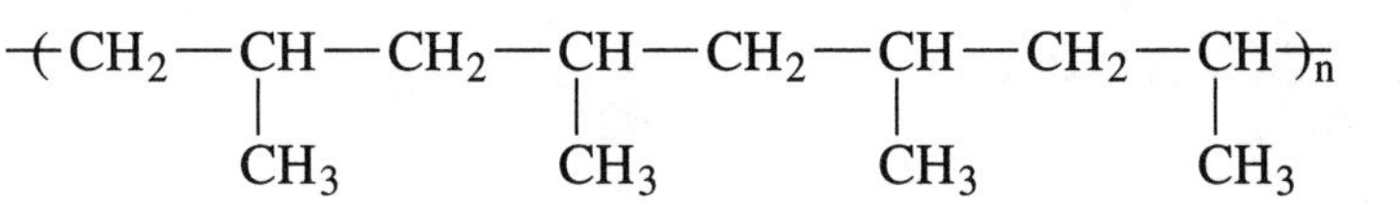

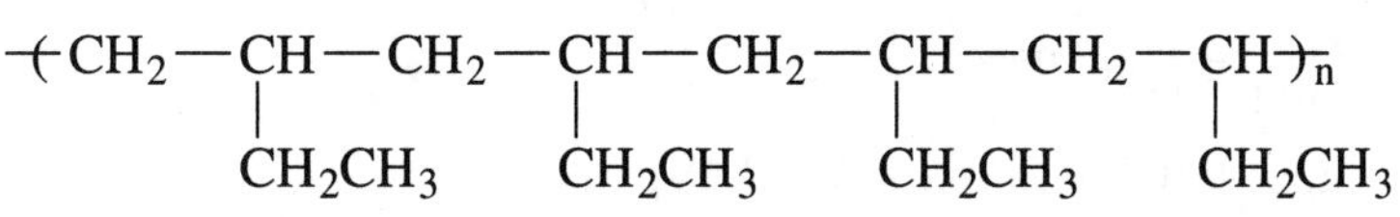

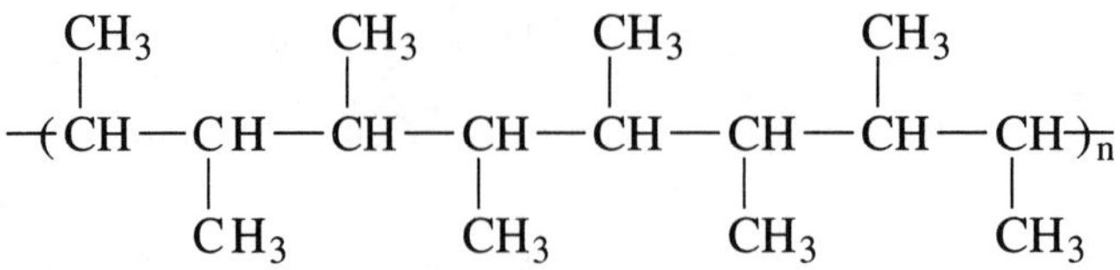

poly-2-butene

(b) Molar mass of ethylene monomer = 28.05 g/mol

$$\frac{\text{molar mass of polymer}}{\text{molar mass of monomer}} = \text{number of monomer units}$$

$$\frac{35000 \text{ g/mol}}{28.05 \text{ g/mol}} = 1.2 \times 10^3 \text{ ethylene units}$$

89. Dibromobenzenes

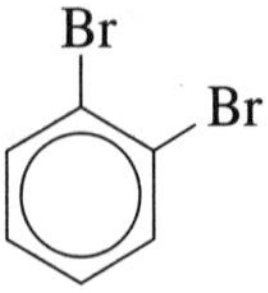

ortho

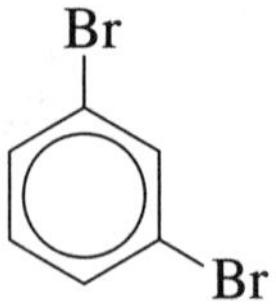

meta

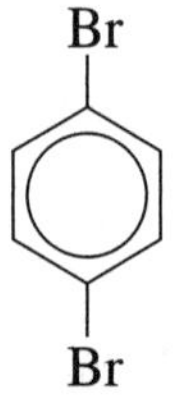

para

Tribromobenzenes

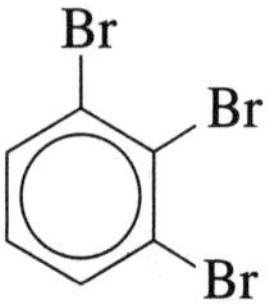

1,2,3-

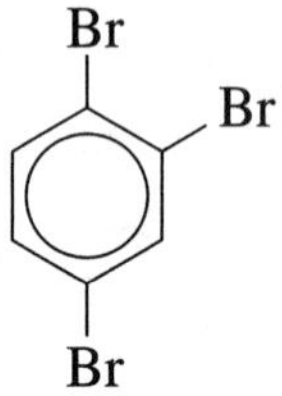

1,2,4-

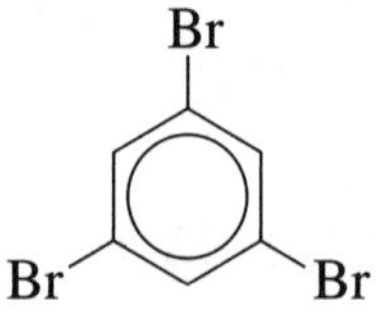

1,3,5-

90. Assume 100. g of material

C $\quad \frac{24.3 \text{ g}}{12.01 \text{ g/mol}} = 2.02 \text{ mol} \qquad \frac{2.02}{2.02} = 1 \text{ mol}$

H $\quad \frac{4.1 \text{ g}}{1.008 \text{ g/mol}} = 4.1 \text{ mol} \qquad \frac{4.1}{2.02} = 2 \text{ mol}$

Cl $\quad \frac{71.7 \text{ g}}{35.45 \text{ g/mol}} = 2.02 \text{ mol} \qquad \frac{2.02}{2.02} = 1 \text{ mol}$

Empirical formula = CH_2Cl (mass = 49.48 g/mol)

$$PV = nRT \qquad n = \frac{PV}{RT}$$

$$n = \left(\frac{740\text{ mm Hg} \times 1\text{ atm}}{760\text{ mm Hg}}\right)\left(\frac{(0.1403\text{ L})}{(0.0821\text{ L atm/mol K})(373\text{ K})}\right) = 4.5 \times 10^{-3}\text{ mol}$$

$$\text{molar mass} = \frac{0.442\text{ g}}{4.5 \times 10^{-3}\text{ mol}} = 98\text{ g/mol}$$

dividing $\frac{98\text{ g/mol}}{49.48\text{ g/mol}} = 2.0$,

therefore the molecular formula is 2 × empirical formula = $C_2H_4Cl_2$

Possible isomers are

```
     H   Cl
     |   |
  H—C—C—Cl
     |   |
     H   H
```

and

```
    Cl  Cl
     |   |
  H—C—C—H
     |   |
     H   H
```

91. $\frac{24\text{ g C}}{12.01\text{ g/mol}} = 2\text{ mol C}$ The empirical formula is CH_2O

$$\frac{4\text{ g H}}{1.008\text{ g/mol}} = 4\text{ mol H}$$

$$\frac{32\text{ g O}}{16.00\text{ g/mol}} = 2\text{ mol O}$$

The only structure given that could match this formula is (e) CH_3COOH

92. (a) CH_3CH_2OH

(b) CH_3I

(c) $CH_3CH_2CH_2CHClCH_3$

(d) $CH_3CH_2CH_2CH_2CH_2CH_2CH_2CH_2CH_2CH_3$

(e)

```
       CH3
       |
   CH3C—OH
       |
       CH3
```

93. (a) $HCOOH + CH_3OH \xrightarrow{H^+} HCOOCH_3 + H_2O$

methyl formate

(b) $CH_3CH_2CH_2COOH + CH_3CH_2CH_2CH_2OH \xrightarrow{H^+}$

$CH_3CH_2CH_2COOCH_2CH_2CH_2CH_3 + H_2O$

butyl butanoate

(c) $CH_3(CH_2)_4COOH + CH_3(CH_2)_4CH_2OH \xrightarrow{H^+}$

$CH_3(CH_2)_4COOCH_2(CH_2)_4CH_3 + H_2O$

hexyl hexanoate

94. (a)

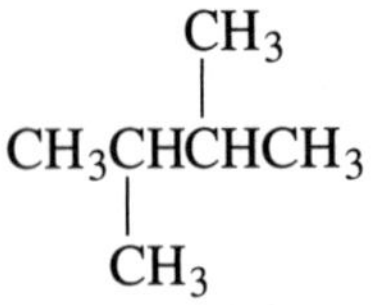

$$\begin{array}{c} \quad CH_3 \\ \quad | \\ CH_3CHCHCH_3 \\ | \quad\quad \\ CH_3 \quad\quad \end{array}$$

(b)

$$\begin{array}{c} CH_3CH_2CHCH_2CH_2CH_3 \\ | \\ OH \end{array}$$

(c)

$$\begin{array}{c} O \\ \| \\ CH_3CH_2C-OH \end{array}$$

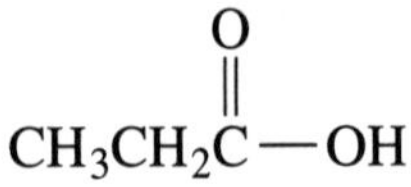

(d)

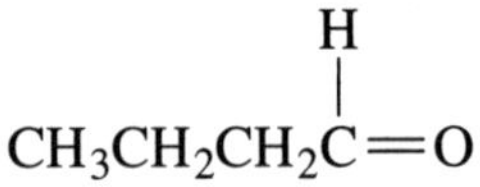

$$\begin{array}{c} H \\ | \\ CH_3CH_2CH_2C=O \end{array}$$

(e)

$$\begin{array}{c} O \\ \| \\ CH_3CH_2C-O-CH_2CH_2CH_2CH_3 \end{array}$$

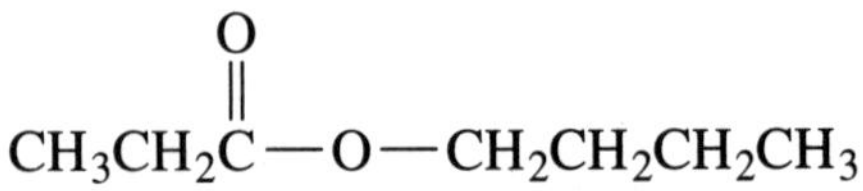

CHAPTER 20

INTRODUCTION TO BIOCHEMISTRY

1. The three principal classes of animal food are carbohydrates, lipids and proteins.

2. Chromosomes are composed of proteins and nucleic acids.

3. The N-terminal residue is the amino-group ($—NH_2$) end of a linear peptide. The C-terminal residue is the carboxyl-group ($—COOH$) end of a linear peptide.

4. Saturated fats are composed of single bonds between carbon atoms. Unsaturated fats are composed of both single and double bonds between carbon atoms. Saturated fats are solids and unsaturated fats are liquids at room temperature.

5. In Table 20.1, the sweetest disaccharide is sucrose. The sweetest monosaccharide is fructose.

6. Fatty acids in vegetable oils are more unsaturated than fatty acids in animal fats. This is because vegetable oils contain higher percentages of oleic and linoleic (unsaturated) acids than animal fats.

7. Of the common amino acids listed in Table 20.3, two, aspartic acid and glutamic acid, have more than one carboxyl group. The following amino acids have more than one amino group: arginine, histidine, lysine, and tryptophan.

8. There are three disulfide linkages in each molecule of beef insulin.

9. In DNA, the nitrogen bases are off to the side while the deoxyribose and phosphoric acid are part of the backbone chain.

10. In the double stranded helix structure of DNA (Figure 20.6), the hydrogen bonding of the nitrogen bases is as follows: guanine and cytosine are mutually bonded to each other as are adenine and thymine.

11. The four major classes of biomolecules are carbohydrates, lipids, proteins, and nucleic acids.

12. Monosaccharides, disaccharides, and polysaccharides. The simplest type of carbohydrate is the monosaccharide.

13. An aldose is a monosaccharide containing an aldehyde group on one carbon atom and a hydroxyl group on each of the other carbon atoms. An aldotetrose is an aldose containing four carbon atoms. A ketose is a monosaccharide containing a ketone group on one carbon atom and a hydroxyl group on each of the other carbon atoms. A ketohexose is a ketone containing six carbon atoms. Examples: aldose, glucose; aldotetrose, erythrose; ketose, fructose; ketohexose, fructose.

14. Classifications of saccharides.

    Monosaccharides: Glucose, Fructose, Galactose, Ribose
    Disaccharides: Sucrose, Maltose, Lactose
    Polysaccharides: Cellulose, Glycogen, Starch

15. Open chain formulas

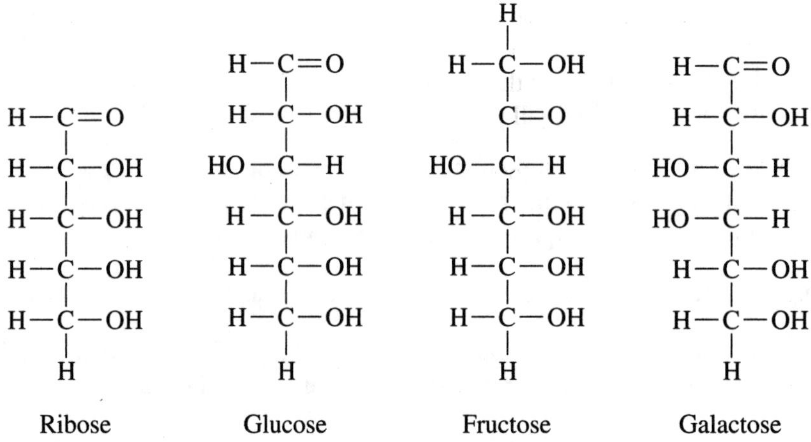

16. Cyclic structural formulas

Ribose

Glucose

Fructose

Galactose

17. Properties and sources

    Ribose: Ribose is a white, crystalline, water soluble pentose sugar, present in adenosine triphosphate (ATP), one of the chemical energy carriers in the body. Ribose and one of its derivatives, deoxyribose, are also important components of the nucleic acids, DNA and RNA, the genetic information carriers in the body.

    Glucose: Glucose is an aldohexose and is found in the free state in plant and animal tissues. Glucose is commonly known as dextrose. It is a component of the disaccharides sucrose, maltose, and lactose, and it is also the monomer of the polysaccharides starch, glycogen, and cellulose. Glucose is the key sugar of the body and is circulated by the blood stream to provide energy to all parts of the body.

    Fructose: Fructose, also called levulose, is a ketohexose and occurs in fruit juices as well as in honey. Fructose is also a constituent of sucrose. Fructose is the sweetest of all sugars, being about twice as sweet as glucose. This accounts for the sweetness of honey. The enzyme invertase, present in bees, splits sucrose into glucose and fructose. Fructose is metabolized directly, but it is also readily converted to glucose in the liver.

    Galactose: Galactose is an aldohexose and occurs, along with glucose, in lactose and in many oligo- and polysaccharides, such as pectin, gums and mucilages. Galactose is synthesized in the mammary glands to make the lactose of milk.

18. Lactic acid has a hydroxyl group and an acid group. It is not a carbohydrate, because it has neither an aldehyde nor a ketone group, and will not yield one upon hydrolysis.

19. The monosaccharide composition of:
    (a) Sucrose; a disaccharide made from one unit of glucose and one unit of fructose.
    (b) Maltose; a disaccharide made from two units of glucose.
    (c) Lactose; a disaccharide made from one unit of galactose and one unit of glucose.
    (d) Starch; a polysaccharide made from many units of glucose.

20. Cyclic structural formulas for sucrose and maltose.

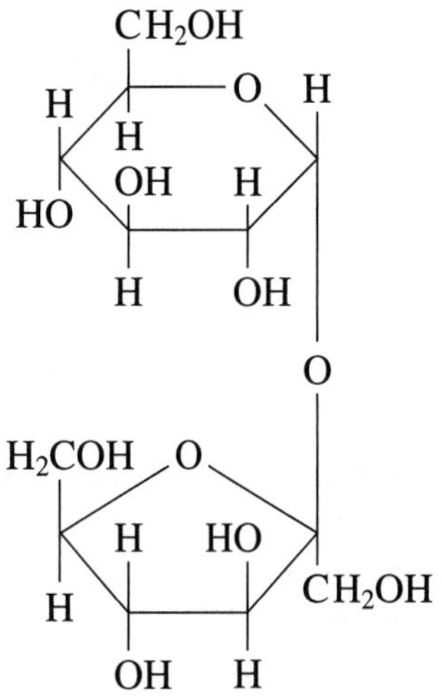

Sucrose

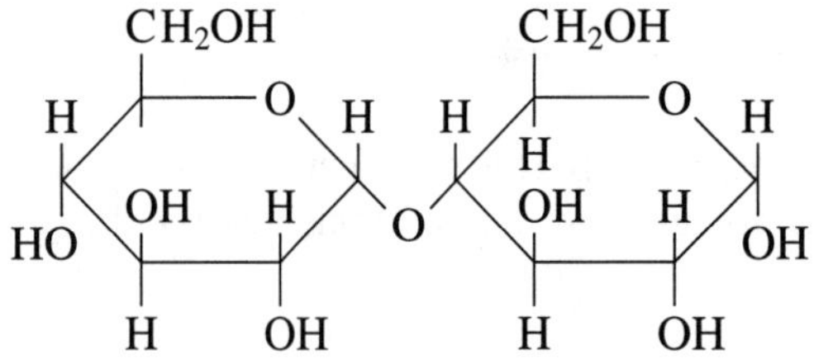

Maltose

21. The formation of a disacharide, $C_{12}H_{22}O_{11}$, from a monosaccharide, $C_6H_{12}O_6$, involves combining two monosaccharide units with a molecule of water splitting out between them.

22. (a) Sucrose + H_2O $\xrightarrow{\text{sucrase}}$ Glucose + Fructose

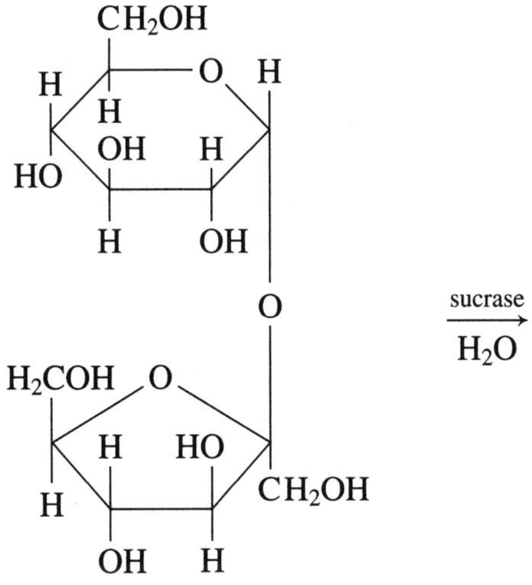

CH₂OH

Sucrose $\xrightarrow[H_2O]{\text{sucrase}}$ Glucose + Fructose

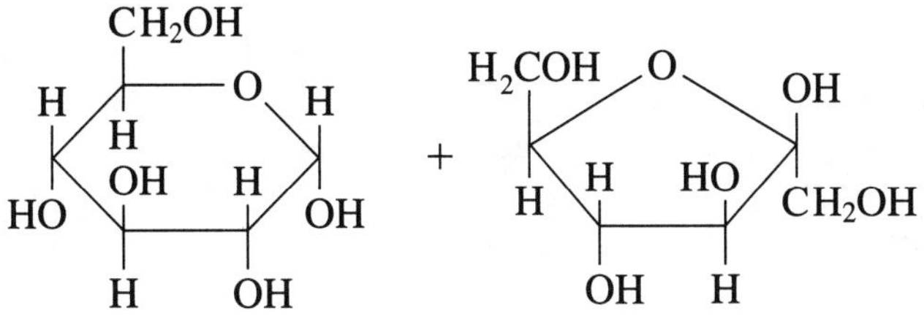

Sucrase catalyzes the hyrolysis of sucrose while maltase catalyzes the hydrolysis of maltose.

(b) Maltose + $H_2O \xrightarrow{\text{maltase}}$ Glucose + Glucose

Maltose $\xrightarrow[H_2O]{\text{maltase}}$ Glucose + Glucose

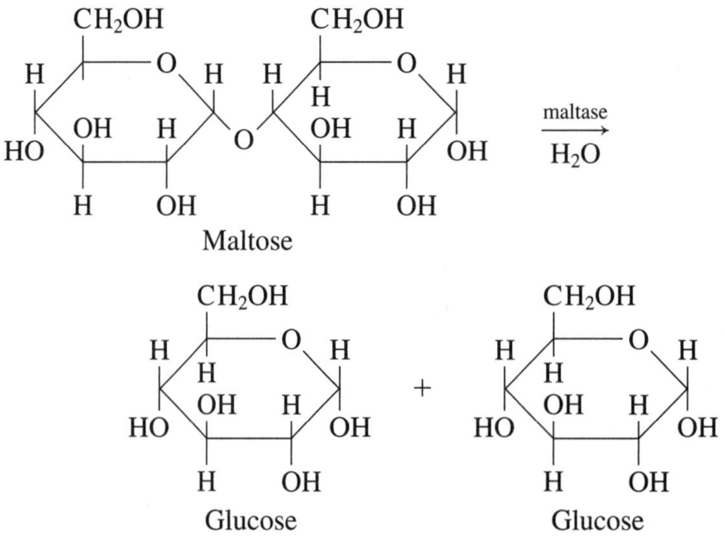

23. Starch and cellulose have their basic composition in common. They contain many glucose units joined in long chains forming polysaccharide molecules with a higher molar mass. The difference in properties, which are highly significant, are due to the different manner in which the glucose units are attached to each other. The primary use of starch is for food. Man cannot utilize cellulose as food due to a lack of the necessary enzymes to hydrolzye cellulose to usable glucose.

24. Carbohydrates are stored in the human body as glycogen, a polysaccharide of glucose.

25. Carbohydrate metabolism

$$\text{Starch} \xrightarrow[\text{mouth}]{\text{amylase}} \text{Dextrins} \xrightarrow[\text{intestines}]{\text{amylase}} \text{Maltose} \xrightarrow[\text{intestines}]{\text{maltase}} \text{Glucose}$$

Glucose is absorbed through the intestinal walls into the blood stream. From there the glucose may be stored in the liver as glycogen or utilized as energy by oxidation to carbon dioxide and water.

Sugars, such as sucrose and lactose, are converted to monosaccharides by specific enzymes in the intestines.

26. Natural sources of sucrose, maltose, lactose, and starch are:

Sucrose: Sucrose is found in the free state throughout the plant kingdom. Sugar cane contains 15% to 20% sucrose, while sugar beets contain 10% to 17% sucrose. Maple syrup and sorghum are also good sources of sucrose.

Maltose: Maltose is found in sprouting grain, but occurs much less commonly in nature than either sucrose or lactose.

Lactose: Lactose, also known as milk sugar, is found free in nature mainly in the milk of mammals. Human milk contains about 6.7% lactose and cow milk about 4.5% of this sugar.

Starch: Starch is a polymer of glucose. It is found mainly in the seeds, roots, and tubers of plants. Corn, wheat, potatoes, rice, and cassava are the chief sources of starch.

27. Invert sugar is sweeter than sucrose, from which it comes, because it contains free frucose which is far sweeter than sucrose. The relative sweetnesses are: fructose, 100; sucrose, 58; glucose, 43.

28. Galactosemia is the inability of infants to metabolize galactose to glucose. It is usually caused by a deficiency of an enzyme, resulting in a build-up of galactose in the blood and urine. Galactosemia causes vomiting, diarrhea, enlargement of the liver, and often mental retardation. Newborn infants are relatively checked for galactosemia.

29. Substances are classified as lipids on the basis of their solubility in nonpolar solvents such as diethyl ether, benzene, and chloroform, their insolubility in water, and their greasy feeling.

30. Structural formulas

Glycerol

$$\begin{array}{l} CH_2OH \\ | \\ CHOH \\ | \\ CH_2OH \end{array}$$

Palmitic acid $CH_3(CH_2)_{14}COOH$

Oleic acid $CH_3(CH_2)_7CH{=}CH(CH_2)_7COOH$

Stearic acid $CH_3(CH_2)_{16}COOH$

Linoleic acid $CH_3(CH_2)_4CH{=}CHCH_2CH{=}CH(CH_2)_7COOH$

31. Fats are solid and vegetable oils are liquid at room temperature. Fats contain higher amounts of saturated fatty acids and vegetable oils contain higher amounts of unsaturated fatty acids.

32. A triacylglycerol (triglyceride) is a triester of glycerol. Most animal fats are triacylgycerols. Tristearin, in the next exercise, is a triacylglycerol.

33. Tristearin

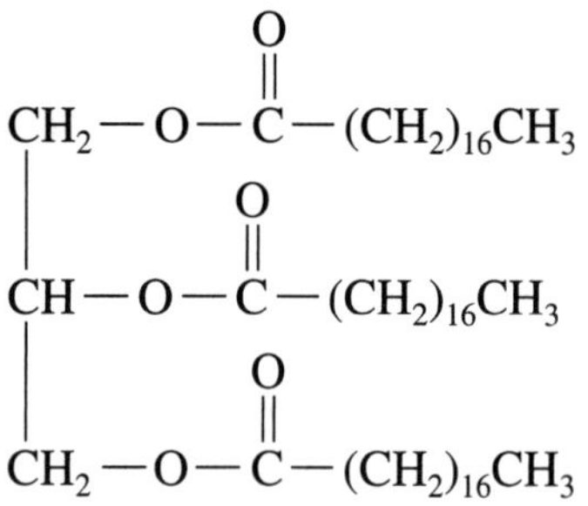

$$
\begin{array}{l}
CH_2-O-\overset{\overset{\displaystyle O}{\|}}{C}-(CH_2)_{16}CH_3 \\
| \\
CH-O-\overset{\overset{\displaystyle O}{\|}}{C}-(CH_2)_{16}CH_3 \\
| \\
CH_2-O-\overset{\overset{\displaystyle O}{\|}}{C}-(CH_2)_{16}CH_3
\end{array}
$$

34. Triacylglycerol

$$
\begin{array}{ll}
CH_2-O-\overset{\overset{\displaystyle O}{\|}}{C}-(CH_2)_7CH{=}CHCH_2CH{=}CH(CH_2)_4CH_3 & \text{(linoleic)} \\
| & \\
CH-O-\overset{\overset{\displaystyle O}{\|}}{C}-(CH_2)_{16}CH_3 & \text{(stearic)} \\
| & \\
CH_2-O-\overset{\overset{\displaystyle O}{\|}}{C}-(CH_2)_7CH{=}CH(CH_2)_7CH_3 & \text{(oleic)}
\end{array}
$$

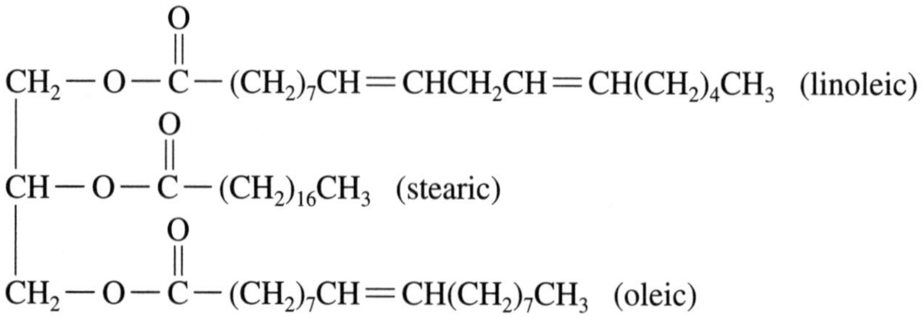

There are two other formulas using the same three acids. In one, the linoleic acid would be in the middle and in the other, oleic acid would be in the middle. The top and bottom positions are equivalent.

35. (a) Tripalmitin is a fat in which all the fatty acid units are palmitic acid.

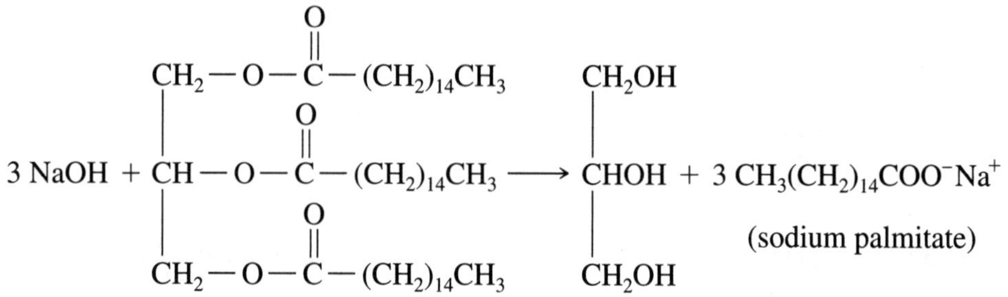

$$
3\,NaOH + \begin{array}{l}
CH_2-O-\overset{\overset{\displaystyle O}{\|}}{C}-(CH_2)_{14}CH_3 \\
| \\
CH-O-\overset{\overset{\displaystyle O}{\|}}{C}-(CH_2)_{14}CH_3 \\
| \\
CH_2-O-\overset{\overset{\displaystyle O}{\|}}{C}-(CH_2)_{14}CH_3
\end{array}
\longrightarrow
\begin{array}{l}
CH_2OH \\
| \\
CHOH \\
| \\
CH_2OH
\end{array}
+ 3\,CH_3(CH_2)_{14}COO^-Na^+ \ \text{(sodium palmitate)}
$$

(b)

$$\begin{array}{l} CH_2-O-\overset{O}{\overset{\|}{C}}-(CH_2)_7CH{=}CHCH_2CH{=}CH(CH_2)_4CH_3 \\ | \\ CH-O-\overset{O}{\overset{\|}{C}}-(CH_2)_{16}CH_3 \qquad\qquad + 3\ NaOH \longrightarrow \\ | \\ CH_2-O-\overset{O}{\overset{\|}{C}}-(CH_2)_7CH{=}CH(CH_2)_7CH_3 \end{array}$$

$$\begin{array}{lll} CH_2OH & & CH_3(CH_2)_4CH{=}CHCH_2CH{=}CH(CH_2)_7COO^-Na^+ \\ | & & \\ CHOH & + & CH_3(CH_2)_{16}COO^-Na^+ \\ | & & \\ CH_2OH & & CH_3(CH_2)_7CH{=}CH(CH_2)_7COO^-Na^+ \end{array}$$

The soaps top to bottom are: sodium linoleate, sodium stearate, and sodium oleate.

36. Vegetable oils can be solidified by hydrogenation, which adds hydrogen to the double bonds, to saturate the double bonds and form fats. Solid fats are preferable to oils for the manufacture of soaps and for certain food products. Hydrogenation extends the shelf-life of oils, because it is oxidation at the points of unsaturation that leads to rancidity of fats and oils.

37. Fats are an important food source for man. They normally account for 25 to 50 percent of caloric intake. Fats are the major constituent of adipose tissue, which is distributed throughout the body. In addition to being a source of reserve energy, fat deposits function to insulate the body against loss of heat and protect vital organs against mechanical injury.

38. The three essential fatty acids are linoleic, linolenic, and arachidonic acids. Diets lacking these fatty acids lead to impaired growth and reproduction and skin disorders, such as eczema and dermatitis.

39. The structural formula of cholesterol is

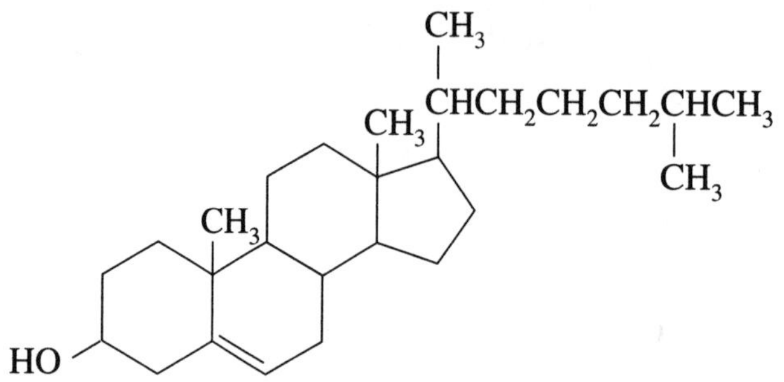

CH_3 | $CHCH_2CH_2CH_2CHCH_3$ | CH_3; CH_3; CH_3; HO

40. The ring structure common to all steroids is

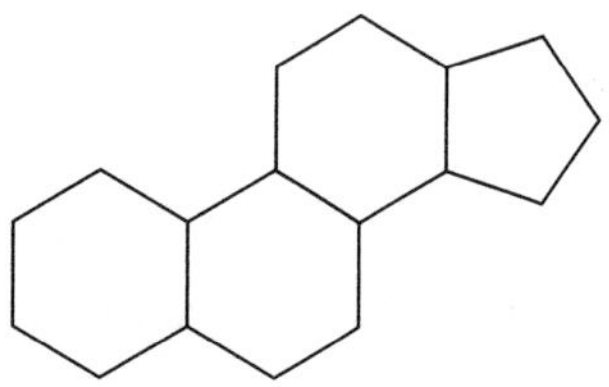

41. Some common foods with high (over 10%) protein content are gelatin, fish, beans, nuts, cheese, eggs, poultry and meat of all kinds.

42. Amino acids contain a carboxylic acid group and an amino acid group.

43. The amino acids in proteins are called α-amino acids because the amine group is always attached in the α position, that is, the first carbon next to the carboxylic acid group.

44. (a) phenylalanine, tryptophan, tyrosine

    (b) cysteine, methionine

    (c) serine, threonine, tyrosine

45. Dipeptides of glycine and phenylalanine

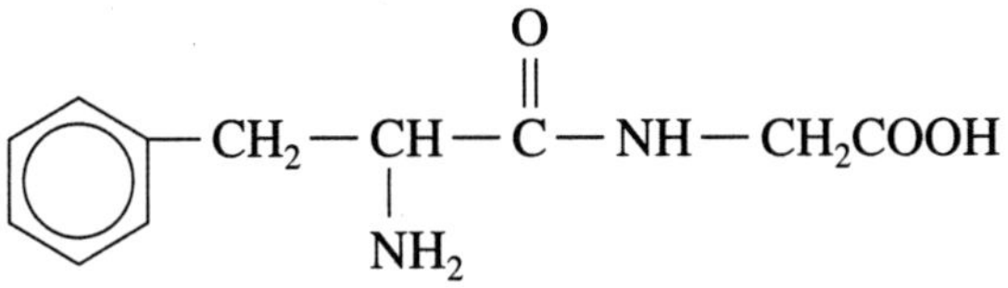

$$C_6H_5-CH_2-\underset{\displaystyle NH_2}{\underset{|}{CH}}-\overset{\displaystyle O}{\overset{\|}{C}}-NH-CH_2COOH \qquad \text{phe-gly}$$

$$H_2N-CH_2-\overset{\displaystyle O}{\overset{\|}{C}}-NH-\underset{\displaystyle CH_2C_6H_5}{\underset{|}{CHCOOH}} \qquad \text{gly-phe}$$

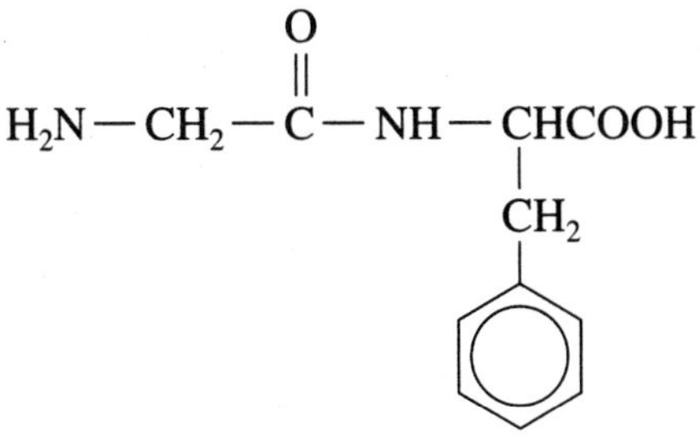

46. (a) Glycylglycine

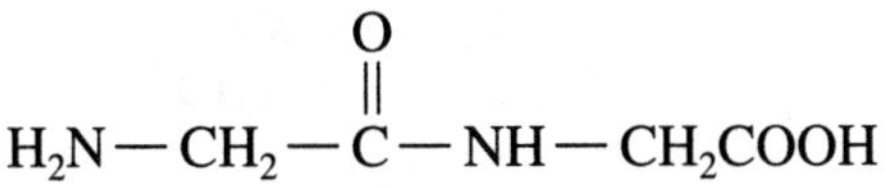

$$H_2N-CH_2-\overset{\displaystyle O}{\overset{\|}{C}}-NH-CH_2COOH$$

(b) Glycylglycylalanine

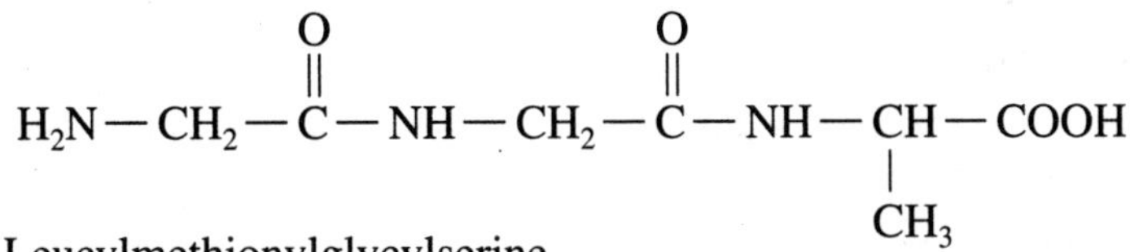

$$H_2N-CH_2-\overset{\displaystyle O}{\overset{\|}{C}}-NH-CH_2-\overset{\displaystyle O}{\overset{\|}{C}}-NH-\underset{\displaystyle CH_3}{\underset{|}{CH}}-COOH$$

(c) Leucylmethionylglycylserine

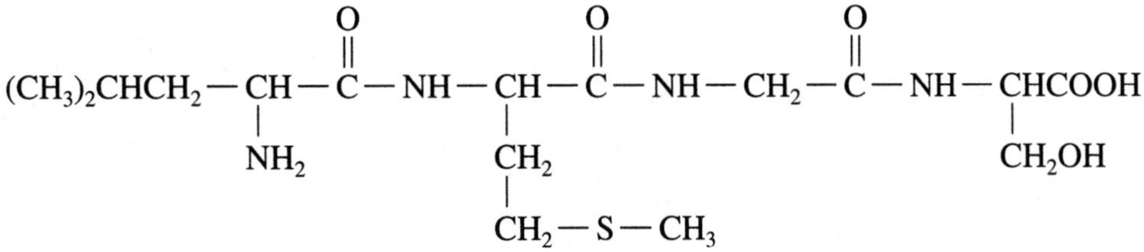

$$(CH_3)_2CHCH_2-\underset{\displaystyle NH_2}{\underset{|}{CH}}-\overset{\displaystyle O}{\overset{\|}{C}}-NH-\underset{\displaystyle CH_2-CH_2-S-CH_3}{\underset{|}{CH}}-\overset{\displaystyle O}{\overset{\|}{C}}-NH-CH_2-\overset{\displaystyle O}{\overset{\|}{C}}-NH-\underset{\displaystyle CH_2OH}{\underset{|}{CHCOOH}}$$

47. All possible tripeptides of glycine (gly), phenylalanine (phe), and leucine (leu).

gly - phe - leu	leu - phe - gly	gly - leu - phe
phe - gly - leu	leu - gly - phe	phe - leu - gly

48. Aspartame: two possible structures

$$C_6H_5-CH_2\underset{\displaystyle NH_2}{\underset{|}{C}}HCOOH \qquad HOOCCH_2\underset{\displaystyle NH_2}{\underset{|}{C}}HCOOH$$

phenylalanine   aspartic acid

$$HOOCCH_2\underset{\displaystyle COOH}{\underset{|}{C}H}-NH-\underset{\displaystyle O}{\underset{\|}{C}}-\underset{\displaystyle NH_2}{\underset{|}{C}H}CH_2-C_6H_5 \qquad \text{asp-phe}$$

$$C_6H_5-CH_2\underset{\displaystyle COO\textcircled{H}}{\underset{|}{C}H}-NH-\underset{\displaystyle O}{\underset{\|}{C}}-\underset{\displaystyle NH_2}{\underset{|}{C}H}CH_2COOH \qquad \text{phe-asp}$$

The circled H is replaced by CH_3 in aspartame.

49. Essential amino acids are those which are needed by the human body but cannot be synthesized by the body. Therefore, it is essential that they be included in the diet. They are: isoleucine, leucine, lysine, methionine, phenylalanine, threonine, tryptophan, and valine.

50. The proteins consumed by a human are converted by digestive enzymes into smaller peptides and amino acids. These smaller units are utilized in many ways:
    (a) to replace and repair body tissues
    (b) to synthesize new proteins
    (c) to synthesize other nitrogen-containing substances, such as enzymes, certain hormones, and heme molecules
    (d) to synthesize nucleic acids
    (e) to synthesize other necessary foods, such as carbohydrates and fats.

    Proteins are catabolized (degraded) to carbon dioxide, water, and urea. Urea, containing nitrogen, is eliminated from the body in the urine.

51. Tissue proteins are continuously being broken down and resynthesized. Protein is continually needed in a balanced diet because the body does not store free amino acids. They are needed to:
    (a) replace and repair body tissue
    (b) synthesize new proteins
    (c) synthesize other nitrogen-containing substances, such as enzymes, some hormones, and bone
    (d) synthesize nucleic acids
    (e) synthesize other necessary foods, such as carbohydrates and fats

52. The component parts that make up DNA

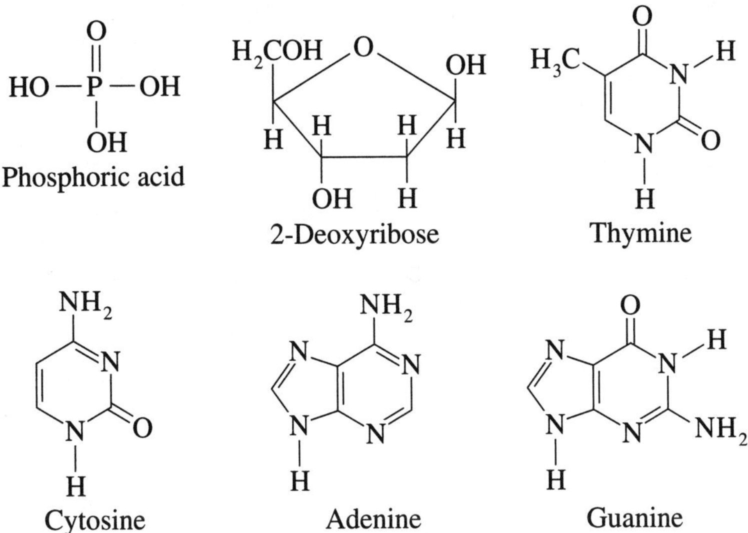

Phosphoric acid 2-Deoxyribose Thymine

Cytosine Adenine Guanine

53. (a) The three units that make up a nucleotide are a phosphate, deoxyribose, and one of the four nitrogen-containing bases. (A, T, G, C)

(b) In DNA, the four types of nucleotides are
phosphate - deoxyribose - thymine
phosphate - deoxyribose - cytosine
phosphate - deoxyribose - adenine
phosphate - deoxyribose - guanine

(c) Structure and name of one of the nucleotides
phosphate - deoxyribose - cytosine

Cytosine deoxyribonucleotide

54. The stucture of DNA, as proposed by Watson and Crick, is in the form of a double helix with both strands coiled around the same axis. Along each strand, phosphate and deoxyribose units alternate. Each deoxyribose unit has one of the four bases attached, which in turn, is hydrogen-bonded to a complementary base on the other strand. Thus, the two strands are linked at each deoxyribose unit by two bases.

55. The two helices of the double helix are joined together by hydrogen bonds between bases. The structure of the bases is such that one base will hydrogen bond to only one other specific base. That is, adenine is always hydrogen-bonded to thymine and cytosine is always bonded to guanine. Therefore, the hydrogen bonding requires a specific structure on the adjoining helix.

56. Complementary bases are the pairs that "fit" to each other by hydrogen bonds between the two helices of DNA. For DNA, the complementary pairs are thymine with adenine and cytosine with guanine, or T-A and C-G.

57. If a segment of a DNA strand has a base sequence C-G-A-T-T-G-C-A, the other strand of the double helix will have the base sequence G-C-T-A-A-C-G-T.

58. Replication of DNA begins with the unwinding of the double helix by breaking the hydrogen bonds between the bases to form two separate strands. Each strand then combines with the proper free nucleotides to produce two identical replicas of the original double helix. This replication of DNA occurs just before the cell divides, giving each daughter cell the full genetic code of the parent cell.

59. DNA contains the genetic code of life. For any individual, the sequence of bases and the length of the nucleotide chains in the DNA molecules contain the coded messages that determine all the characteristics of the individual, including the reproduction of that species. Because of the mechanics of human reproduction, the offspring is a combination of the chromosomes of each parent, thus will not be a carbon copy of either parent.

60. The structural differences between DNA and RNA are:
    (a) RNA exists in the form of a single-stranded helix, whereas DNA is a double helix.
    (b) RNA contains the pentose sugar ribose, whereas DNA contains deoxyribose.
    (c) RNA contains the base uracil, whereas DNA contains thymine.

61. In ordinary cell divisions, known as mitosis, each DNA molecule forms a duplicate by uncoiling to single strands. Each strand then assembles the complementary portion from available free nucleotides to form duplicates of the original DNA molecule.

    In most higher forms of life, reproduction takes place by the union of the sperm with the egg. Cell splitting to form the sperm cell and the egg cell occurs by a different and more complicated process called meiosis. In meiosis, the sperm cell carries only one half of the chromosomes from its original cell, and the egg cell also carries one half of its original chromosomes. Between them, they form a new cell that once again contains the correct number of chromosomes and all the hereditary characteristics of the species.

62. The location of protein synthesis is a ribosome.

63. Proteins are polymers of amino acids.

64. The red dotted lines in Figure 20.6 represent hydrogen bonds between complementary base pairs.

65. Enzymes are proteins that act as catalysts by generally lowering the activation energy of specific biochemical reactions. With the assistance of enzymes, these chemical reactions proceed at high speed at normal body temperature.

66. Enzymes are usually specific for one particular reaction because the substrate (substance acted upon by the enzyme) usually fits exactly into a small part of the enzyme (known as the "active site") to form an intermediate enzyme-substrate complex. Most substrates do not fit any other enzyme.

67. Polypeptides are numbered starting with the N-terminal amino acid.

    N-terminal tyr C-terminal val

68. 1 2 3 4 5
    tyr - gly - his - phe - val

    Polypeptides are numbered starting with the N-terminal amino acid.

69. The bond that connects one amino acid to another in a protein is called a peptide bond.

70. In the lock and key hypothesis the active site of an enzyme exactly fits the complementary-shaped part of a substrate to form an enzyme-substrate reaction complex on the way to forming the products. In the flexible site hypothesis (induced-fit) the enzyme changes its shape to fit the shape of the substrate to form the enzyme-substrate reaction complex.

71. Enzyme specificity is due to the particular shape of a small segment of the enzyme known as the "active site". This site fits a complementary shape of the substrate with which the enzyme is reacting.

72. Enzymes act as catalysts for biochemical reactions. Their function is to increase the rate of biochemical reactions by lowering the activation energy of these biochemical reactions.

73. (a) Fructose contains a C=O group in its carbon chain.

```
                        CH2OH
                          |
ketone group ——→ C=O
                          |
               HO—C—H
                          |
                 H—C—OH
                          |
                 H—C—OH
                          |
                        CH2OH
```

(b) Glucose contains a H—C=O group at the beginning of its carbon chain.

```
aldehyde group ——→ H—C=O
                               |
                      H—C—OH
                               |
                    HO—C—H
                               |
                      H—C—OH
                               |
                      H—C—OH
                               |
                             CH2OH
```

74. The simplest empirical formula for a carbohydrate is CH_2O.

75. An amino acid contains an amino group (—NH_2) on the carbon chain in addition to the carboxylic acid group. The amino group cannot be bonded to a >C=O group to be an amino acid. For example

CH_3CH_2COOH is a carboxylic acid

$CH_3CH(NH_2)COOH$ is an amino acid

76. $C_{12}H_{22}O_{11}$ $\quad 22\,H = +22 \quad 11\,O = -22 \quad 12\,C = 0$ $\qquad CO_2$ $\quad 2\,O = -4 \quad 1\,C = +4$

$$C_{12}H_{22}O_{11} + 12\,O_2 \longrightarrow 12\,CO_2 + 11\,H_2O$$

The change in oxidation state of carbon is from 0 to +4.

77. (a)

$$\begin{array}{l} CH_2{-}O{-}\overset{\displaystyle O}{\overset{\|}{C}}{-}CH_2CH_2CH_3 \\ | \\ CH{-}O{-}\overset{\displaystyle O}{\overset{\|}{C}}{-}CH_2CH_2CH_3 \\ | \\ CH_2{-}O{-}\overset{\displaystyle O}{\overset{\|}{C}}{-}CH_2CH_2CH_3 \end{array}$$

(b)

$$\begin{array}{l} CH_2{-}O{-}\overset{\displaystyle O}{\overset{\|}{C}}{-}C_{17}H_{35} \\ | \\ CH{-}O{-}\overset{\displaystyle O}{\overset{\|}{C}}{-}C_{15}H_{31} \\ | \\ CH_2{-}O{-}\overset{\displaystyle O}{\overset{\|}{C}}{-}C_{13}H_{27} \end{array}$$

78. Molar mass of $C_6H_{10}O_5 = 162.1$ g/mol

Cellulose: $\dfrac{6.0 \times 10^5 \text{ g/mol}}{162.1 \text{ g/mol}} = 3.7 \times 10^3$ monomer units

Starch: $\dfrac{4.0 \times 10^3 \text{ g/mol}}{162.1 \text{ g/mol}} = 25$ monomer units

Celulose: $(3.7 \times 10^3 \text{ units})(5.0 \times 10^{-10} \text{ m/unit}) = 1.9 \times 10^{-6}$ m long

Starch: $(25 \text{ units})(5.0 \times 10^{-10} \text{ m/unit}) = 1.3 \times 10^{-8}$ m long